Larval Fish and Shellfish Transport through Inlets

Cover photo: Oregon Inlet, North Carolina (U.S. Army Corps of Engineers.)

Publication of these proceedings was made possible by the

U.S. Army Corps of Engineers

Waterways Experiment Station

Larval Fish and Shellfish Transport through Inlets

Edited by

Michael P. Weinstein

Proceedings of a Workshop Held in
Ocean Springs, Mississippi, USA
August 19–20, 1985

American Fisheries Society Symposium 3

Bethesda, Maryland
1988

The American Fisheries Society Symposium series is a registered serial. Suggested citation formats follow.

Entire book

Weinstein, M. P., editor. 1988. Fish and shellfish transport through inlets. American Fisheries Society Symposium 3.

Article within the book

Seabergh, W. C. 1988. Observations on inlet flow patterns derived from numerical and physical modeling studies. American Fisheries Symposium 3:16–25.

Library of Congress Catalog Card Number: 88-70511
ISSN 0892-2284 ISBN 0-913235-46-6

Address orders to

American Fisheries Society
5410 Grosvenor Lane, Suite 110
Bethesda, Maryland 20814, USA

CONTENTS

Preface

The collection of papers in this volume represents one product of a workshop held at the Gulf Coast Research Laboratory in Ocean Springs, Mississippi, on 19 and 20 August 1985. Sponsorship for the workshop was provided by the Coastal Ecology Group of the U.S. Army Corps of Engineers Waterways Experiment Station, and funding was provided under the Environmental Impact Research Program. The moderator and sponsors gratefully acknowledge Dr. Harold Howse, Director, and the staff of the Gulf Coast Research Laboratory for hosting the activities of this workshop.

The intent of the workshop was to address long-standing questions regarding the effects of coastal engineering projects, particularly construction of jetties at coastal inlets, on the recruitment of egg and larval stages of fishes and shellfishes to estuarine nursery areas. These questions had originally surfaced in connection with the proposed jetty construction project at Oregon Inlet, North Carolina.

During the planning phase for the Oregon Inlet jetties, a technical advisory board was charged with the tasks of reviewing and evaluating several plans of study to monitor transport phenomena at the inlet proper. Preliminary physical modeling studies, although inconclusive, had indicated that changes in the circulation pattern induced by jetty construction at the inlet might delay or impede passage of passive buoyant or semibuoyant "particles" into the estuary. Later, however, the advisory board determined that the complexity of the transport issue was such that multiyear intensive field sampling would be required to overcome the masking effect of natural variability in the system; the probability that these efforts would lead to definitive conclusions would be low. Costs for such studies were found to be prohibitive, and a decision was made not to pursue field investigations. Thus, the basic questions remained unresolved.

The Larval Transport Workshop was convened to reexamine the technical foundation of the transport issue as it pertains to jetty projects. Rather than limiting the scope of discussions, a generic approach to the Oregon Inlet case was taken. In essence, experts from both physical and biological disciplines were brought together; the question posed was: "Is the null hypothesis that transport processes at jettied inlets are not different from those at unaltered inlets a testable one?" Formal presentations (upon which the contributed papers contained herein were based) led off the workshop. The objective was to summarize the present state of knowledge regarding transport processes, because there were no existing data sets obtained specifically to address the topic. Several authors were assigned the explicit task of producing broad topic overviews; others focused on specific case studies. Working discussions followed in an attempt to reach a consensus on the feasibility and direction of future studies on the transport issue. A condensation of the participants' recommendations is contained in the epilogue appended to this volume.

Thirteen peer-reviewed papers constitute the proceedings of this workshop. The topics considered include transport of water masses (shelf–estuarine exchange) and larvae from the continental shelf to inlets and ultimately (for the larvae) to estuarine nurturing grounds. Examples of cross-shelf transport of water and of larvae are provided by Wiseman et al. and by Shaw et al., respectively, for the Gulf of Mexico; they concentrate on species of commercial importance. Several papers deal with events immediately in and around inlets, including transport processes and mechanisms (Kjerfve and Wolaver; Seabergh; Wang), and with the general physical oceanographic processes affecting larval fish transport through North Carolina inlets (Pietrafesa and Janowitz). One of two workshop presentations on models of the recruitment process for larvae passing through the Cape Fear River inlet in North Carolina and into primary nurseries is included herein (Lawler et al.). Behavioral tactics and physical factors that cue larvae during their passage from spawning to nurturing ground are independently reviewed by Boehlert and Mundy and by Miller. The relative roles of passive and active transport were discussed and hypotheses presented. Discussions of invertebrate larval transport, by McConaugha and by Epifanio, concentrate primarily on decapod crustacea. Within-estuary retention and circulation of larval fish are considered by Bourne and Govoni. Finally, an elegant consideration of null hypothesis testing and statistical design criteria for measurements of larval fish and shellfish transport is provided by Colby.

MICHAEL P. WEINSTEIN
Lawler, Matusky & Skelly Engineers
One Blue Hill Plaza
Pearl River, New York 10965, USA

DOUGLAS G. CLARKE AND EDWARD J. PULLEN
Coastal Ecology Group
U.S. Army Corps of Engineers
Waterways Experiment Station
Post Office Box 631
Vicksburg, Mississippi 39180, USA

American Fisheries Society Symposium 3:1–8, 1988

Shelf–Estuarine Water Exchanges between the Gulf of Mexico and Mobile Bay, Alabama

WILLIAM J. WISEMAN, JR.

Coastal Studies Institute and Department of Geology and Geophysics, Louisiana State University Baton Rouge, Louisiana 70803, USA

WILLIAM W. SCHROEDER

Marine Sciences Program, The University of Alabama, Dauphin Island, Alabama 36528, USA

SCOTT P. DINNEL

Coastal Studies Institute and Department of Marine Sciences, Louisiana State University

Abstract.—One month of current meter data from Main Pass, which connects lower Mobile Bay with the Gulf of Mexico, demonstrated shelf–estuarine exchange driven by north–south wind stress at periodicities longer than the tide. Riverine discharge fluctuations may modulate the gravitational circulation over time scales shorter than seasons. Tidal diffusion, long-period advection, and the mean circulation are of equal importance to dispersion of water through Main Pass.

The exchange of water between estuaries and the inner continental shelf through tidal passes has been studied for many years. Coastal engineers have focused attention on the stability of the inlets and concentrated on tidal flows (e.g., Bruun and Gerritsen 1960; Brunn 1966; O'Brien 1969). Oceanographers have been more concerned with the net nontidal exchange of water between estuary and ocean and the gravitational circulation driven by the density distribution in the estuary (Pritchard 1956; Rattray and Hansen 1962). In recent years, the importance of flow variations at time scales longer than the tides (subtidal) has been recognized and studied extensively (Carter et al. 1979; Wiseman 1986). Tidal, subtidal, wind-driven, and gravitational motions advect a variety of dissolved and particulate matter including nutrients, suspended sediment, planktonic larvae, and marsh detritus.

Several studies of estuaries along the Atlantic coast of the USA have demonstrated a variety of forcings and responses possible in coastal inlets, but few such studies have been carried out in the estuaries and inlets of the Gulf of Mexico coast. These shallow estuaries are traditionally thought of as wind driven (Ward 1980). This paper presents measurements of subtidal flow variability in Main Pass, Mobile Bay, Alabama, and estimates the relative importance of tidal, wind-driven, and gravitational flows.

Main Pass connects Mobile Bay, a broad, shallow estuary, with the Gulf of Mexico (Figure 1). The pass is 5.4 km wide. Its western two-thirds is only 4 m deep, but depths as great as 15 m occur in the eastern section. A ship channel through the pass is maintained to a depth of 13 m. Main Pass carries approximately 85% of the mass flux between Mobile Bay and the adjacent continental shelf waters (Schroeder 1978). The remaining 15% flows through Pass-aux-Herons, which connects Mobile Bay to east Mississippi Sound.

An extensive study of the hydrography of Main Pass was carried out by Lysinger (1982). Waters in the deeper eastern side of the pass always show salinity stratification; the shallow western waters can be either stratified or well mixed. The magnitude of salinity and the degree of stratification vary during the tidal period. Salinities of the upper water layers within the pass are significantly related to the river discharge into Mobile Bay.

Schroeder and Wiseman (1986) analyzed 2 years of data to determine relationships among wind stress, river runoff, barometrically corrected water level, and water flow through Main Pass and within Mobile Bay. They inferred that at least three forcing functions caused significant flow variations over three different time scales. Strong north–south winds, particularly those accompanying winter frontal passages, were extremely effective in forcing water exchange through Main Pass at periods between 2 and 10 d. This response appears to be due directly to friction at the air–water interface. At periods longer than 3 d, the east–west wind stress also forced an exchange because of Ekman convergence and divergence at the coastline. Over periods longer than 40 d, river

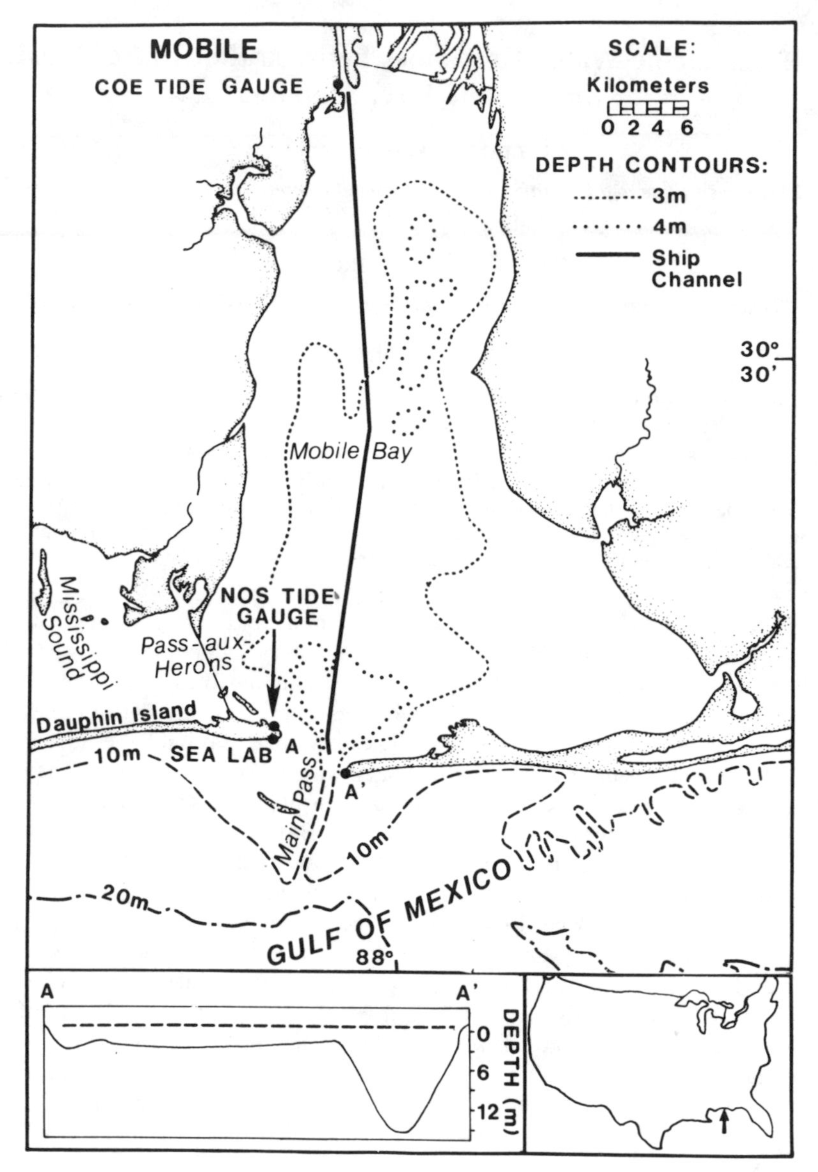

FIGURE 1.—Bathymetric chart of Main Pass, Mobile Bay, Alabama, and a cross-section of Main Pass (A–A'); NOS is National Oceanographic Survey; COE is U.S. Army Corps of Engineers.

discharge at the northern end of Mobile Bay was coherent with water levels within the bay and thus with the total mass flow through Main Pass. It was expected that gravitational circulation also would be modulated by variability in river flow, but this could not be confirmed without direct current measurements.

Thirty-five days (6 November–11 December 1979) of simultaneous current meter records from Main Pass and lower Mobile Bay have recently been made available to us. The analysis and interpretation of this data set form the main subject of this paper.

Current Data

The current records were made by five Endeco type 105 meters moored at three locations in and near Main Pass (moorings DI, TR, and RA: Figure 2). Mooring DI (water depth, 2.8 m) had one current meter at middepth (station DIM). Mooring

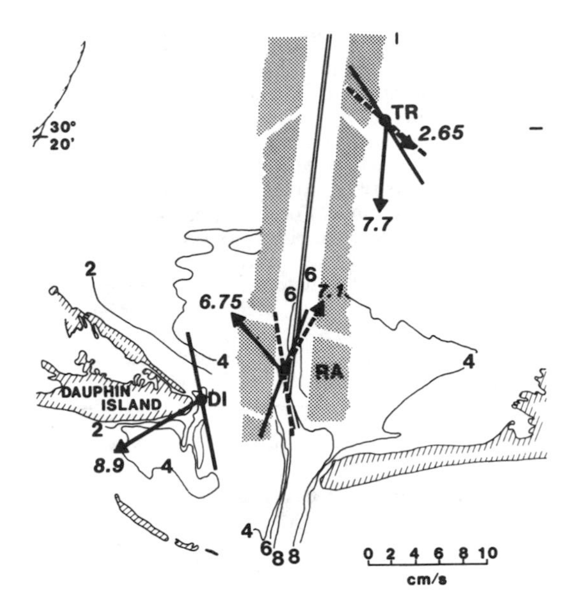

FIGURE 2.—Location of current meter moorings (DI, RA, and TR) in and near Main Pass, Mobile Bay. The stippled area represents spoil banks on either side of the shipping channel. Bottom contours are in meters. Arrows represent mean flows. Straight lines are the major principal axes of the low-pass-filtered currents. Their lengths are proportional to the standard deviations. Dashed lines represent currents 1 m off the bottom. Solid lines represent middepth currents.

TR (water depth, 3.2 m) had two meters; one was 1 m off the bottom (station TRB) and the other was at middepth, 2 m off the bottom (station TRM). Mooring RA (depth, 6.4 m) also had two meters, 1 m (station RAB) and about 3.2 m (station RAM) above the bottom.

Current data were resolved into a coordinate system with a u positive to the east and v positive to the north. They were then low-pass filtered with a 40-h cutoff to eliminate all fluctuations with periods shorter than 40 h. Figure 2 gives mean flows and major principal axes of the low-pass-filtered data from each meter. Table 1 presents statistics of the flow.

Mean wind velocities, $\mathbf{U}_w = (u_w, v_w)$, at Dauphin Island Sea Laboratory were converted to a parameter related to stress, $\mathbf{U}_w|\mathbf{U}_w|$. Water levels at Dauphin Island (National Ocean Survey) and Port of Mobile (U.S. Army Corps of Engineers) were corrected for barometric pressure. All records were then low-pass filtered with a 40-h cutoff for comparison with the currents (Figure 3).

Variance spectra (Bendat and Piersol 1986) for winds and currents and phases and coherences

TABLE 1.—Statistics of low-passed-filtered water currents measured by middepth (M) and near-bottom (B) meters at moorings DI, RA, and TR in and near Main Pass, Mobile Bay (Figure 2). Current coordinates u and v are positive to the east and north, respectively.

Station	Mean velocity (cm/s) u	Mean velocity (cm/s) v	Variance (cm²/s²) u	Variance (cm²/s²) v	Covariance of u and v (cm²/s²)
DIM	−7.75	−4.30	4.71	34.73	5.40
RAB	3.41	6.17	6.70	27.78	2.88
RAM	−4.37	5.12	13.68	33.71	8.60
TRB	1.93	−1.82	20.56	16.46	12.21
TRM	−0.68	−7.68	17.15	35.53	16.16

(Bendat and Piersol 1986) between records were estimated. Complex empirical orthogonal function analyses (Legler 1983) of the winds and currents were also performed.

Daily river discharge data from the Mobile River system were calculated from U.S. Geological Survey records. The record was too short to accurately estimate the coherence of river flow with currents. Correlations, though, were run with the current records after these had been low-pass filtered with a 72-h cutoff to eliminate a strong signal near 2–2.5 d, which would only be marginally resolvable if it were present in the daily discharge data.

Results

Three of the current meters (stations TRB, TRM, and DIM, all less than 3 m deep) showed seaward mean flow indicative of the upper layer of a normal estuarine circulation. The two meters at mooring RA (stations RAB and RAM, both deeper than 3 m) exhibited landward mean flows that would be found in the lower layer of a gravitational estuarine flow. Most of the variance in the low-pass-filtered current records from all stations except TRB was confined to the north–south direction. The following analysis concentrates on these north–south components.

When the wind-stress and current vectors were analyzed with complex empirical orthogonal functions (Legler 1983), three important modes (patterns of variability) emerged (Table 2). A first mode involved the wind stress and currents at TRM and DIM. A second mode of nearly equal importance involved the wind stress and currents at RAB and RAM. The third mode involved only the currents at TRB, which, together with the similarity of the u and v vector variances at TRB (Table 1), convinced us to ignore this record in further analyses. The raw data from this meter

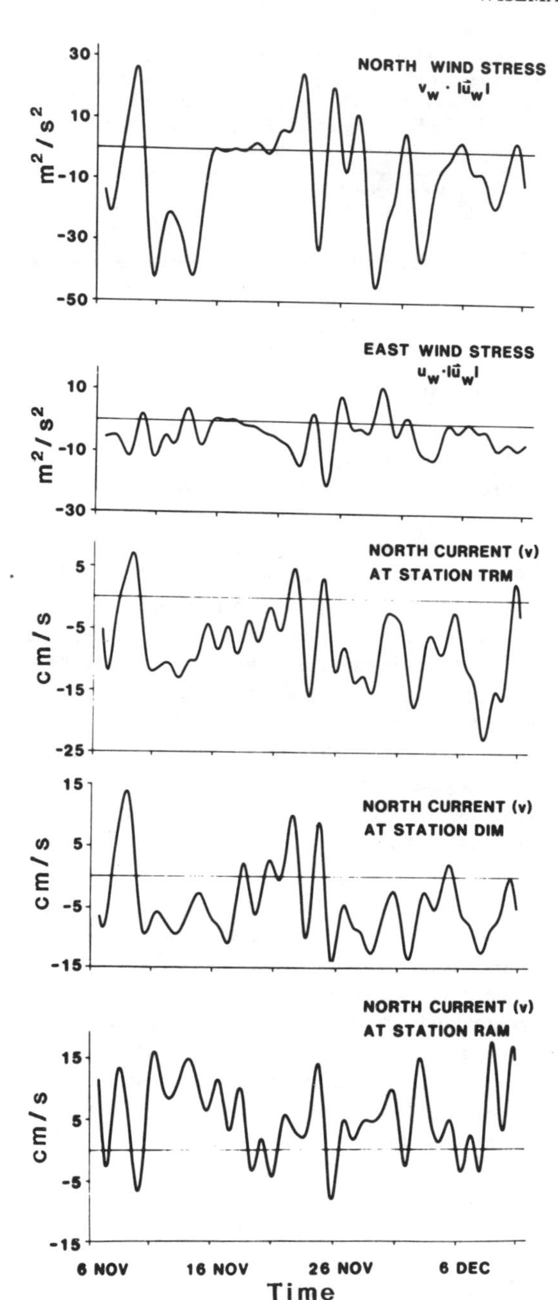

FIGURE 3.—Low-pass-filtered records of north and east wind stresses and north currents at middepth stations TRM, DIM, and RAM near Main Pass, Mobile Bay.

showed no anomalies, suggesting that the meter malfunctioned; we can only assume that currents there were steered by local bottom topography.

Variance spectra indicated that the weak east–west wind stress had energy peaks near 10 and 3-d periods, while the stronger north–south stress exhibited peaks in a 10–20 d band and a 4–5 d band with only a minor trough between the two (Figure 4). At TRM, the north–south currents had maximum energy in the 5–7 d band. The north–south currents at DIM showed maximum variance at periods near 10 d and a minor peak near 3–4 d. The north–south currents at RAM exhibited peaks at 3 and 10-d periods. The resemblance of the spectrum shapes suggest that flow at RAM was associated with the east–west wind stress, whereas the flows at TRM and DIM were associated with the north–south wind stress. These relationships are consistent with the conclusions of Schroeder and Wiseman (1986).

We attempted to further investigate these relationships through cross-spectrum analysis. Many difficulties were encountered because of the short record lengths. At the longest periods, the number

TABLE 2.—Complex analysis of winds and currents for three dominant modes (eigenvectors) as defined by their percent normalized variances (eigenvalues) of the correlation matrix of wind and current vectors. Percent of total normalized variances refers to the total variance of the input set of six vectors. Percent variance of individual records refers to the variance of each record explained by each mode. For example, mode 1 explains 41% of the total variance of the six vector records and 77.8% of the variance at DIM, but only 3% of the variance at RAB. Amplitude refers to relative magnitude of the eigenvector at each station within each mode. The phase relates the orientations of the eigenvectors at each station within each mode. The origin is arbitrary but consistent within each mode.

Wind or station	Amplitude	Phase (°)	% Variance of individual records
Mode 1 (41% of total normalized variance)			
Wind stress	0.457	22.8	51.2
DIM	0.563	1.8	77.8
RAB	0.110	46.2	3.0
RAM	0.148	13.5	5.4
TRB	0.358	−31.0	31.6
TRM	0.559	−12.9	76.8
Mode 2 (30.2% of total normalized variance)			
Wind stress	0.376	−15.5	25.6
DIM	0.111	142.2	2.2
RAB	0.654	−179.9	77.5
RAM	0.630	165.9	71.9
TRB	0.141	−42.8	3.6
TRM	0.048	−149.6	0.4
Mode 3 (13.8% of total normalized variance)			
Wind stress	0.332	−102.8	9.1
DIM	0.255	−130.9	5.4
RAB	0.171	15.4	2.4
RAM	0.145	163.0	1.7
TRB	0.850	43.8	59.8
TRM	0.226	−44.5	4.2

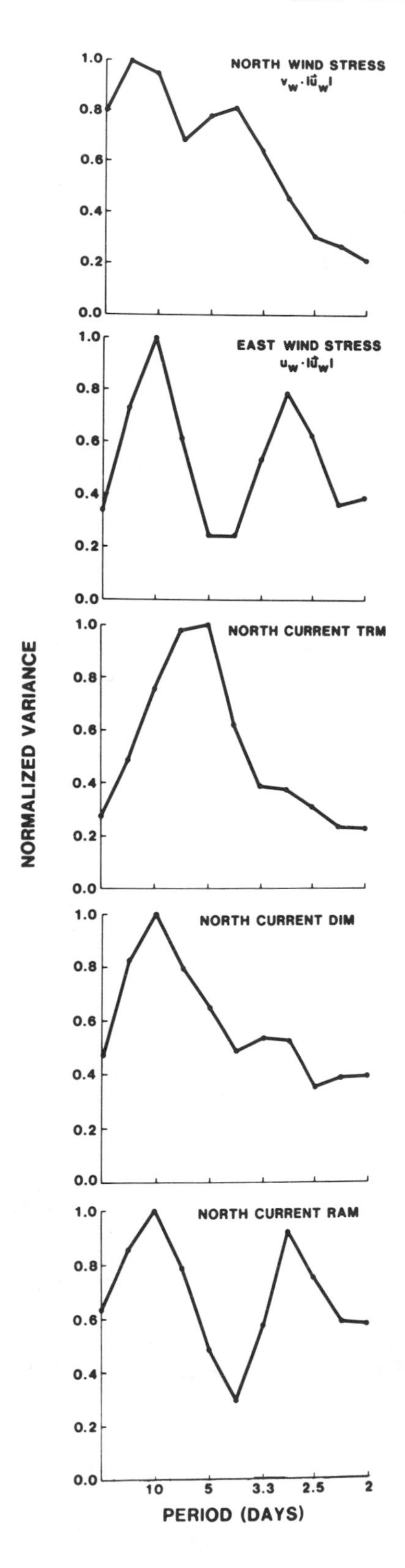

of repetitions of a phenomenon in our records is small (near one) and statistical reliability suffers. At the 2- and 4-d periods, when strong relationships between wind and current appear to occur, the two wind stress components are coherent and interpretation of phase relationships is made obscure. In an effort to resolve this, we calculated the multiple coherence (Jenkins and Watts 1968) between both wind stress components and the north–south currents at each middepth site (Figure 5).

It is clear that the currents were coherent with the wind at all sites. Much of the strong coherence at periods from 2 to 5 d, though, cannot be attributed definitively to one stress component or the other. When we removed from the east–west stress and the currents that part of the records that is coherent with the north–south stress, the resultant time series were generally incoherent at the 95% significance level of 0.58. When the portions of the current and north–south stress coherent with the east–west stress were removed, though, the resultant time series were usually coherent at the very longest periods and at periods shorter than 4 d. We ignored the long–period band for the reasons mentioned above.

Schroeder and Wiseman (1986) discussed the importance of water level slope to the dynamics of Mobile Bay. We estimated this parameter by comparing the barometrically corrected low-pass-filtered water levels at the Port of Mobile (50 km north of Main Pass) and Dauphin Island. At the high subtidal frequencies of interest, though, this record was coherent with both the east–west and north–south wind stress, and relationships could not be unequivocally interpreted.

River discharge was sampled daily, and the record (Figure 6) was too short to allow reliable cross-spectrum analysis. Correlations between this record and other observed variables (not shown), though, were suggestive. At lags of 9 d, the northward flow at DIM, RAM, and TRM showed negative, positive, and negative peaks in their correlations with discharge, respectively, indicating an increase in the intensity of the gravitational circulation with increased discharge. These peaks compared well with the lag of 5–9 d for water to reach the bay from the river gauging stations (Schroeder 1979).

←

FIGURE 4.—Normalized variance spectra with 9 degrees of freedom for north wind stress and east wind stress and for north currents at middepth stations TRM, DIM, and RAM.

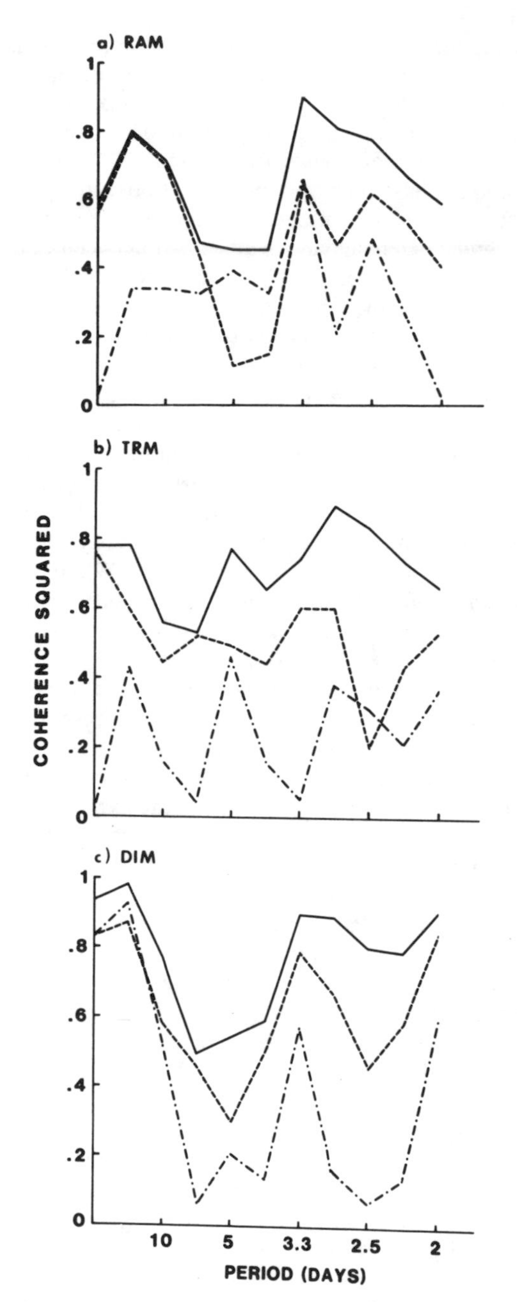

FIGURE 5.—Multiple coherence between wind stress and north current (solid line) at middepth stations RAM, TRM, and DIM, partial coherence between north wind stress and north current after the influence of east wind stress (dashed line) is removed, and partial coherence between east wind stress and north current after the influence of north wind stress (dot-dashed line) is removed. Significance level is 0.58.

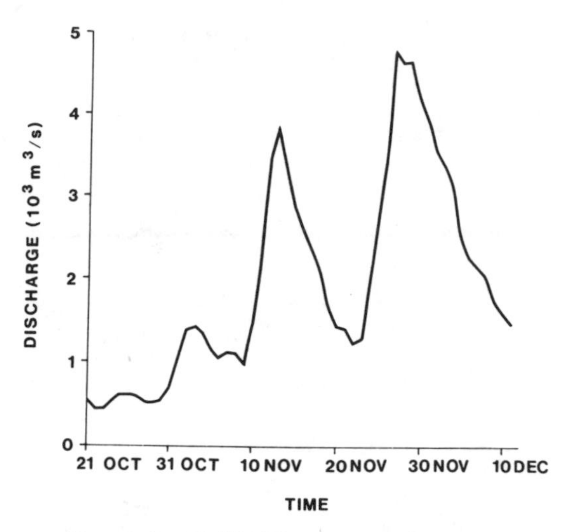

FIGURE 6.—Mobile River system discharge.

Discussion

Wang (1979) discussed the seasonal variability of the low-frequency dynamics of Chesapeake Bay. Similar variability occurs in Mobile Bay. Figure 7 shows the low-pass-filtered record of a parameter which is proportional to the volume flux of water to Mobile Bay. The record was constructed by averaging the instantaneous water level readings at the Port of Mobile and Dauphin Island and then taking the time derivative of the resultant time series. Exchanges are clearly larger in the winter season than during the summer. The 1 month of current meter data we have available is from early winter.

The usual concept of estuarine circulation is a two-layered flow with fresher water overlying saltier water. Such is the case within the Mobile Bay Ship Channel and the deeper portions of lower Mobile Bay and Main Pass. Two-thirds of Main Pass, though, and most of the Bay is very shallow. Bottom friction in these regions will be much different than the friction between a fresh-

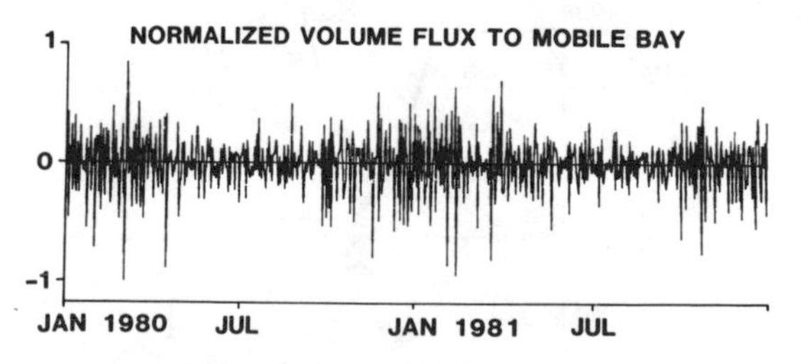

FIGURE 7.—Normalized, low-pass-filtered volume flux to Mobile Bay, 1980–1981.

and saltwater layer; therefore, the response of near-surface currents to wind stress applied over the shoals may be very different from what one would find in the deeper channels. Given these two restrictions on the available data, the results support the conclusions reached by Schroeder and Wiseman (1986): (1) that the north–south wind stress frictionally drives surface water in a coherent manner through Main Pass (the associated time delay has not been clearly determined); and (2) that river discharge appears to readily modulate the strength of the gravitational circulation.

Shaw et al. (1985) discussed the importance of wind-induced shelf–estuary exchange processes to the recruitment of gulf menhaden larvae *Brevoortia patronus*. They suggested that the larvae, which are spawned over the midshelf, are transported to the coastal boundary layer by the mean shelf circulation patterns. Once within the coastal boundary layer, they are rapidly advected parallel to the coast. If they pass the mouth of an estuary when a major wind-induced filling event is occurring, their probability of being recruited to the estuary is greatly increased. The wind-driven shelf–estuarine exchanges we have described above are precisely the type of process Shaw et al. (1985) invoked to account for recruitment success along the western Louisiana coast. We anticipate that these events will be equally important to the recruitment success of shelf-spawned larvae to Mobile Bay.

To properly evaluate the importance of these long-period flow variations as transport mechanisms, the variations should be compared to both the mean gravitational circulation and diffusive transports. The standard deviation of the north–south speed fluctuations (root-mean-square speed) at DIM, RAM, and TRM all lie between 0.5 and 1 times the corresponding mean speeds. Peak fluctuation speeds, though, can exceed the means by a factor of 3.

We estimated the mean advection due to the gravitational circulation from the mean flows at the current meters. A mean velocity through the upper layers of Main Pass similar to that observed at DIM would empty Mobile Bay in about 20 d. Austin (1954) estimated the flushing time for Mobile Bay at about 50 d using a modified tidal prism technique. Monthly longitudinal sections of salinity along the center line of Mobile Bay (Schroeder and Wiseman 1986) also suggested time scales of the order of a month for major exchanges of water between the bay and the shelf.

The effects of tidal dispersion (as estimated from Austin 1954), subtidal advection, and mean advection appear, then, to be of the same order of magnitude within Main Pass. These processes are additive. Which, if any, will be dominant for any given transport problem depends upon the season and the time scales of interest.

Acknowledgments

Funding for this work was provided by the University of Alabama Marine Science Program, the Dauphin Island Sea Laboratory, and the Mississippi–Alabama Sea Grant Consortium and the Louisiana Universities Sea Grant Program (U.S. Department of Commerce). The current records were supplied by Mobil Oil Exploration and Producing Southeast, Incorporated. The Port of Mobile tide data were provided by Geary McDonald, Mobile District, U.S. Army Corps of Engineers. Celia Harrod prepared the figures. This publication is contribution 99 from the Aquatic Biology Program, University of Alabama, and contribution 117 from the Marine Environmental Sciences Consortium, Dauphin Island, Alabama.

References

Austin, G. B. 1954. On the circulation and tidal flushing of Mobile Bay, Alabama. Part 1. Texas A&M College Research Foundation Project 24, Technical Report 12, College Station.

Bendat, J. S., and A. G. Piersol. 1968. Random data analysis and measurement procedures. Wiley, New York.

Bruun, P. 1966. Tidal inlets and littoral drift. H. Skipnes Offsettrykkeri, Trondheim, Norway.

Bruun, P., and F. Gerritsen. 1960. Stability of coastal inlets. North Holland, Amsterdam.

Carter, H. H., T. O. Najarian, D. W. Pritchard, and R. E. Wilson. 1979. The dynamics of motion in estuaries and other coastal water bodies. Reviews of Geophysics and Space Physics 17:1585–1590.

Jenkins, G. M., and D. G. Watts. 1968. Spectral analysis and its applications. Holden-Day, San Francisco.

Legler, D. M. 1983. Empirical orthogonal function analysis of wind vectors over the tropical Pacific region. American Meteorological Society Bulletin 64:234–241.

Lysinger, W. R. 1982. An analysis of the hydrographic conditions found in the Main Pass of Mobile Bay, Alabama. Masters' thesis. University of Alabama, Tuscaloosa.

O'Brien, M. P. 1969. Dynamics of tidal inlets. Pages 397–406 *in* A. Ayala-Castaneres and F. B. Phleger, editors. Memoir, symposium on coastal lagoons. UNAM-UNESCO, Universidad Nacional Autonoma de Mexico, Mexico City.

Pritchard, D. W. 1956. The dynamic structure of a

coastal plain estuary. Journal of Marine Research 15:33–42.
Rattray, M. J., Jr., and D. V. Hansen. 1962. A similarity solution for circulation in an estuary. Journal of Marine Research 20:121–133.
Schroeder, W. W. 1978. Riverine influence on estuaries: a case study. Pages 347–364 *in* M. S. Wiley, editor. Estuarine interaction. Academic Press, New York.
Schroeder, W. W. 1979. The dispersion and impact of Mobile River system waters in Mobile Bay, Alabama. WRRI (Water Resources Research Institute) Auburn University Bulletin 37.
Schroeder, W. W., and W. J. Wiseman, Jr. 1986. Low-frequency shelf–estuarine exchange processes in Mobile Bay and other estuarine systems on the northern Gulf of Mexico. Pages 355–367 *in* D. A. Wolfe, editor. Estuarine variability. Academic Press, New York.
Shaw, R. F., W. J. Wiseman, Jr., R. E. Turner, L. J. Rouse, Jr., R. E. Condrey, and F. J. Kelly, Jr. 1985. Transport of larval Gulf menhaden *Brevoortia patronus* in continental shelf waters of western Louisiana: a hypothesis. Transactions of the American Fisheries Society 114:452–460.
Wang, D.-P. 1979. Sub-tidal sea level variations in Chesapeake Bay and relations to atmospheric forcing. Journal of Physical Oceanography 9:413–421.
Ward, G. H., Jr. 1980. Hydrography and circulation processes of Gulf estuaries. Pages 183–215 *in* P. Hamilton and K. B. MacDonald, editors. Estuarine and wetland processes with emphasis on modelling. Plenum, New York.
Wiseman, W. J., Jr. 1986. Estuarine–shelf interactions. Pages 109–115 *in* C. N. K. Mooers, editor. Coastal and estuarine sciences, volume 3. Baroclinic processes on continental shelves. American Geophysical Union, Washington, D.C.

American Fisheries Society Symposium 3:9–15, 1988

Transport Model for Water Exchange between Coastal Inlet and the Open Ocean

DONG-PING WANG

Marine Sciences Research Center, State University of New York
Stony Brook, New York 11794, USA

Abstract.—A numerical model capable of predicting interactions between coastal inlets and the open ocean is a useful tool for assessment of environmental impacts. Most previous estuarine models have been based on the assumption of a two-dimensional flow field, which is quite restrictive for studies of water exchange between inlets and open ocean. An example of using a generalized, three-dimensional, density-stratified model to simulate the inlet–ocean interaction is presented. Results indicate that dispersal of the outflow plume is the combined effect of tidal, gravity, and wind-driven currents. The model may be used to assess effects of jetty construction on the transport of larvae.

Estuaries are semienclosed water bodies that have free connection with the ocean and contain water from land drainage and the sea. The circulation in estuaries is very complex, depending on river runoffs, winds, tides, and conditions in the adjacent ocean. Because the external forcings change continually, physical processes in an estuary are better described as a succession of episodic events. Rapidly changing physical conditions can have drastic effects on estuarine environment. For example, Taft and Wang (1982) reported a rapid buildup of water stratification and subsequently an anoxic lower layer within a day in the Potomac River. Prolonged anoxia can have catastrophic effects on the benthic species (Seliger et al. 1985).

Because of the transient nature of an estuary, it is difficult to define a "baseline" state against which human influences can be measured. Management decisions, however, must be made with regard to the uses of estuaries. Environmental impacts can result from dredging, water intake, and disposal of waste heat and sewage. How human activities will interfere with estuaries generally is too complex a matter to be decided by trial and error. Predictive models are needed to isolate and identify the impacts. Monitoring strategies can be better designed if they are based on information derived from model predictions.

Ideally, a predictive model should be verified against observations for a particular site of interest. In practice, collection of ocean data is still such a formidable task that a meaningful model verification generally will not be feasible. The credibility of a predictive model, therefore, must depend on model's ability to incorporate the proper physical processes. In this paper, limitations of the two-dimensional-model approach are briefly reviewed, and a generic study of the interactions between coastal inlets and ocean based on a three-dimensional model is described. The study gives a basic description of the important physical processes in inlets and their effects on the transport of larvae. Effects of jetty construction on transport processes are also discussed.

Two-Dimensional Models

Difficulties with numerical models of estuarine circulation mainly arise from limited computer resources and an inadequate understanding of physical processes. Most estuarine models use a two-dimensional representation of the water body in order to reduce computation effort and to simplify model physics. A two-dimensional vertically averaged model (Butler 1980) assumes a homogeneous estuary. This approach is most useful for prediction of sea levels (storm surges) and tidal currents in which the stratification effect is minimal. It also is being used in efforts to predict residual tidal circulation in shallow estuaries (Cheng and Casulli 1982). Because the vertically averaged model does not consider variation in water column properties, it is less useful for studies of particle transport and water quality in estuaries of appreciable salinity stratification (such as partially mixed estuaries).

A two-dimensional laterally averaged model (Wang and Kravitz 1980) assumes a narrow estuary. This approach considers the vertical exchange of momentum and material; therefore, it is suitable for use in partially mixed estuaries. The laterally averaged model has been used in investigations of salt intrusion, flushing, and water quality. Because the laterally averaged model does not consider variation in properties across the flow, it is not applicable to estuaries with complex geometries (multiple channels and large flats). Efforts are being made to develop multiple channels (Wang 1983).

Three-Dimensional Model

A two-dimensional estuarine model is computationally simple. However, it is too restrictive for study of the water exchange between an estuary and its adjacent coastal ocean. Spreading of a freshwater plume into the coastal ocean is a three-dimensional process that cannot be approximated by vertical or lateral averaging. Thus, a three-dimensional model will be required for study of the estuary–ocean interaction. A three-dimensional model solves a set of partial differential equations that describe the force balance (momentum equation), the mass conservation (continuity equation), and the salt conservation (buoyancy equation). Mixing and dissipation are sub grid processes, that is, they occur in spatial scales much smaller than those resolved by a circulation model, which must be parameterized. A numerical model also requires initial and boundary conditions and a description of external forcings.

Physical processes that determine the estuary–

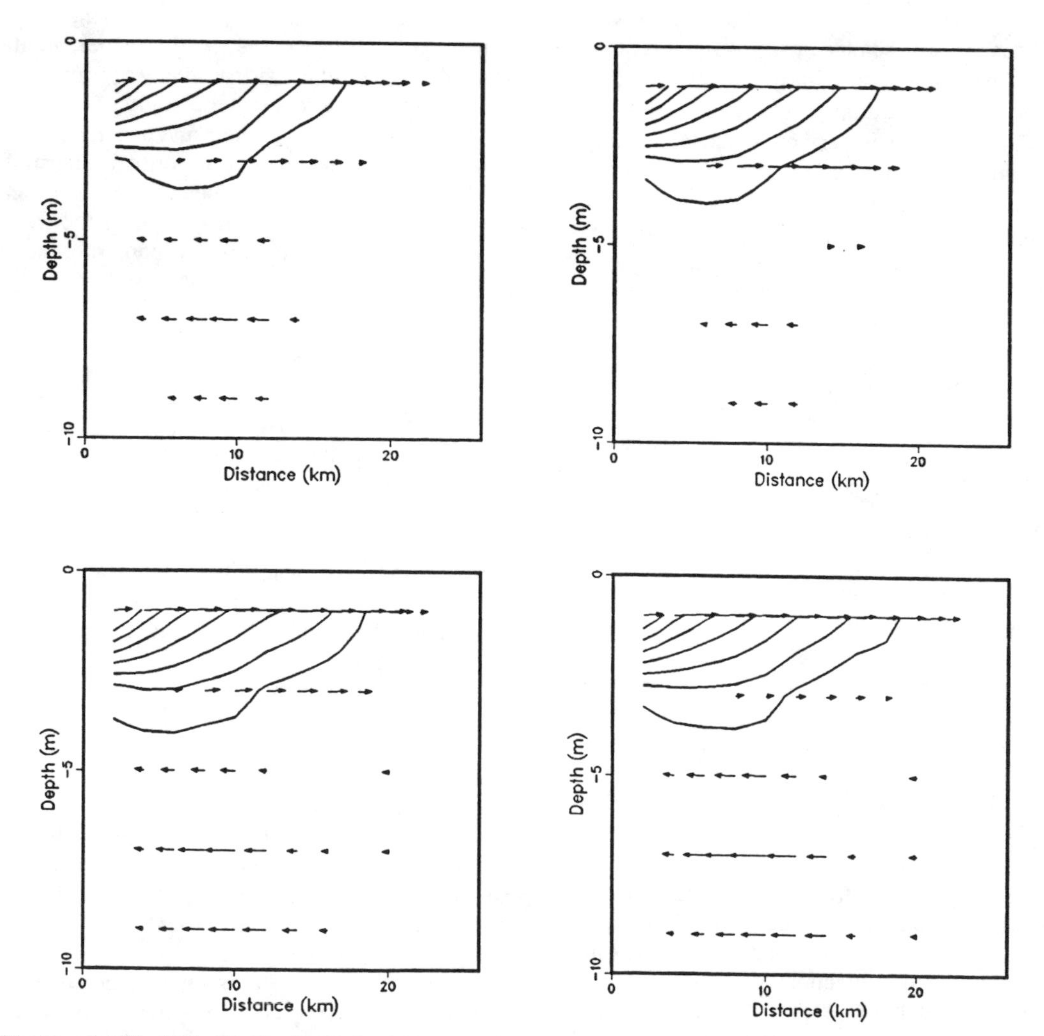

FIGURE 1.—Vertical distributions of salinity (2‰ isopleth interval, continuous lines) and longitudinal water velocity (vector length between two grids, 20 cm/s) along the central axis of a model single-inlet system during four tidal phases. The inlet mouth is at kilometer 10, and the open ocean extends to the right.

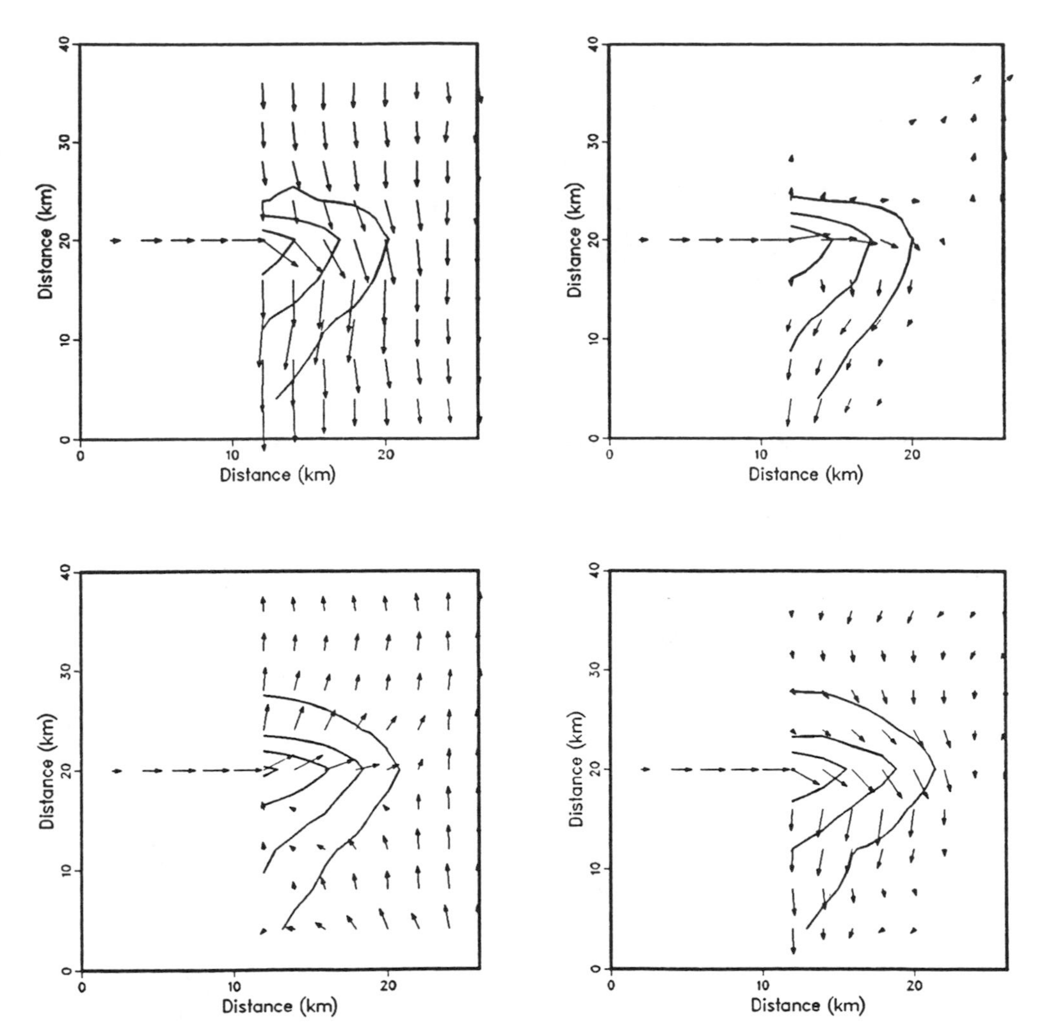

FIGURE 2.—Horizontal surface distributions of salinity (2‰ isopleth interval, continuous lines) and water velocity (vector length between two grids, 20 cm/s) in a model single-inlet system during four tidal phases. The inlet mouth is at kilometer 10 along the horizontal axis, and the open ocean extends to the right.

ocean interaction include the effects of tide, wind, and density gradient. Tidal currents are responsible for mixing in partially mixed estuaries, and they also are important for transport in shallow, well-mixed estuaries. Wind stirring can rapidly mix an entire water column. Alongshore winds also drive strong coastal currents that rapidly redistribute the outflow plume. Density gradients caused by the salinity difference between estuarine and oceanic waters drive a freshwater outflow in the surface layer and a saltwater intrusion in the bottom layer. The total flow field in the vicinity of inlet entrance reflects the combined effects of tide, wind, and density gradient.

Simulations from a three-dimensional stratified circulation model illustrate predicted estuary–ocean interactions. The three-dimensional model previously has been used in studies of coastal ocean circulation (Wang 1982), exchange of waters in a lock (Wang 1985b), and far-field plume from a buoyant discharge (Wang 1985a). Parameterization of the vertical mixing is based on the Munk–Anderson formula, which relates the degree of mixing to the stability of the water column. The model ocean consists of a narrow coastal inlet, 10 km long, 2 km wide, and 10 m deep, and a coastal ocean having a bottom that slopes from a 10-m depth at the coast to a 50-m depth 15 km offshore. The model domain is 20 km in each alongshore direction and 25 km in the offshore

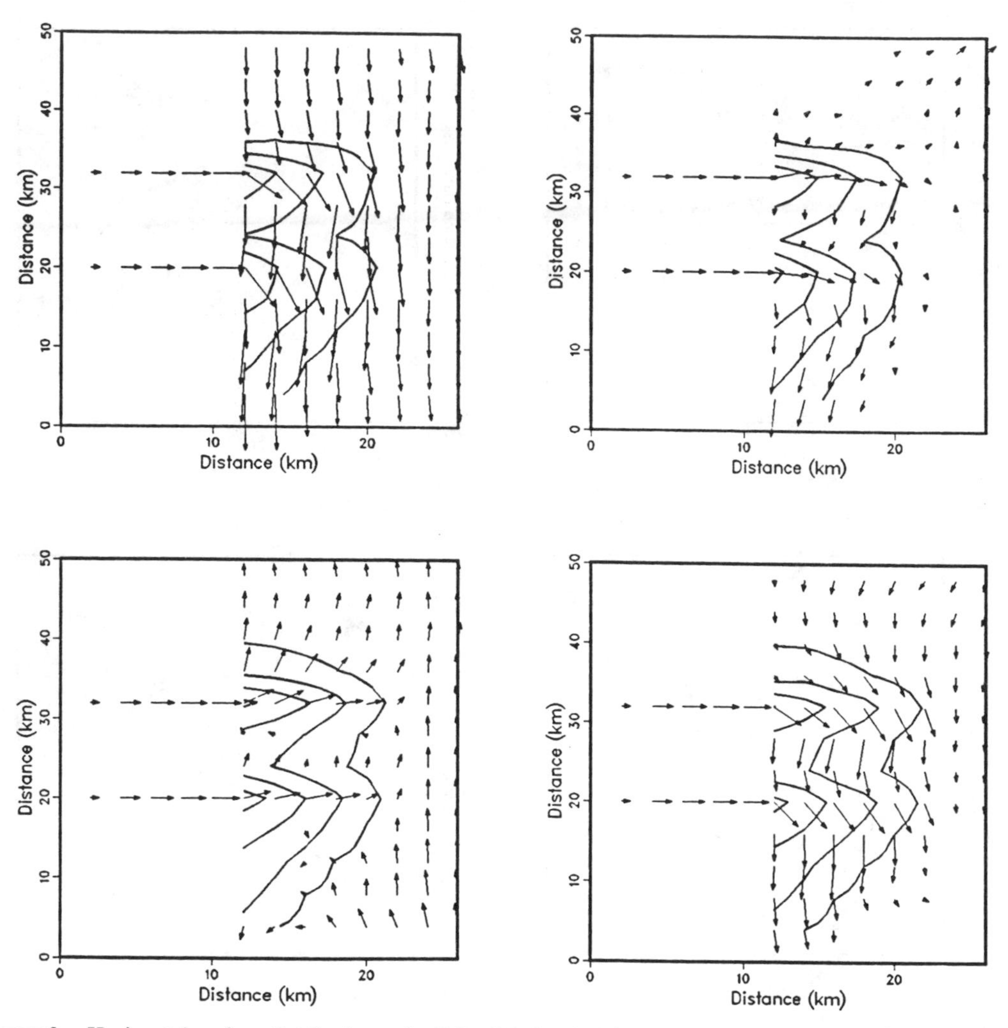

FIGURE 3.—Horizontal surface distributions of salinity (2‰ isopleth interval, continuous lines) and water velocity (vector length between two grids, 20 cm/s) in a model two-inlet system during four tidal phases. Inlet mouths are 12 km apart and enter the ocean at kilometer 10 along the horizontal axis; the open ocean extends to the right.

direction from the head of the estuary. Because the ocean is not bounded by land, it is essential to specify conditions at the "open" boundaries. At model's offshore open boundary, an alongshore-propagating semidiurnal tide of 20-cm amplitude is specified. At the model's two cross-shelf open boundaries, a radiation condition is specified. A constant river runoff of 200 m^3/s, which is the source of buoyant water for the system, is also specified at the head of the estuary. Initially, the system is at rest, and the entire model domain is filled with ocean water of 33‰ salinity. The model results at the 7th tidal cycle, after the initial adjustment is completed, are described.

The model coastal inlet is small but deep; consequently, the tidal current in the inlet is small, and the vertical mixing is weak. The river discharge is confined to the upper 3 m, forming a thin surface plume. Figure 1 shows the vertical distributions of salinity and longitudinal water velocity along the axis of the inlet during four tidal phases. The flow in the inlet is a classical two-layer gravitational circulation. Because of the small amount of mixing, the surface velocity is considerably larger than the subsurface velocity. The gravitational flow also is much larger than the tidal flow, so the flow variation over a tidal cycle is small.

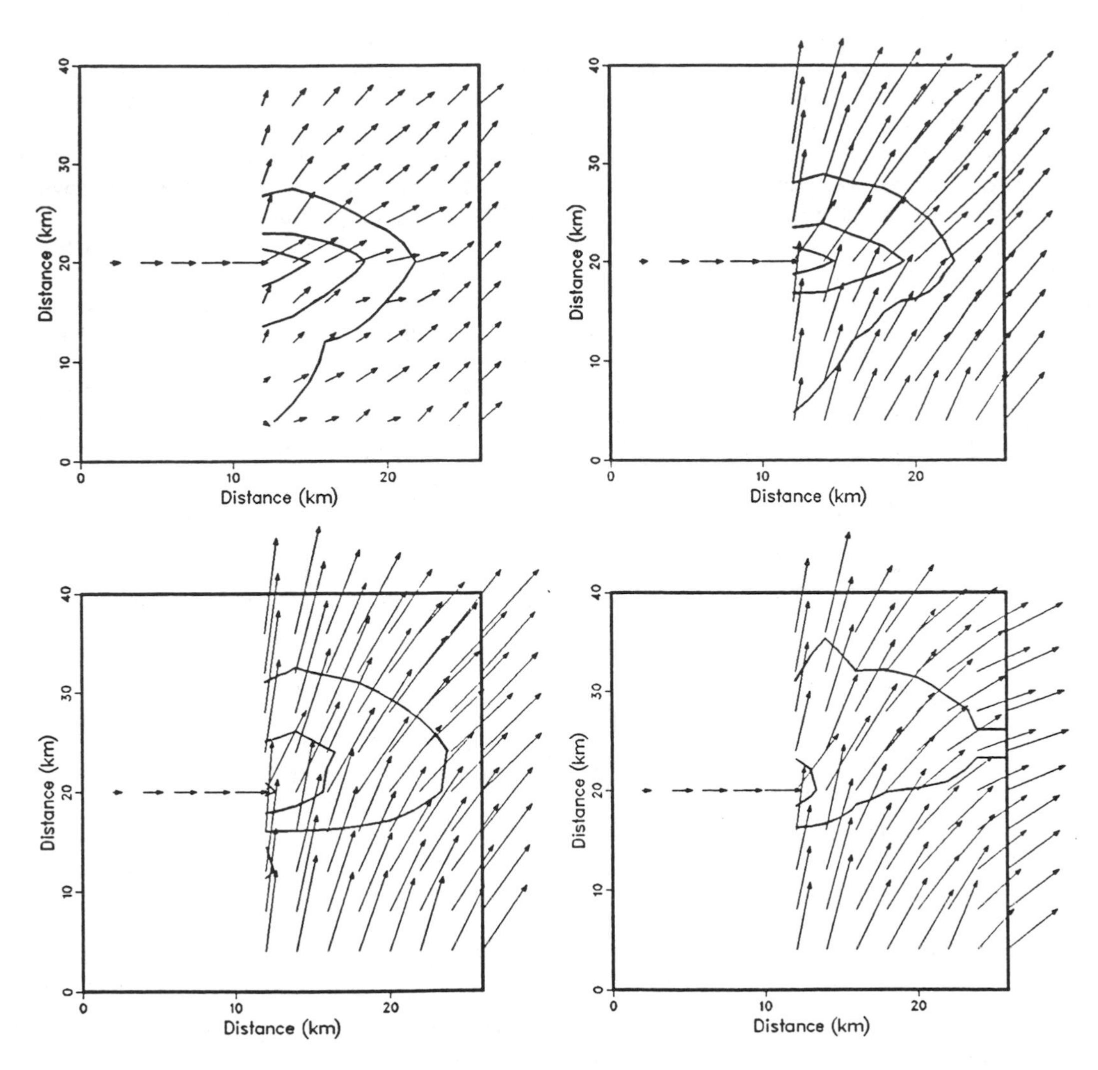

FIGURE 4.—Horizontal surface distributions of salinity (2‰ isopleth interval, continuous lines) and water velocity (vector length between two grids, 20 cm/s) in a model single-inlet system subjected to a constant alongshore wind stress of 1 dyne/cm^2 during four tidal phases. The inlet mouth is at kilometer 10 along the horizontal axis, and the open ocean is to the right.

Figure 2 shows the surface distributions of salinity and horizontal velocity during four tidal phases. Upon exiting the inlet, the fresh-water plume turns to the right (in the northern hemisphere) because of the Earth's rotation. Tidal currents in the coastal ocean are about 20 cm/s, which are similar to currents of the surface plume. Consequently, transport of the outflow plume shows the combined effects of self- and passive advections. To the right of the inlet entrance, the plume interferes with the tide, producing a strong asymmetry of the flow field over a tidal cycle. To the left of the inlet entrance, the plume forms a sharp front, beyond which the flow field is entirely tidal.

Small coastal inlets usually occur in multiples. Hence, plumes from one coastal inlet may interact with plumes from another one. However, because the plume is not merely a passive tracer, the net effect will not be a linear superposition of the two plumes. To study the interaction between multiple inlets and the ocean, the model simulation was repeated with two inlets separated by 12 km, each inlet identical to the one used in the previous simulation. Figure 3 shows the surface distributions of salinity and horizontal velocity during four tidal phases. The two plumes tend to overlap, forming a band of low-salinity water along the coast. Thus, when two inlets are near each other, there is no clear distinction between their respective plumes; rather, the two point sources act as if a line source is distributed uniformly along the coast.

Alongshore winds can drastically change circulation in the shallow coastal zone. To demonstrate the effect of wind, the previous model simulation of a single inlet was continued with a constant alongshore wind of 1 dyne/cm^2 starting with the 8th tidal cycle. Figure 4 shows the surface distributions of salinity and horizontal velocity at four tidal phases; the wind starts at the beginning of the tidal cycle. The surface flow follows the direction of wind, but it also has an offshore component (an Ekman flow) due to the Earth's rotation. The wind-driven flow is considerably larger than the tidal flow so there is no flow reversal over a tidal cycle. The surface salinity distribution is strongly affected by the Ekman flow so the plume spreads out offshore and its concentration decreases considerably.

Discussion

Transport processes in an inlet–ocean system are complex and variable. This study showed examples of the interaction between inlet discharge plumes and tidal currents, the interference of multiple inlets, and the wind-driven flow superposed on a discharge plume. No attempt was made to quantify the model simulation. Rather, the objective was to demonstrate the physical processes that are important for the transport of particles. Given the transient nature of transport processes, a reliable estimate of particle exchanges between inlets and ocean must derive from a large number of flow simulations with real external forcings.

Coastal inlets are important migratory pathways for fish larvae. Because of their low swimming ability, small larvae essentially move with the current as passive particles. The transport of larvae into the inlet depends on the circulation near the inlet entrance and the local density of larvae. The water movement through the inlet entrance should be nearly random when averaged over many episodes. On the other hand, the spatial density of larvae will not be homogeneous, because of the asymmetrical flow pattern of the discharge plume. The plume forms a strong convergence zone ("front"), which will concentrate larvae near the inlet entrance. A well-developed plume causes a higher density of larvae, which makes transport of larvae into the inlet more efficient. In contrast, a diluted plume reduces the transport efficiency for larvae.

Jetty construction near the inlet entrance will displace the discharge plume off the coast. If the seaward extension of the jetty is much smaller than the dimensions of the plume, the jetty will not have important effects on the plume. On the other hand, if the jetty and the plume have a similar seaward dimension, the jetty will distort and enlarge the plume. Most likely, the discharge plume will become more diluted when there is a jetty. Consequently, the efficiency of transport of larvae may be reduced by jetty construction. This suggestion is tentative; nevertheless, the need for a better understanding of the plume circulation and of the relation between density of larvae and plume circulation is clearly demonstrated.

Acknowledgments

The manuscript was prepared while I was a visiting Chair in the Department of Oceanography, Naval Postgraduate School. M. Weinstein and R. Kendall offered useful suggestions during the revision of this manuscript.

References

Butler, H. L. 1980. Evolution of a numerical model for simulating long-period wave behavior in ocean–estuarine systems. Pages 147–182 *in* P. Hamilton and K. B. Macdonald, editors. Estuarine and wetland processes. Plenum, New York.

Cheng, R. T., and V. Casulli. 1982. On Lagrangian residual currents with application in south San Francisco Bay, California. Water Resources Research 18:1652–1662.

Seliger, H. H., J. A. Boggs, and W. H. Biggley. 1985. Catastrophic anoxia in the Chesapeake Bay in 1984. Science (Washington, D.C.) 228:70–73.

Taft, J. L., and D.-P. Wang. 1982. Vertical mixing and nutrient transport in the Chesapeake Bay. Maryland Power Plant Siting Program, RRRP-58, Annapolis.

Wang, D.-P. 1982. Development of a three-dimensional limited-area (island) shelf circulation model. Journal of Physical Oceanography 12:605–617.

Wang, D.-P. 1983. Two-dimensional salt intrusion model. American Society of Civil Engineers, Journal of Waterway, Port, Coastal and Ocean Division 109:103–112.

Wang, D.-P. 1985a. A far-field model of the regional influence of effluent plumes from ocean thermal

conversion (OTEC) plants. Argonne National Laboratory, ANL/OTEC-EV-3, Argonne, Illinois.
Wang, D.-P. 1985b. Numerical study of gravity currents in a channel. Journal of Physical Oceanography 15:299–305.
Wang, D.-P., and D. W. Kravitz. 1980. A semi-implicit two-dimensional model of estuarine circulation. Journal of Physical Oceanography 10:441–454.

American Fisheries Society Symposium 3:16–25, 1988

Observations on Inlet Flow Patterns Derived from Numerical and Physical Modeling Studies

WILLIAM C. SEABERGH

Coastal Engineering Research Center, U. S. Army Engineer Waterways Experiment Station Post Office Box 631, Vicksburg, Mississippi 39180, USA

Abstract.—Due to its responsibility to provide navigable entrance channels through coastal inlets, the U.S. Army Corps of Engineers has conducted numerous site-specific model studies of tidal inlets and has performed generalized research on tidal inlets. This paper presents some insights about flow patterns at tidal inlets, and the factors that influence those patterns, that have arisen from the model studies.

The U.S. Army Corps of Engineers has responsibility for providing functional and structural design criteria at many tidal inlets on the coasts of the United States. If entrance channels at tidal inlets cannot be maintained economically at depths suitable for the boats and ships that must use the inlet, jetty structures may be proposed in order to remove or minimize the effects of coastal sediment filling; jetties also protect boats from wave action as they navigate the entrance. Many Corps of Engineers projects involve relatively shallow sandy inlets, such as Oregon Inlet, North Carolina (Figure 1), where wave action and tidal currents create shallow shifting channels surrounded by breaking wave conditions, making navigation difficult. In order to aid in the design of inlet projects, the U.S. Army Engineer Waterways Experiment Station (WES) conducts numerical and physical model studies of tidal inlets. In this paper, I discuss flow patterns in and near some of those modeled inlets, patterns that may be relevant to the transport of fish larvae through inlets.

Inlet Characteristics

Tidal inlets generally have a relatively short, narrow channel passing through a sandy barrier island and connect the ocean to a relatively large (or long) bay. The bay is usually small enough (on the order of tens of kilometers or less) for the water surface to rise and fall (co-oscillate) relatively uniformly in response to the forcing ocean tide. Larger estuaries sometimes have broader junctions with the sea and may be long enough (hundreds of kilometers) to contain nearly an entire tidal wave length.

Hydrodynamic conditions at tidal inlets can vary from a relatively simple ebb-and-flood tidal system to a very complex one involving tides, wind stresses, freshwater influxes, and surface gravity waves (4- to 25-s periods). Flows through inlets associated with large open bays and small tidal amplitudes can be dominated by wind stress; ebb conditions can last for days when winds pile up water near the bay side of the inlet, or long floods can occur when winds force bay water away from the inlet. Most inlet bays, however, are small and some are covered with vegetation, so wind stress does not dominate there. Although many bays do not receive much fresh water relative to the volume of tidal flow, substantial inputs of fresh water can create vertically stratified flows through the tidal inlet. Convergence of flows from several directions at either side of the inlet can create strong turbulence that scours the channel deeply through the narrowest part of the inlet and silts in the channel on bay and ocean sides. Maximum depths generally in the range of 6–15 m may occur in such channels, whereas seaward channel depths may diminish to 1.5–3 m. Waves breaking on ocean beaches can generate alongshore currents; when these intersect with currents caused by waves breaking over sand shoals near the inlet, flow patterns can be very complex on the exposed ocean side of the inlet. Inside the inlet, water may diverge into one or more channels among shoal areas created by the deposition of sand from the ocean beaches. The natural inlet can vary in location over time (Figure 2), and natural changes can be greater than human-made changes.

Inlet Variables

Key parameters are the cross-sectional area at the minimum inlet width (A_c), channel length (L), and channel-controlling depth (DCC; Figure 3). The ebb tidal delta area (AED) is the area seaward of the inlet bounded by the shoreline, the contour

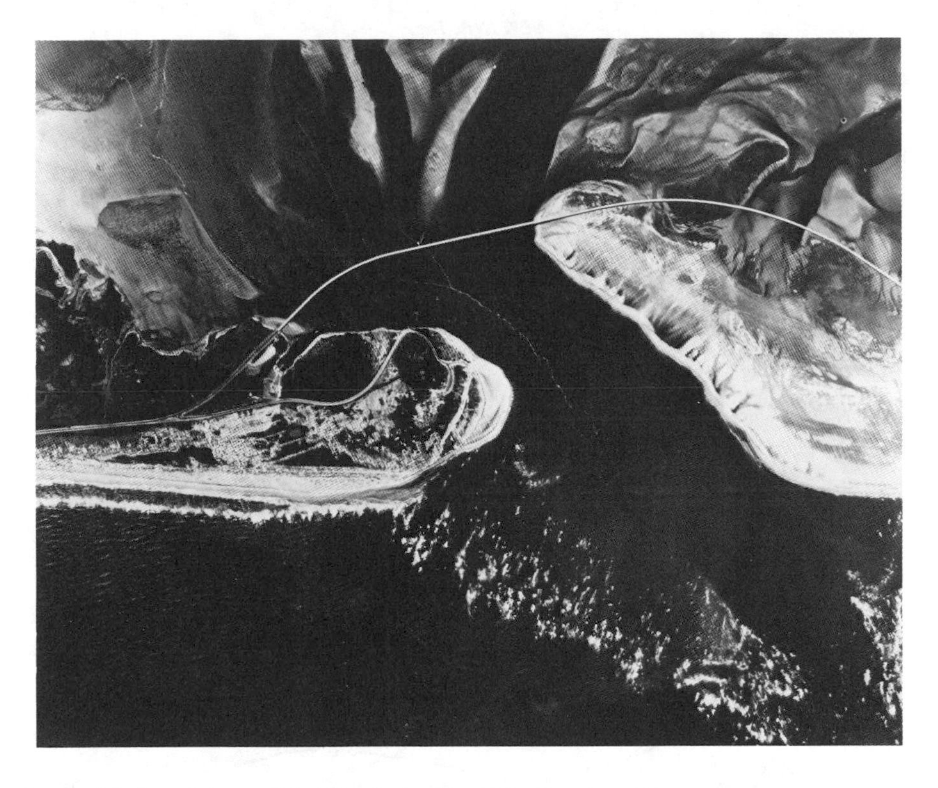

FIGURE 1.—Aerial view of Oregon Inlet, North Carolina, February 12, 1976, an example of a continuously shifting tidal inlet. Oregon Inlet connects Pamlico Sound (top, west) with the Atlantic Ocean (bottom, east); it presently is crossed by a traffic causeway and bridge.

depth at the crest of the outer bar in the channel (DCC) to a point where it parallels the shoreline, and the line of the inlet minimum width. Both AED and L vary in log-linear relation to A_c over several orders of magnitude (Figure 4), based on 67 inlets examined by Vincent and Corson (1980) that had been subjected to little or no human modification.

The addition of jetty structures to natural inlets can be expected to cause some modification of the inlet's morphology. Careful engineering design can reduce the amount of induced change. For example, jetties are placed so that the minimum cross-sectional area of the inlet is maintained, leaving the volume of water exchanged over a tidal cycle (i.e., the tidal prism) unchanged from the natural state. This is based on O'Brien's (1931) observation that there is a direct relationship between the inlet's minimum cross-sectional flow area, A (this is the minimum area, not necessarily the area at the location of the minimum width A_c), and the tidal prism (P) filling the bay (Figure 5). If the same minimum area is maintained between the entrance channel's jetties that existed for the natural inlet, the tidal prism will be the same, and tides will flush out the bay behind the inlet as well as they did in the natural state. Actually, a more hydraulically efficient channel usually will exist at a jettied inlet because sediment influx is reduced and there are fewer shoals. Also, the relationship between cross-sectional area and tidal prism can be combined with the integrated flow continuity to determine that the maximum velocity (due to astronomical tides, no freshwater flow) through the minimum inlet cross-section will be about 1.0 m/s.

Several other principles of tidal inlet geomorphology, determined by Dean and Walton (1973) and Walton and Adams (1976), should be recognized when inlets are to be modified for navigation. For example, outer ocean bars increase in size as wave exposure (or average wave energy) decreases. Also, the outer ocean bar volume increases for increasing tidal prism. Finally, bay shoals are larger in regions of high wave energy.

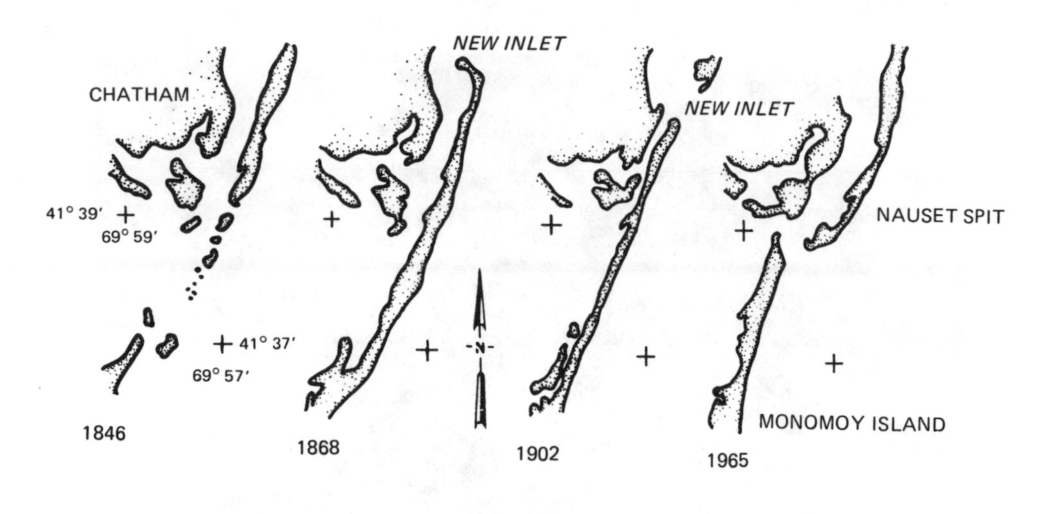

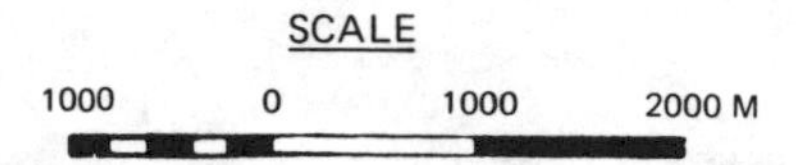

FIGURE 2.—Historical changes of inlet bathymetry in the Monomay–Nauset Inlet, Cape Cod, Massachusetts, 1846–1965. (After Hayes 1971.)

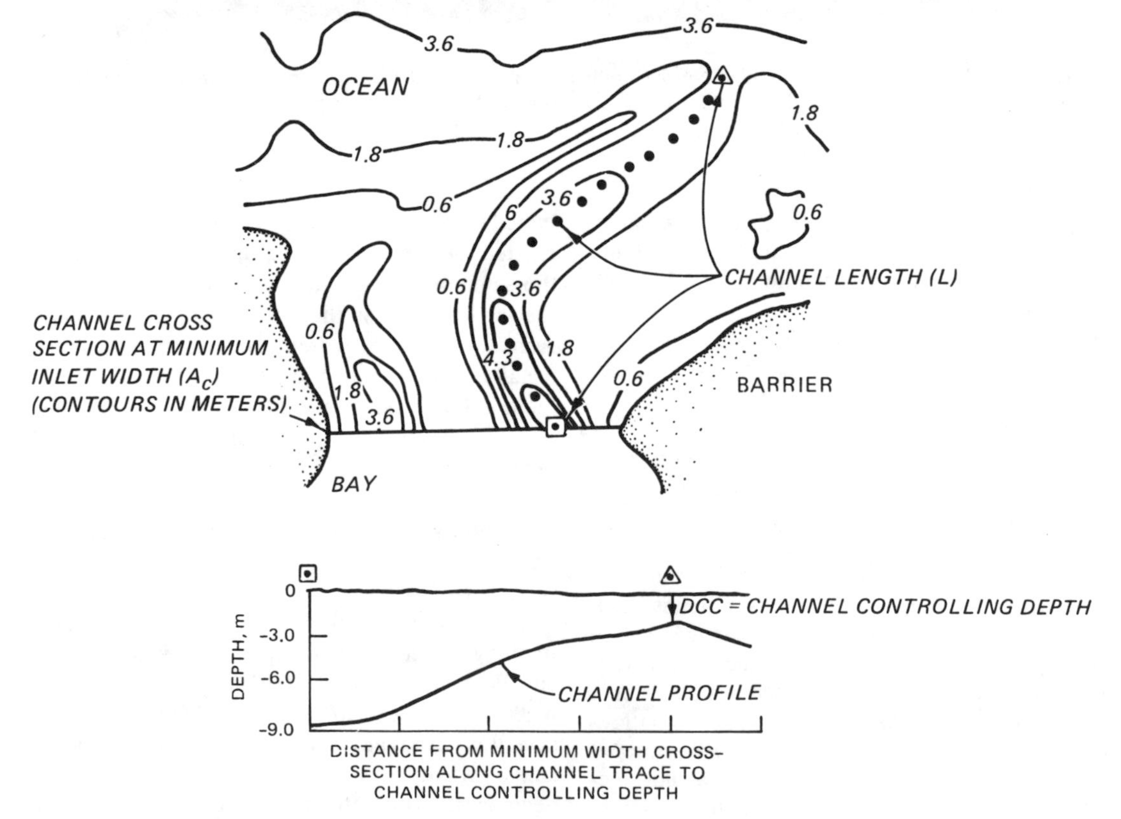

FIGURE 3.—Inlet channel parameters: A_c (m^2) channel cross-section at minimum inlet width; L (m), channel length; DCC (m), channel-controlling depth. (From Vincent and Corson 1980.)

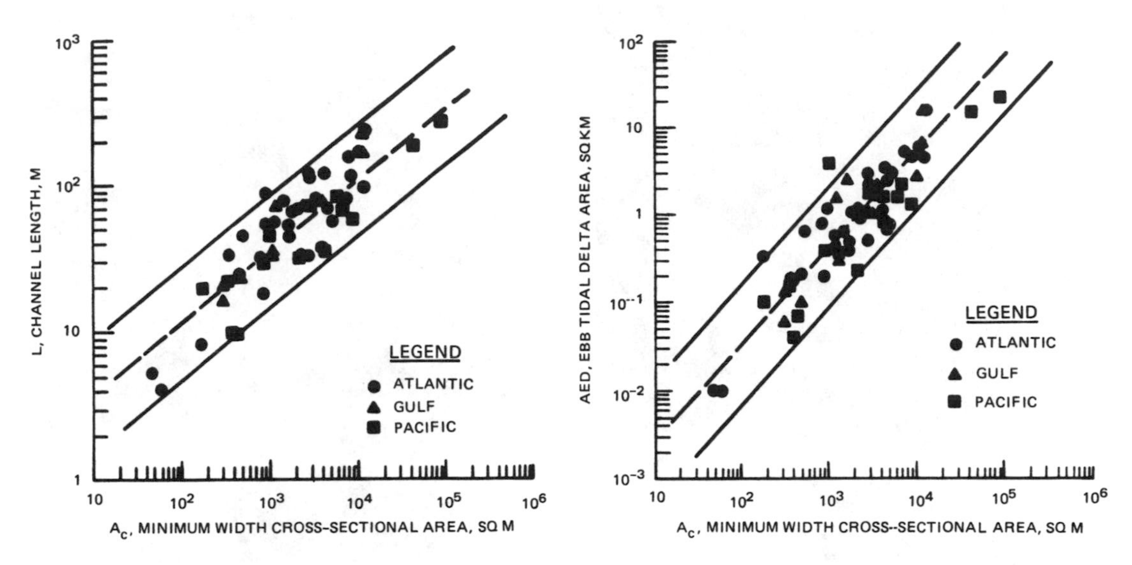

FIGURE 4.—Cross-sectional area at the minimum inlet width (A_c, m) versus channel length (L, m) and the ebb tidal delta area (AED, km^2) The dashed line is a linear best-fit curve and the solid lines are 95% confidence bands. (From Vincent and Corson 1980.)

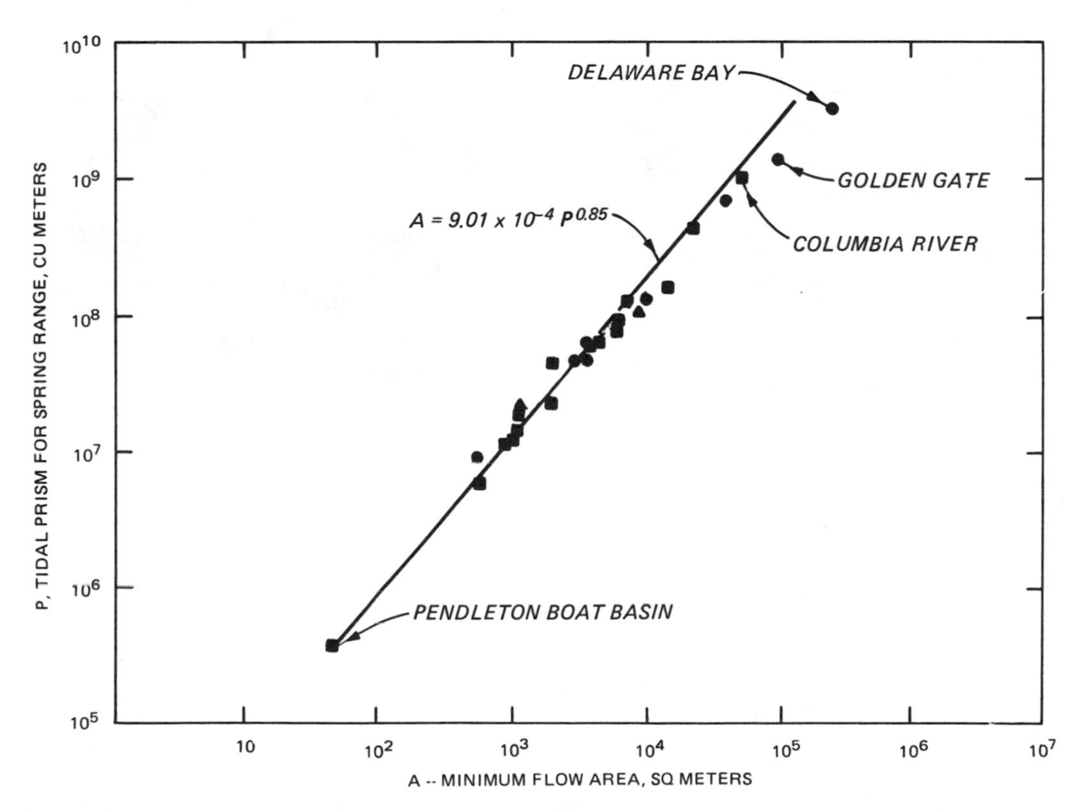

FIGURE 5.—A, the minimum cross-sectional flow area (m^2) of an inlet versus P, the tidal prism (m^3) of the spring tidal range.

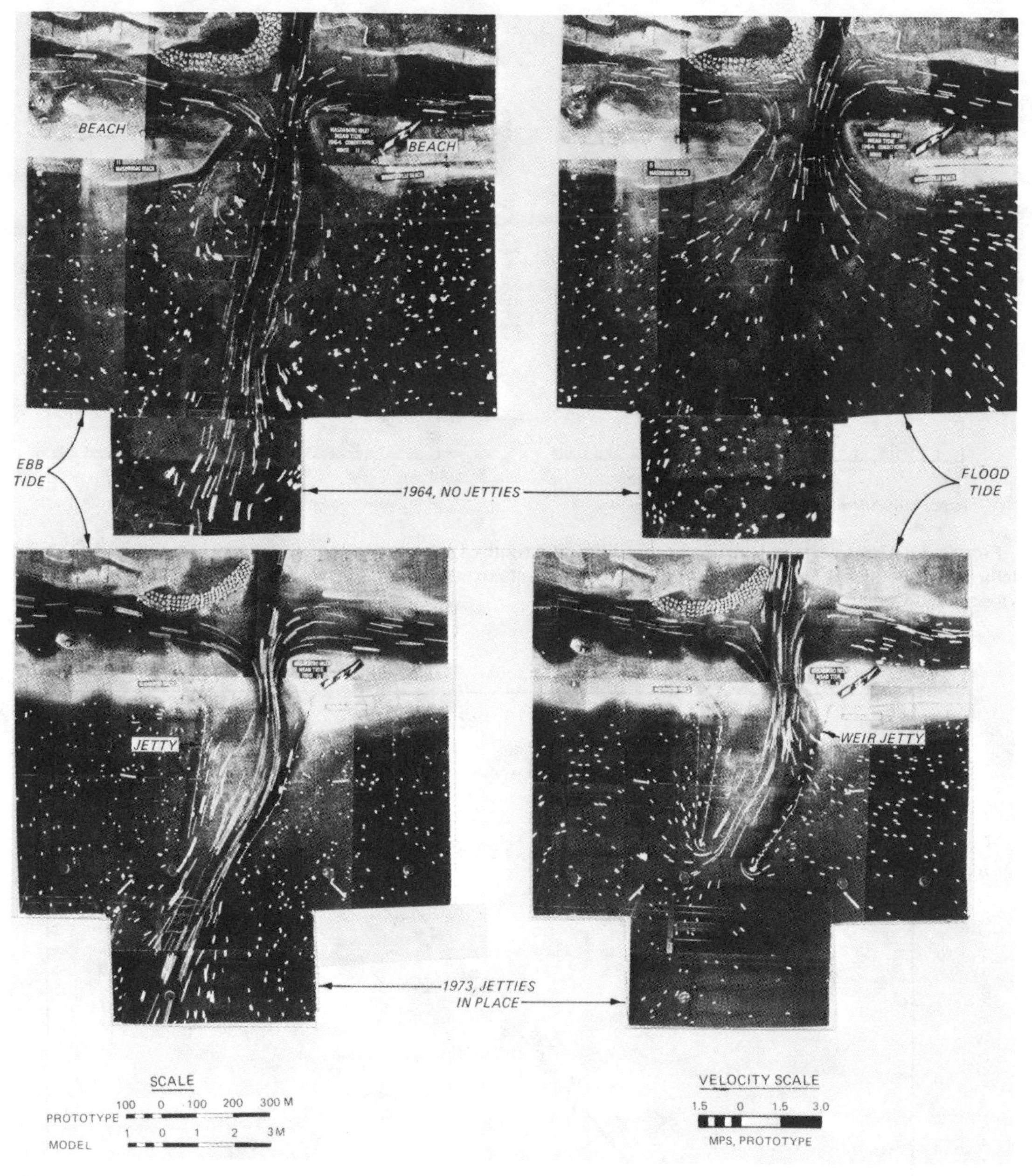

FIGURE 6.—Surface current patterns at Masonboro Inlet, North Carolina, during maximum ebb (left) and flood (right) tides before (1964) and after (1973) jetties were installed at either side of the channel on the ocean side of the inlet. The streaks are surface floats (4-s exposures) on aerial photographs; streak length is proportional to the speed of a float's motion.

Inlet Flow Patterns

An inlet has a "gorge" where flows converge before they expand again on the opposite side. Shoal (shallow) areas that extend bayward and oceanward from the gorge depend on inlet hydraulics, wave conditions, and general geomorphology. All these interact to determine flow patterns in and around the inlet and where main flow channels occur. Some analytical approaches to flow patterns at inlets treat parameters affecting the "plume" of water exiting the inlet (i.e., ebb flow). Joshi (1982) examined the effect of bottom

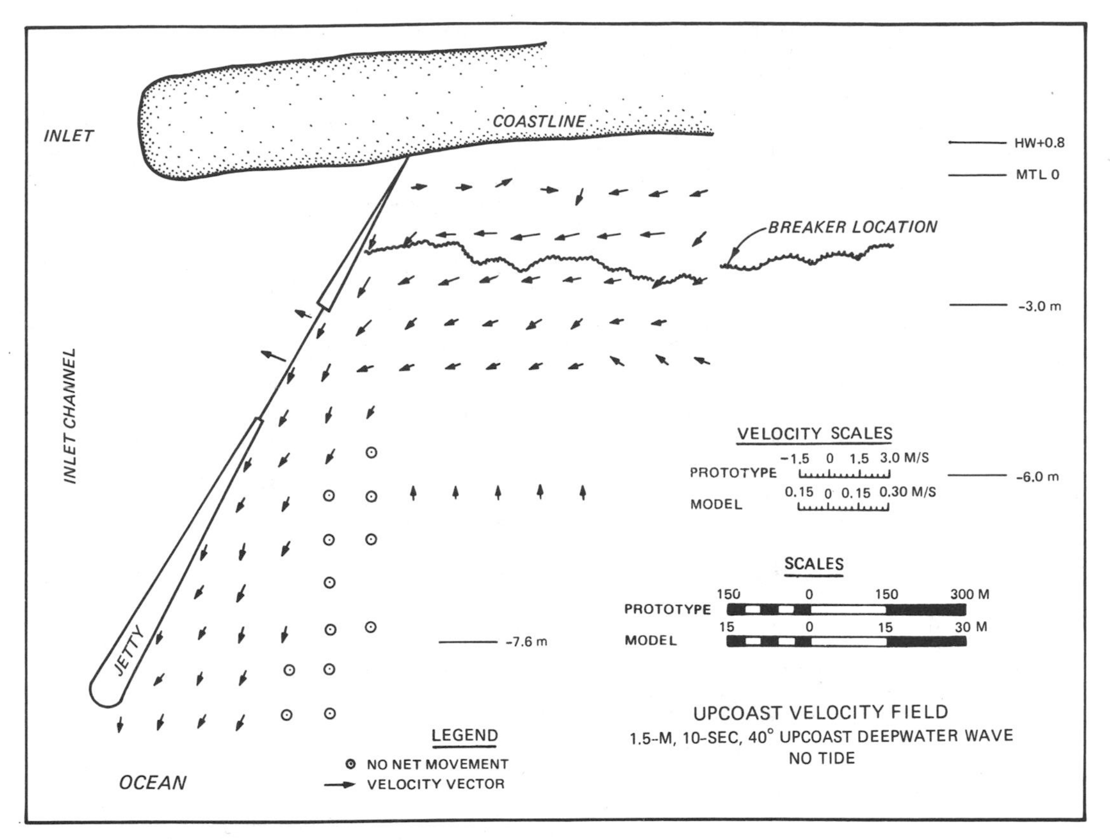

FIGURE 7.—Model of current patterns resulting from waves breaking along the shoreline. A strong current is moving oceanward along the jetty structure. (This is for wave-generated currents only; no tidal currents were reproduced in the model for this test.) For the depth scale at the right margin, HW is high water, and MTL is mean tide level.

friction, bathymetric changes, and lateral entrainment of water due to turbulent mixing. Generally, as bottom friction increases, the plume becomes wider and water velocity along the center line decreases. As bottom slope increases, the plume contracts. Joshi and Taylor (1983) examined the alongshore circulation created by tidal jets and predicted that currents toward the inlet along the shoreline (induced by the ebb tidal jet or plume) can range from 0.01 to 0.1 m/s within 10 half-widths (of the minimum inlet width) of the natural inlet entrance. With a jetty in place, the range of induced currents along the shoreline is slightly less than this from six inlet-widths away and further. Within six inlet widths, the current along the shoreline is reduced, finally becoming zero at the junction of the shoreline and the jetty. The above work is based on a simplified bathymetry and does not include wave-induced longshore currents, which can be an order of magnitude larger than the jet-induced longshore currents. Ismail and Wiegel (1983) found that the waves increased the rate at which the ebb jet spreads.

Surface flow patterns during physical model studies are obtained from aerial 4-s time exposures of surface floats; Figure 6 shows examples for Masonboro Inlet, North Carolina, representing natural and jettied conditions during maximum ebb and flood flows (Seabergh 1976). Ebb flow through the natural inlet was channelized between the shallow shoal areas on each side of the channel, whereas flooding waters moved in broad current patterns toward the inlet. The channel incised in the shoals reduced entrainment of nearshore water, resulting in slower flooding flows along the shoreline toward the inlet than predicted by the analytical model's smooth offshore profiles. Most of the shear force of the ebbing jet was expended on the sides of the channel rather than in the entire water column. Two jetties were then installed; the south one (left in Figure 6) is solid, but the north one is a "weir" jetty that permits

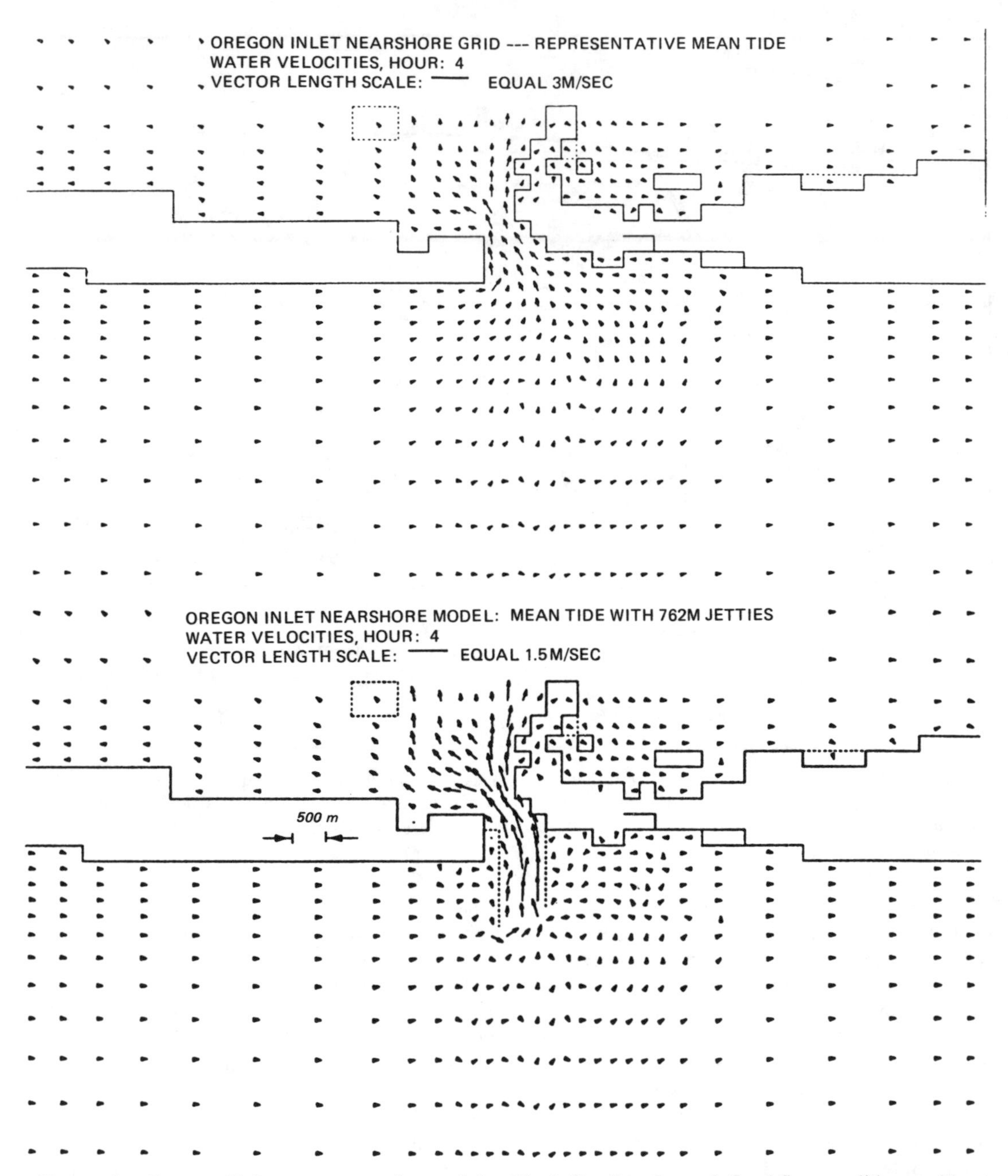

FIGURE 8.—Coarse-grid flow patterns at Oregon Inlet, North Carolina, for peak flood flow conditions, without (top) and with jetties (bottom).

flow over its shoreward third when the water level reaches mid tide. Flood tides with the jetties in place showed a fairly strong current effect on the south side. Flow over the north weir created a slightly slower flow toward the inlet on the shoreward portion of the flow field.

Figure 7 shows physical model velocity data (Seabergh 1983) on the oceanward side of a weir jetty for wave-generated currents. It illustrates the strength and extent of influence of the velocity field associated with 1.5-m-high waves breaking on a smoothly sloping beach. For these tests, no tides were reproduced in the model. The tests indicate that substantial current effects can occur

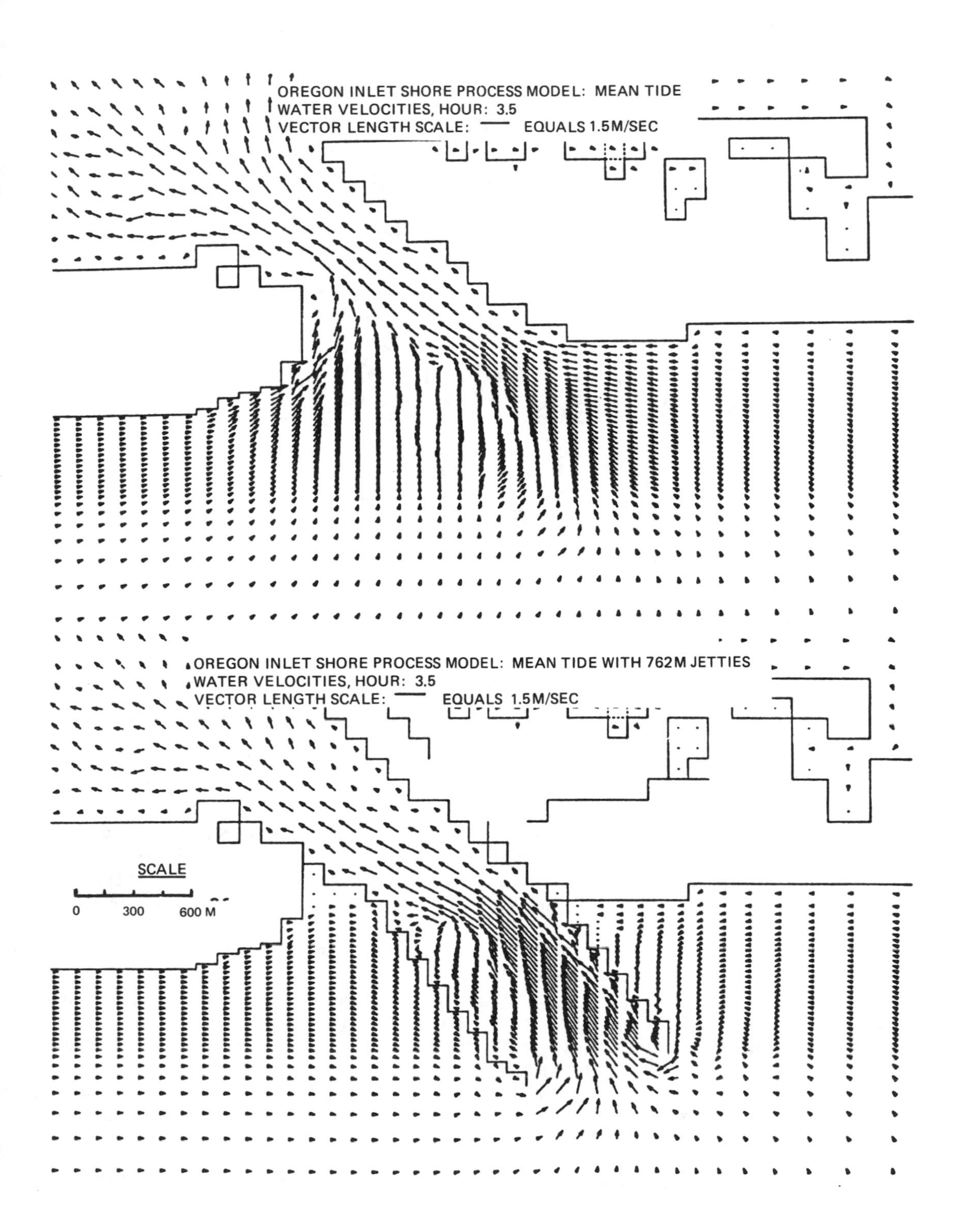

FIGURE 9.—Fine-grid flow patterns at Oregon Inlet, North Carolina, for peak flood flow conditions, without (top) and with jetties (bottom).

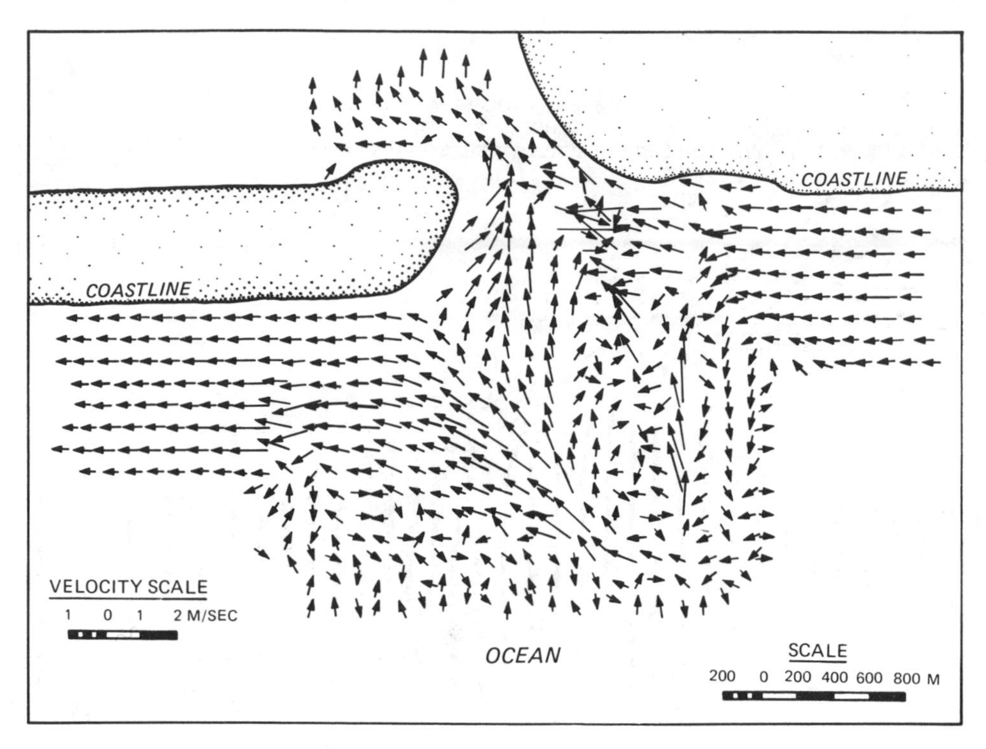

FIGURE 10.—Wave-generated currents at Oregon Inlet, North Carolina as determined by the CURRENT numerical model.

outside the jetties when nearby wave energies are moderate to large.

In recent years, the Corps of Engineers has used numerical models to study flow patterns in tests of jetty designs. Figure 8 shows flow patterns at Oregon Inlet (Leenknecht et al. 1984) for existing flood tide conditions and for a postulated jettied condition. Velocity vector plots were created with a two-dimensional depth-integrated model known as the WES implicit flooding model (WIFM). There are two velocity scales on the illustration, but the effects of jetties on water velocity and direction can be inferred easily. A rough estimate is that the jetties will influence flood flow, relative to flow through the natural inlet, for about one inlet width around the jetty system, i.e., upcoast and downcoast of the jetties and offshore from the tip of the jetties. This model grid was used to provide boundary conditions for a fine-scale grid of inlet shore processes. Figure 9 provides a closer look at flow patterns near the inlet with the use of the fine-scale model. There are variations in currents adjacent to the shore between unjettied and jettied conditions. Longshore flow patterns adjacent to the upcoast and downcoast shoreline for flood flow conditions are similar except in the immediate vicinity of the inlet entrance. Figure 10 shows wave-induced currents occurring at the natural inlet determined from a numerical model known as CURRENT (Vemulakonda 1984). The currents result from 3.5-m-high waves approaching the shoreline obliquely and breaking at 8-s intervals due to shallow depths. As noted previously in the discussion of physical models, wave-generated currents can markedly affect flow patterns in shallow water near inlets.

The Corps of Engineers also has a three-dimensional numerical model known as CELC3D to simulate coastal currents and sediment dispersion (Sheng and Butler 1982). This model resolves currents driven by tide, wind, and density gradients. It has not been used in a tidal inlet study because most inlets recently studied were well-mixed systems without important variations in velocity through the water column.

Modeling Approaches to Larval Transport

Numerical and physical models have been used to study tidal inlets for many years. A first try at examining larval transport at a tidal inlet could involve a two-dimensional depth-averaged numerical model applied to the immediate vicinity of the inlet. Two somewhat idealized inlets could be examined: the natural inlet and the same inlet with jetties and a shoreline that has adjusted to them. The bathymetry should be realistic. Sand bar forms and channels incised through shoals should be reproduced instead of a smoothly sloping offshore bathymetry. If models such as WIFM were used for numerous setups, varying grid blocks could be tagged with a conservative dye to represent locations of larvae, and appropriate dispersion coefficients could be chosen in order to follow movements of larval patches for several tidal cycles. Three-dimensional situations where salinity stratification (for example) may occur could be modeled in a similar manner with CELC3D. As various grid cells are examined, probabilities that a larva will be successfully transported through the inlet could be calculated for the jettied and nonjettied inlet to determine if significant differences occur.

The next step in a numerical approach would be to impose short-period wave effects, which would have substantial effects on grid cells adjacent to the shoreline. Input to these finely gridded models may have to come from a larger regional grid that accounts for wind effects on water movement.

Conclusions

Tidal inlet systems can be viewed hydrodynamically as varying from relatively simple flow orifices to rather complex flow systems as the number of influencing parameters increases. Wind waves, freshwater inflow, and wind stress on the water surface can cause significant effects. The scale of inlets extends from the very small to the very large and the extent of their influence along the coast varies proportionately to their size. As structures (jetties) are added to an inlet for producing channel stability and navigation safety, local changes to the flow patterns occur. At this stage of research, it would be difficult to determine the net effect that hydrodynamics would have on larval recruitment into the estuary.

Acknowledgments

The Office of Chief of Engineers authorized publication of this paper. Helpful comments by H. L. Butler and C. E. Chatham are appreciated. Janie Daughtry typed the manuscript.

References

Dean, R. G., and T. L. Walton, Jr. 1973. Sediment transport processes in the vicinity of inlets with special references to sand trapping. Estuarine Research 2:129–149.

Hayes, M. O. 1971. Lecture notes for course on inlet mechanics and design 10–20 May 1971. (Unpublished; available from U.S. Army Engineer Waterways Experiment Station, Vicksburg, Mississippi.)

Ismail, N. M., and R. L. Wiegel. 1983. Opposing wave effect on momentum jets spreading rate. ASCE (American Society of Civil Engineers) Journal of Waterway, Port, Coastal and Ocean Engineering 109:465–483.

Joshi, P. B. 1982. Hydromechanics of tidal jets. ASCE (American Society of Civil Engineers) Journal of Waterway, Port, Coastal and Ocean Engineering 108:239–253.

Joshi, P. B., and R. B. Taylor. 1983. Circulation induced by tidal jets. ASCE (American Society of Civil Engineers) Journal of Waterway, Port, Coastal and Ocean Engineering 109:445–464.

Leenknecht, D. H., J. A. Earickson, and H. L. Butler. 1984. Numerical simulation of Oregon Inlet control structures' effects on storm and tide elevations in Pamlico Sound. U.S. Army Engineer Waterways Experiment Station, Technical Report CERC-84-2, Vicksburg, Mississippi.

O'Brien, M. P. 1931. Estuary tidal prisms related to entrance areas. Civil Engineering 1:738–739.

Seabergh, W. C. 1976. Improvements for Masonboro Inlet, North Carolina, volumes 1 and 2. U.S. Army Engineer Waterways Experiment Station, Technical Report H-76-4, Vicksburg, Mississippi.

Seabergh, W. C. 1983. Weir jetty performance: hydraulic and sedimentary considerations. U.S. Army Engineer Waterways Experiment Station, Technical Report HL-83-5, Vicksburg, Mississippi.

Sheng, Y. P., and H. L. Butler. 1982. A three-dimensional hydrodynamic model for coastal, estuarine and lake currents. Pages 531–574 *in* Proceedings of the 1982 Army numerical analysis and computer conference. U.S. Army Research Office, Report 82-3, Research Triangle Park, North Carolina.

Vemulakonda, S. R. 1984. Erosion control of scour during construction, report 7. CURRENT—a wave induced current model. U.S. Army Engineer Waterways Experiment Station, Technical Report HL-80-3, Vicksburg, Mississippi.

Vincent, C. L., and W. D. Corson. 1980. The geometry of selected U.S. tidal inlets. Report 20 *in* General investigation of tidal inlets. U.S. Army Coastal Engineering Research Center, Fort Belvoir, Virginia, and U.S. Army Engineer Waterways Experiment Station, Vicksburg, Mississippi.

Walton, T. L., and W. D. Adams. 1976. Capacity of inlet outer bars to store sand. Fourteenth Coastal Engineering Conference, Honolulu, Hawaii.

American Fisheries Society Symposium 3:26–33, 1988

Sampling Optimization for Studies of Tidal Transport in Estuaries[1]

BJÖRN KJERFVE[2] AND T. G. WOLAVER

Belle W. Baruch Institute for Marine Biology and Coastal Research
University of South Carolina, Columbia, South Carolina 29208, USA

Abstract.—Measurements of material and water transport in estuarine cross sections are costly and require a substantial effort. It is desirable to optimize the sampling design by reducing the effort and cost without losing important information in the process. To illustrate how to optimize a sampling design in a tidal transport study, we calculated and analyzed fluxes of water, nitrogen (nitrate plus nitrite), and particulate organic carbon (POC) through a cross section of a South Carolina marsh creek during two tidal cycles. Discharge explained 95% of the variation in nitrate plus nitrite flux and 92% of the variation in POC flux. The remaining fluxes were presumed to result from biogeochemical marsh–estuary processes. Because nontidal transport is usually of great interest yet is often masked by tidal dynamics, it is essential to make careful flow measurements to assess biogeochemical processes from direct flux measurements.

The direct measurement of transports in tidal creeks and estuaries is not a trivial undertaking. Nixon (1980) reviewed several projects aimed at determining estuary–coastal exchanges and pointed out the lack of well-designed long-term studies. Indirect transport estimation is not a simple matter either, and it is affected by simplifying assumptions and much uncertainty. In general, water and material transport measurements in tidal creeks are made difficult by the reversing tidal currents that are orders of magnitude greater than the underlying net flow. Gravitational flow driven by freshwater input, residual tidal flow, wind drift, and estuarine storage and emptying due to dynamic coupling to the coastal ocean are physical mechanisms causing net water movements. Material deposition or erosion from adjacent wetlands is related to the time–velocity asymmetry of the currents (Bayliss-Smith et al. 1979), which can control the overall concentrations of dissolved and suspended materials in the adjacent estuarine waters. The change in distribution of these constituents in an estuary over one or more tidal cycles depends on advection, dispersion due to tidal flow oscillations, lateral and vertical shear effects, and turbulence (e.g., Dyer 1974). The end result is that net water movements and net movements of each dissolved or suspended constituent seldom coincide in magnitude and direction. To obtain reliable net transport estimates from direct measurements requires a careful design to account for flow and concentration variability with respect to time as well as to the depth and width of the water body (e.g., Kjerfve et al. 1981).

Bly Creek is an arm of the North Inlet marsh–estuarine system located on the northeastern coast of South Carolina (Figure 1). The Bly Creek Ecosystem Study was designed to evaluate the effects of a vegetated marsh, an oyster reef community, and terrestrial freshwater runoff on seasonal and annual material transport in a tidal salt marsh estuarine basin (Dame et al. 1984, 1985; Wolaver et al. 1985; and Chrzanowski and Zingmark 1986). The cumulative effects of internal material processing were determined from outputs by the system and inputs to the system. These fluxes were measured directly along a single cross-sectional transect of Bly Creek.

An important consideration in studies of this sort is the trade-off between sampling effort, which can be very costly when dynamic processes are investigated, and sampling error, which can be unacceptably high if sampling effort is too low. Many studies have been conducted on material transport through tidal creeks (Nixon 1980), but only a few have addressed the critical problem of spatial and temporal sampling errors (Boon 1978; Kjerfve et al. 1981). The main purpose of this paper is to describe an intensive calibration program used to formulate a long-term sampling design that is optimal in terms of cost, feasibility, and errors.

Methods

The calibration study was conducted along a transect across the mouth of the Bly Creek basin

---

[1]Contribution 654 from the Belle W. Baruch Institute for Marine Biology and Coastal Research.

[2]Also Department of Geology and Marine Science Program, University of South Carolina.

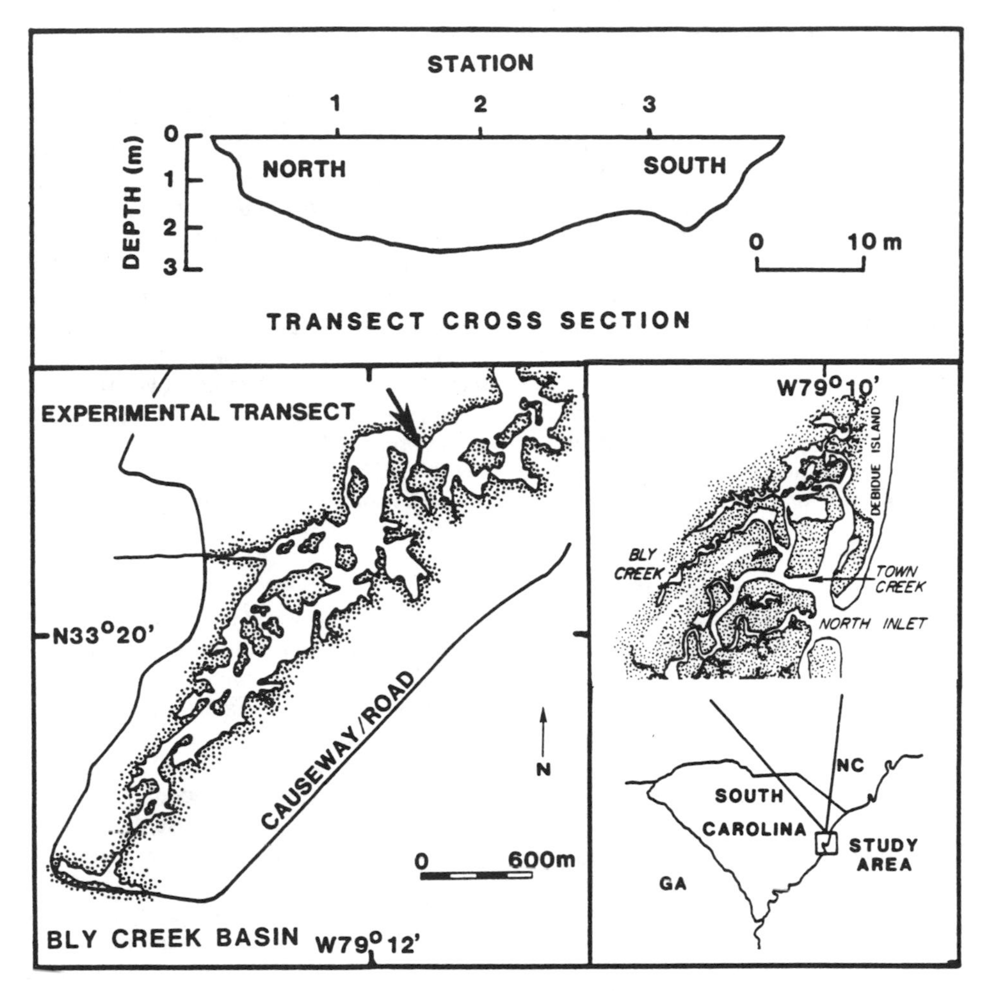

FIGURE 1.—Area map of Bly Creek and North Inlet, showing the location and cross section of the Bly Creek transect.

(Figure 1). The basin is rectangular and approximately 450 m wide and 1,600 m long. It is closed at the upper end by a causeway and bounded on two sides by Pleistocene beach ridges vegetated with loblolly pine *Pinus taeda*. The interior of the basin is largely covered by the cord grass *Spartina alterniflora*.

Because of the enclosed nature of the basin, the tidal prism, which is the total water volume between consecutive high and low tides, is exchanged only through the experimental transect. Fresh water enters the upper basin from groundwater or intermittent stream flow, but it only amounts to about 1% of the tidal prism. At mean low water, the water volume remaining in the basin is approximately 0.7% of the water volume of the full basin (Eiser and Kjerfve 1986). Flow dynamics in the Bly Creek system are essentially tidally driven.

Calibration measurements were made in the Bly Creek transect over two tidal cycles (from 1200 hours on 15 October to 1240 hours on 16 October 1982). The experimental procedures were similar to those of the North Inlet calibration experiment (Kjerfve et al. 1981). The variables measured were water discharge and fluxes of particulate organic carbon (POC) and nitrogen in the form of nitrate plus nitrite (NO_3^- + NO_2^-). Both POC and NO_3^- + NO_2^- are nonconservative constituents, i.e, their concentrations in the water can be altered through biogeochemical processes. We intended to measure salinity as a conservative constituent whose concentration in the water can change only as a result of mixing and diffusion processes, but problems with the salinometers prevented this.

The Bly Creek transect (Figure 1) was 53 m wide and an average of 1.33 m deep at mean tide height. The tidal range measured 1.53 m and did not vary appreciably between the two tidal cycles. Based on tidal measurements every 20 min, the mean (SD) cross-sectional area was 70.3 (23.1) m^2.

We selected three stations along the transect to be sampled every 20 min over the two tidal cycles. Measurements were taken at three depths (surface, mid, and bottom) when the station depth was greater than 1.5 m, at two depths (surface and bottom) when station depth was 0.5–1.5 m, and at one depth (surface) when station depth was less than 0.5 m. This sampling density represented the longest duration, the most number of stations and depths, and the fastest sampling rate logistically feasible for the calibration experiment.

Water speeds were measured with current crosses (Kjerfve 1982), and flow directions were estimated with a surface streamer and compass. The velocity vectors were decomposed into an along-channel u-component (positive for ebb) and an across-channel v-component (positive towards the northern bank). Because the transect was near a channel bend, the least-squares optimized creek flow-axis differed by 16° between ebbing and flooding tides. The flow was towards 107° true on ebbing tides and towards 271° true on flooding tides. Thus, different flow decomposition angles were chosen for ebb and flood tides. Only the along-creek flows (u) were needed for flux computations; they (and their root-mean-square deviations both in centimeters per second) were 1.7 (31.4) at station 1, 4.9 (41.8) at station 2, and 6.3 (48.2) at station 3. The positive values indicate net flow out of the basin.

Instantaneous water discharge through the transect was calculated according to Kjerfve (1979). A cubic spline was fitted to each vertical profile of the u-component velocity, 11 equidistant interpolated values were calculated for each profile, each value was cross multiplied by the assigned station width and time-varying depth, and the resulting product was summed over the total cross section to yield instantaneous discharge (Figure 2).

Instantaneous fluxes of $NO_3^- + NO_2^-$ and POC were calculated by spline-fitting the concentration values the same way as for velocities, cross-multiplying velocity, concentration, width, and depth and then summing over the cross section. Nitrate plus nitrite nitrogen was determined with an autoanalyzer system (Armstrong et al. 1967). Particulate organic carbon was determined by the dry combustion technique with CuO as an oxidant; the resultant CO_2 was measured by infrared absorption on an Oceanographic International Carbon Analyzer 523 C.

To illustrate the variability in flux intensity, concentrations, and velocity, we constructed cross-sectional maps of time-averaged (net) values and root-mean-square deviations (Figure 3). These maps incorporate the flux due to the time-varying water depth (Stokes' drift) in such a way that the net fluxes integrated over the mean area are identical to the corresponding time averages of the previously computed instantaneous fluxes. The technique for computing the flux-preserving net and root-mean-square maps was detailed by Kjerfve and Seim (1984).

Calibration Procedure

J. D. Spurrier and Kjerfve (unpublished) have calculated that stratified sampling conducted over 34 tidal cycles during 1 year would be an optimum strategy for reliable estimates within logistic constraints. To make measurements every 20 min at three stations for that many tidal cycles is, however, neither feasible nor economical. Thus, we were at first forced to decrease the temporal and spatial sampling density without losing essential information.

With the intensive sampling data in hand, we explored ways in which the sampling effort could be reduced for long-term monitoring without serious loss of information or increase in error. The ideal case is to use data from all three stations. Therefore, we first examined the effects of dropping one or two stations from the analysis. The discharge and material fluxes, both net and root-mean-square deviations from the net values, were then recomputed for the two calibration experiment tidal cycles. The results for the ideal case (three stations), the best case based on two stations, and each case based on a single station are summarized in Table 1.

Paired t-test comparisons of cases 2–5 with case 1 led to rejection of case 5 (station 3 alone). We also used linear regression to evaluate how well cases 2–5 predicted the ideal case. We regressed 38 instantaneous values for the ideal case on the corresponding values for each of cases 2–5 for water, $NO_3^- + NO_2^-$, and POC fluxes. A case was rejected if any of its flux regressions had (1) an r^2 value less than 0.94, (2) an r value outside the range 0.8–1.2, or (3) an absolute value of the

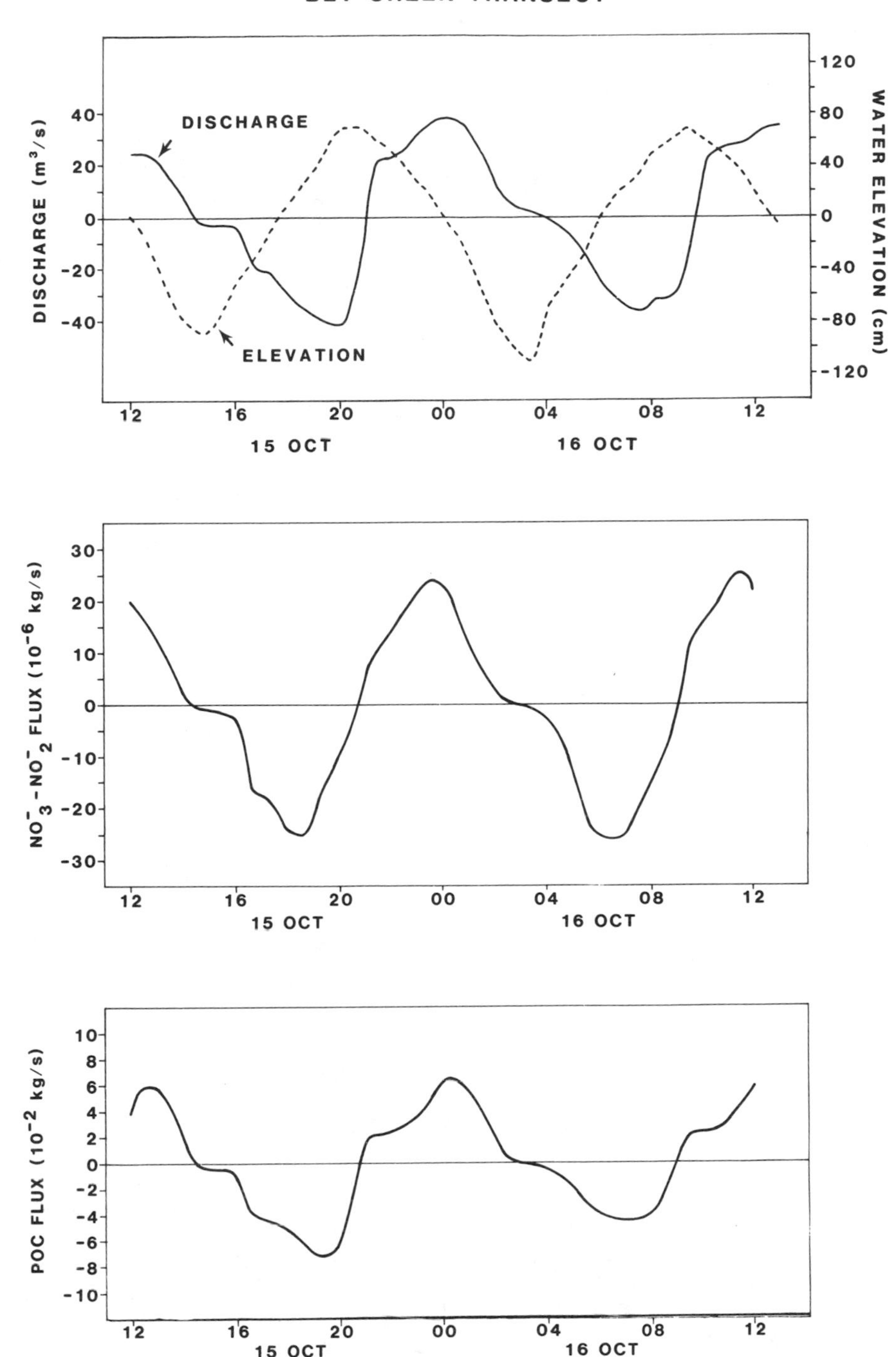

FIGURE 2.—Discharge of water, nitrogen (as NO_3^- + NO_2^-), and particulate organic carbon (POC) through the Bly Creek transect during the calibration experiment in 1982. Positive values indicate flow out of the creek basin. The tidal sequence is superimposed on the discharge graph.

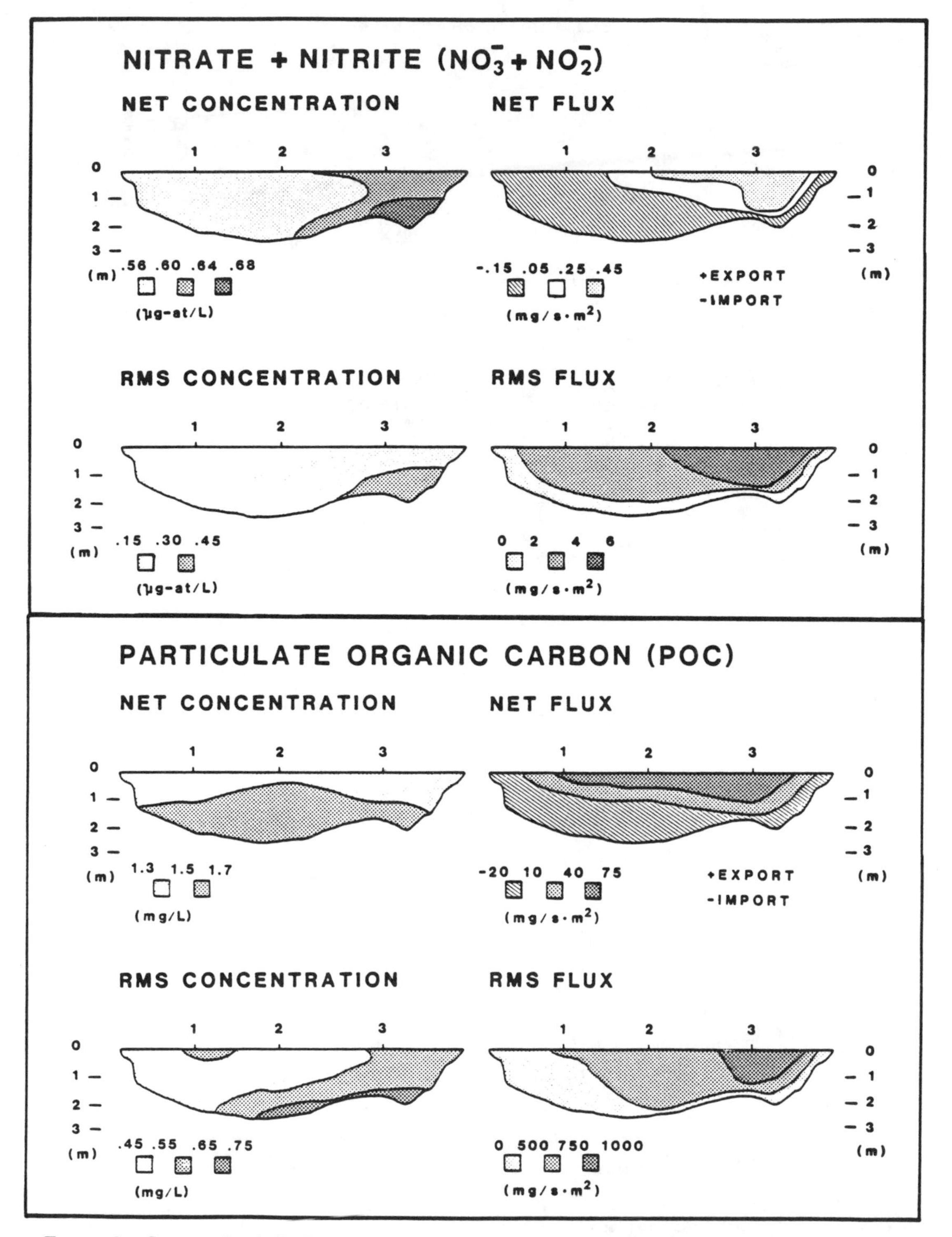

FIGURE 3.—Cross-sectional distribution of the net and root-mean-square (rms) concentrations and fluxes of nitrogen (nitrate plus nitrite, µg-atoms/L) and particulate organic carbon (POC) at the Bly Creek transect during the 1982 calibration experiment.

TABLE 1.—Computed net water discharges and net fluxes of dissolved nitrogen ($NO_3^- + NO_2^-$) and particulate organic carbon (POC) based on various stations along the Bly Creek transect during the 1982 calibration experiment. Positive discharge or flux is ebb-directed (export) and negative flux is flood-directed (import). Root-mean-square deviations are in parentheses. Percentage errors of discharge and fluxes for cases 2–5, relative to case 1, are given also.

Case	Stations included	Discharge (m^3/s)	Nitrogen flux (10^{-6} kg/s)	POC flux (10^{-2} kg/s)	Percentage error, γ[a]		
					Discharge	Nitrogen	POC
1 (ideal)	1, 2, 3	1.2 (26)	0.33 (16)	0.14 (4.0)			
2	1, 3	1.0 (24)	0.50 (15)	0.15 (3.7)	8	9	13
3	1	0.5 (21)	−0.43 (13)	0.06 (3.3)	25	27	34
4	2	1.6 (29)	−0.14 (19)	0.08 (4.8)	15	17	22
5	3	1.4 (27)	1.40 (17)	0.23 (4.2)	14	18	17

[a] $\gamma = 100\ D/P$; D = root-mean-square deviation between case 1 and other cases; P = average tidal prism for the two tidal cycles.

intercept that exceeded the corresponding root-mean-square by more than 2.5%. The regression analyses led to rejection of cases 3 and 5.

Kjerfve et al. (1981) defined a relative percentage error, $\gamma = 100\ D/P$, indicating the error of a particular case relative to the ideal case (case 1); P is the average tidal prism for the two tidal cycles, computed separately for water, $NO_3^- + NO_2^-$, and POC; D is the root-mean-square deviation for the integrated water, $NO_3^- + NO_2^-$, and POC fluxes, respectively, between the ideal case and any other case. The errors were particularly great for case 3 (Table 1).

Because cases 3 and 5 have already been rejected, only cases 2 and 4 are considered further. The average percentage error for all constituents was 10% for case 2 and 18% for case 4, but case 2 required two stations, whereas case 4 required only one. In view of the relatively small information gain from selection of two stations instead of one, we adopted case 4, thereby selecting one station near the middle of the channel in the continued long-term measurement effort for the Bly Creek Ecosystem Study.

The regression analysis indicated that case 4 systematically overestimated the ideal transport. The regression coefficients were 0.86, 0.86, and 0.80 for water, $NO_3^- + NO_2^-$, and POC fluxes, respectively. In the subsequent long-term measurement program, we therefore multiplied each instantaneous flux estimate by a constant 0.84 to unbias the cross-sectional flux computed with only a single station.

Finally, it was necessary to choose an optimum sampling rate for the long-term sampling. To reduce the cost and effort of sampling every 20 min during a tidal cycle, we recalculated fluxes on the basis of 40- and 60-min sampling intervals over the two tidal cycles, as Kjerfve et al. (1982) did for several other North Inlet transects. A sampling rate of once an hour did not alter the calculated fluxes significantly from those based on 20-min intervals. Accordingly, we adopted a sampling interval of 60 min for the continuation of the Bly Creek Ecosystem Study.

Discussion

The purpose of the Bly Creek calibration study was not to estimate long-term net material transport rates, for which measurements over only two tidal cycles are inadequate, but to optimize the sampling design for future long-term measurements. In that respect, our data are more than adequate.

Because field measurements of material transports in tidal creeks are expensive to make and require a large effort to execute well, it is desirable to optimize the sampling design. This implies that the duration between sample times is maximized, the number of measurement points is minimized, and the duration of the study is as short as possible to yield unbiased and adequate information without adversely affecting the quality of the results. We focused on determining the optimal sampling rate and minimizing the number of measurement points. The question of an adequate sampling duration will be addressed in a separate paper by Spurrier and Kjerfve.

Selection of a single transect station and a sampling rate of one measurement set per hour was acceptable for our purposes. This design produced a relative potential error (γ) of 18% compared to sampling three stations every 20 min. The γ value should be interpreted as the maximum possible error, rather than the average error, because of the way it is computed (Kjerfve et al. 1982). Our results are consistent with previous North Inlet area studies (Kjerfve et al. 1981, 1982) and indicate an objective method for optimizing sampling in tidal transport studies.

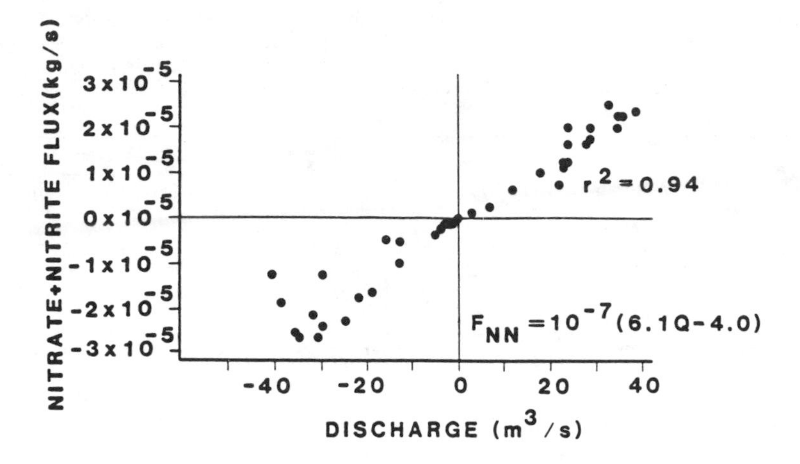

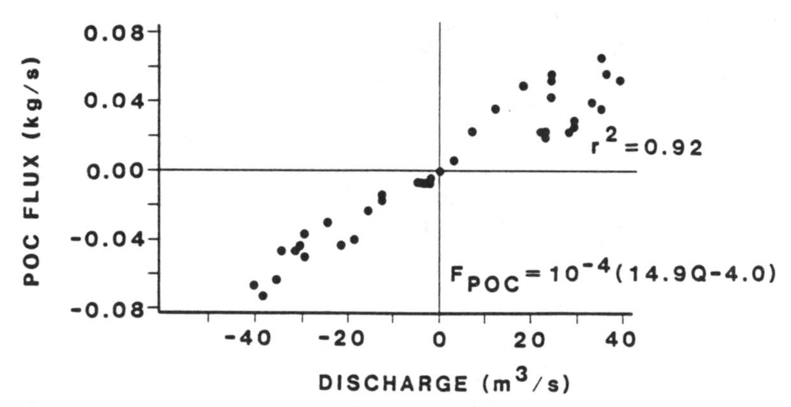

FIGURE 4.—Linear regressions of nitrate plus nitrite flux (F_{NN}) and POC flux (F_{POC}) on discharge (Q) based on data from two experimental tidal cycles at the Bly Creek transect.

Besides serving as a guide to design of optimized long-term mass flux studies, short-term studies like this one can be used for ecological characterization of the system. Within Bly Creek, the NO_3^- + NO_2^- and POC concentrations do not vary systematically over the tidal cycle. For the two experimental tidal cycles examined here, the area-weighted net concentrations (±SD) were 0.59 ± 0.03 μg-atoms N/L (0.0083 ± 0.00045 mg N/L) and 1.50 ± 0.13 mg POC/L, the standard deviations referring to the net spatial deviations from the respective means. The temporal deviations from the tidal averages are considerably larger, on the order of 0.3 μg-atoms N/L (0.004 mg N/L) and 0.6 mg POC/L. The variation in concentrations of these biological constituents is neither large nor systematic. In part, this low variability is probably a function of the time of year of sampling. In the fall, most POC is relatively refractory detrital material and NO_3^- + NO_2^- behaves almost conservatively.

Water flow is the dominant factor controlling the magnitude of both NO_3^- + NO_2^- and POC fluxes (Figure 4). Instantaneous discharge explains 94% of the variability in instantaneous NO_3^- + NO_2^- flux and 92% of instantaneous POC flux. This would not necessarily be the case if the concentration variabilities were greater and more systematic. It is the degree of deviation between a constituent flux and discharge that is of greatest interest. This deviation is an indicator of the magnitude of any biogeochemical process affecting the constituent, but also includes any measurement errors.

On the time scale of two tidal cycles, the instantaneous tidal fluxes dominate and mask concentration changes due to biogeochemical alterations of NO_3^- + NO_2^- and POC (Figure 4).

With data from many tidal cycles, it becomes possible not only to compare net discharge and net constituent flux, but also to evaluate the role of biogeochemical processes within the system relative to net export and import values for individual constituents.

Acknowledgments

We thank Pramot Sojisuporn and H. E. Seim for help with the computations and K. E. Magill for editing the manuscript. In addition, we thank R. D. Dame, T. Williams, R. G. Zingmark, J. D. Spurrier, H. N. McKellar, Jr., T. H. Chrzanowski, and the technicians of the Baruch Institute Coastal Field Laboratory for their help with the sampling. This project was funded by National Science Foundation grant DEB-8119752.

References

Armstrong, F. A. J., C. R. Sterne, and J. D. H. Strickland. 1967. The measurement of upwelling and subsequent biological processes by means of the Technicon autoanalyzer and associated equipment. Deep-Sea Research 14:381–389.

Bayliss-Smith, T. P., R. Healey, R. Lailey, T. Spencer, and D. R. Stoddart. 1979. Tidal flows in salt marsh creeks. Estuarine and Coastal Marine Science 9:235–255.

Boon, J. D., III. 1978. Suspended solids transport in a salt marsh creek—an analysis of errors. Pages 147–175 *in* B. Kjerfve, editor. Estuarine transport processes. University of South Carolina Press, Columbia.

Chrzanowski, T. H., and R. G. Zingmark. 1986. Passive filtering of microbial biomass by *Spartina alterniflora*. Estuarine, Coastal and Shelf Science 22:545–558.

Dame, R. F., T. G. Wolaver, and S. Libes. 1985. The summer uptake and release of nitrogen by an intertidal oyster reef. Netherlands Journal of Sea Research 19:265–268.

Dame, R. F., R. G. Zingmark, and B. Haskin. 1984. Oyster reefs as processors of estuarine materials. Journal of Experimental Marine Biology and Ecology 83:239–247.

Dyer, K. R. 1974. The salt balance in stratified estuaries. Estuarine and Coastal Marine Science 2:273–280.

Eiser, W.C., and B. Kjerfve. 1986. Marsh topography and hypsometric characteristics of a South Carolina salt marsh basin. Estuarine, Coastal and Shelf Science 23:595–605.

Kjerfve, B. 1979. Measurement and analysis of water current, temperature, salinity, and density. Pages 186–226 *in* K. R. Dyer, editor. Hydrography and sedimentation in estuaries. Cambridge University Press, Cambridge, England.

Kjerfve, B. 1982. Calibration of estuarine current crosses. Estuarine, Coastal and Shelf Science 15:553–559.

Kjerfve, B., J. A. Proehl, F. B. Schwing, H. E. Seim, and M. Marozas. 1982. Temporal and spatial considerations in measuring estuarine water fluxes. Pages 37–51 *in* V. S. Kennedy, editor. Estuarine comparisons. Academic Press, New York.

Kjerfve, B., and H. E. Seim. 1984. Construction of net isopleth plots in cross-sections of tidal estuaries. Journal of Marine Research 42:503–508.

Kjerfve, B., L. H. Stevenson, J. A. Proehl, T. H. Chrzanowski, and W. M. Kitchens. 1981. Estimation of material fluxes in an estuarine cross-section: a critical analysis of spatial measurement density and errors. Limnology and Oceanography 26:325–335.

Nixon, S. W. 1980. Between coastal marshes and coastal waters—a review of twenty years of speculation and research in the role of salt marshes in estuarine productivity and water chemistry. Pages 437–525 *in* P. Hamilton and K. B. McDonald, editors. Estuarine and wetland processes. Plenum, New York.

Wolaver, T. G., and eight coauthors. 1985. The flume design—a methodology for evaluating material fluxes between a vegetated salt marsh and the adjacent tidal creek. Journal of Experimental Marine Biology and Ecology 91:281–291.

American Fisheries Society Symposium 3:34–50, 1988

Physical Oceanographic Processes Affecting Larval Transport around and through North Carolina Inlets

LEONARD J. PIETRAFESA AND GERALD S. JANOWITZ

Department of Marine, Earth and Atmospheric Sciences, North Carolina State University Raleigh, North Carolina 27695, USA

Abstract.—Atlantic croaker *Micropogonias undulatus,* flounders *Paralichthys* spp., spot *Leiostomus xanthurus,* and Atlantic menhaden *Brevoortia tyrannus* all spawn in the continental shelf waters of North Carolina during late fall to early winter. The juveniles use the bays and tributaries adjoining estuaries such as Pamlico Sound and the Cape Fear River as nurseries during their first winter and spring. In previous studies of recruitment into the estuaries through barrier island inlets or estuarine mouths, it was assumed that both larvae and juveniles entered the estuaries at the bottom of the water column and moved upstream thereafter. The mechanisms were presumed to be tidal. Larvae can enter Pamlico Sound through Oregon, Hatteras, and Ocracroke inlets not only during flood stages of the tide but also in the presence of favorable ocean-to-estuary sea-level pressure gradients. The Cape Fear River has strong semidiurnal flood and ebb tidal flows and also responds vigorously to one-sided divergences and convergences of the adjacent coastal ocean. Facing seaward, flow at the river mouth is in at the left and out on the right. We conclude that, in addition to flooding tides, nonlocal forcing at the estuary mouths can effect transport of larval fish through the estuary mouths, throughout the entire water column.

In this paper we describe the physical events, and their causes, that are responsible for the transport of larvae into the Pamlico Sound and Cape Fear River estuaries of North Carolina. First we describe the general features of the two systems. We then discuss the circulation within Pamlico Sound and its coupling to coastal Ekman dynamics, which provide the mechanism for larval transport into the sound. Finally we discuss the dynamics of the Cape Fear River estuary.

Estuary Descriptions

Pamlico Sound

Pamlico Sound contains the principal nursery areas for fish juveniles along the mainland periphery of North Carolina. The sound has come under scrutiny in recent years because of increased use of its water and adjacent lands by commercial, municipal, and recreational users and because of its ultimate importance to commercial fishing interests.

Pamlico Sound (Figure 1) is the largest barrier island estuary in the United States. Its approximate dimensions are 140 km in the northeast–southwest direction and 25–55 km in the northwest–southeast direction, and its approximate area is 4,350 km^2 (Roelofs and Bumpus 1953). A maximum depth of about 7 m occurs in the west end of the sound. Shoaling regions are near the mouths of the Neuse and Pamlico rivers and close to the inlets of the Outer Banks. Because of the extensive shoals around the margin and projecting into the sound, the mean depth is about 5 m. The main inlets connecting Pamlico Sound to the Atlantic Ocean are Ocracoke, Hatteras, and Oregon.

Fresh water flows into Pamlico Sound from the Neuse, Tar, Pungo and Pamlico rivers and, via Albemarle, Roanoke, and Croatan sounds to the north, from the Chowan and Roanoke rivers. Evaporation exceeds rainfall in the summer and the converse occurs in winter. Annually, the rainfall into the sound and the evaporation from the sound are nearly equal. The sound is generally isothermal; Roelofs and Bumpus (1953) claimed that vertical temperature differences within the sound do not exceed 2°C. Horizontal temperature gradients occur near the inlets due to temperature differences between the sound and the coastal ocean.

It has been suggested by Roelofs and Bumpus (1953), Posner (1959), and Woods (1967) that wind and freshwater runoff are the factors controlling the horizontal salinity distribution in Pamlico Sound. However, other factors are important. The combination of northerly winds and freshwater inflow from Albemarle Sound drives low-salinity water down into northern Pamlico Sound. Winds from the southwest have the opposite effect. In the southern part of the sound, highest salinities (9–19‰) are found at Ocracoke Inlet. In

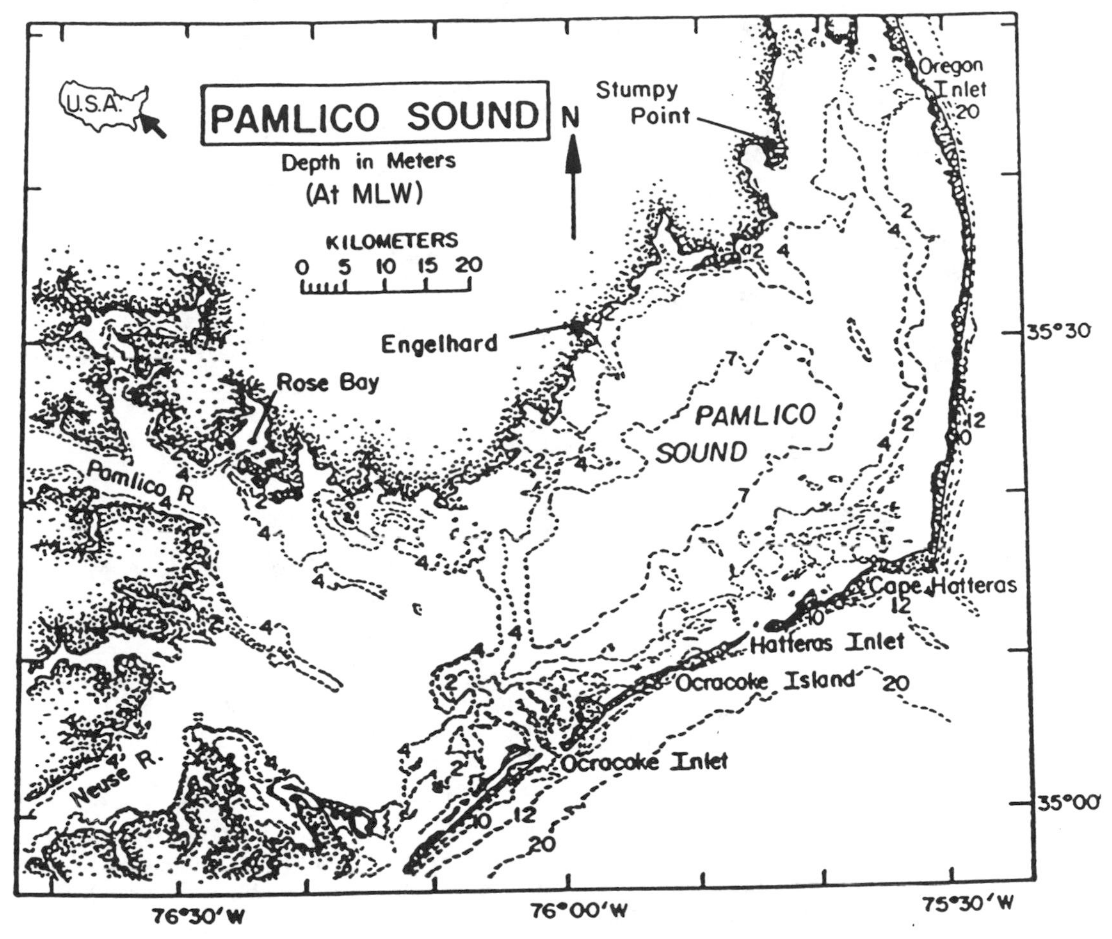

FIGURE 1.—Pamlico Sound, North Carolina. MLW is mean low water.

the northern part of the sound, the lowest salinities occur near Albemarle Sound. The inlets are the source of salt; high-salinity water from the Gulf Stream enters the sound through Ocracoke and Hatteras inlets. As with temperature, salinity tends to be well mixed vertically within the sound. Roelofs and Bumpus (1953) estimated the average surface to bottom salinity differences to be 0.66‰.

Miller et al. (1984) discussed the migratory routes of five species of estuary-dependent fish larvae and juveniles along the North Carolina continental shelf. These five species—Atlantic menhaden *Brevoortia tyrannus,* spot *Leiostomus xanthurus,* Atlantic croaker *Micropogonias undulatus,* southern flounder *Paralichthys lethostigma,* and summer flounder *P. dentatus*—constitute only 10% of fish species that use the sound, but they support 90% of the annual commercial catch in North Carolina coastal waters. All five species spawn during winter near the shelf break at the western wall of the Gulf Stream. Their larvae and juveniles migrate 100 km to major inlets in the barrier islands and then another 25–100 km to nursery areas across Pamlico Sound (Figure 2). Miller et al. (1984) demonstrated the importance of understanding the mechanisms of larval transport through the barrier island inlets. This paper establishes the actual transport mechanisms.

Cape Fear River Estuary

The Cape Fear River (Figure 3) originates in central North Carolina at the junction of the Deep and Haw rivers. It flows generally southeast through the coastal plain province past Wilmington to the Atlantic Ocean, a distance of 684 km. It has a drainage area of about 23,310 km^2. At Wilmington, the upper Cape Fear River and the Northeast Cape Fear River combine to establish the head of the estuary. Below Wilmington, the

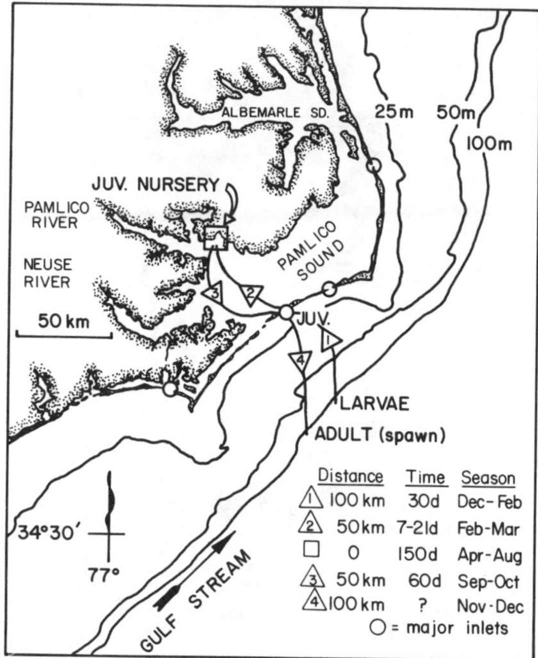

FIGURE 2.—Migration patterns of five dominant fish species off North Carolina. (From Miller et al. 1984.)

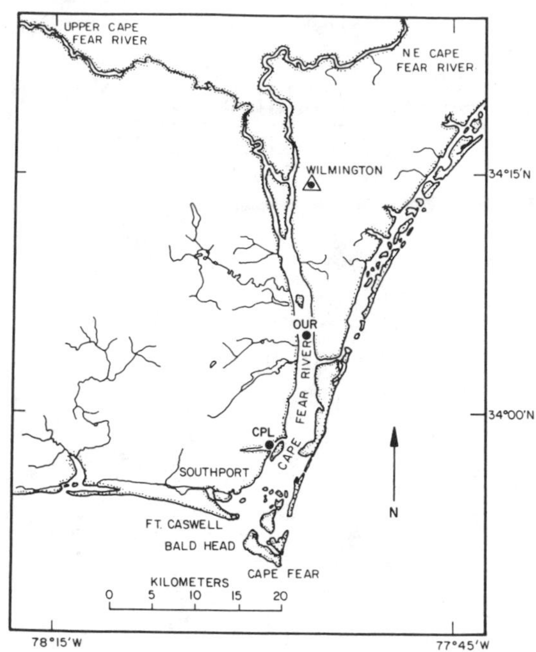

FIGURE 3.—Lower Cape Fear River, North Carolina, showing measurement sites WIL (at Wilmington), OUR, and CPL.

river is a tidal basin about 45 km long with a drainage area of about 906 km^2. The principal feature of the study area is the dredged ship channel, which the U.S. Army Corps of Engineers has dredged and maintained for several years (Welch 1979), the width and depth of which are approximately 90–335 m and 11–12 m, respectively. The ship channel consists of a main channel and extensive shallow areas an order of magnitude wider than the central channel. The average width of the river is about 184 m at Wilmington. Toward the river mouth, the width gradually increases to about 1.6 km at the mouth of the Brunswick River; from this point to the mouth it varies from 1.6 to 4 km. It flows generally south from Wilmington for about 34 km to Federal Point, turns southwest to Southport, and southwest again across the bar to the ocean. The ocean bar is 3.2 km seaward of the river mouth. Most of the width of the estuary is covered by low islands, spoil areas, and tidal flats.

The Cape Fear River estuary is partially mixed and of the coastal plain, drowned river valley type (Dyer 1973). The area affected by the intrusion of salt water varies regularly with the tides. The saline water moves up the estuary during a flood tide in the general shape of a diffuse salt wedge. Due to tidal wind stirring in the river, the wedge is not very sharply defined but is modified to produce a partially mixed condition (Figure 4).

Pamlico Sound

Physical Dynamics

There have been several studies of the physics of Pamlico Sound. One (Pietrafesa et al. 1986) was a study of sea level fluctuations in the basin. Another (Gilliam et al. 1985) was a study of the flow dynamics within and at the mouths of several primary nursery areas and a bay on the southwest coast of the sound. These studies have been used to verify a rudimentary mathematical (actually numerical) model of the subtidal frequency physics of the sound.

Many natural forcing factors could affect water level within Pamlico Sound, including the astronomical tides, changes in water density, river discharge, precipitation and evaporation, atmospheric pressure, and wind stress. It has been suggested that astronomical tides within the Pamlico Sound are negligible except near inlets (Marshall 1951; Posner 1959). Marshall (1951) indicated that the sound was too small to have an appreciable tide generated within it and that any

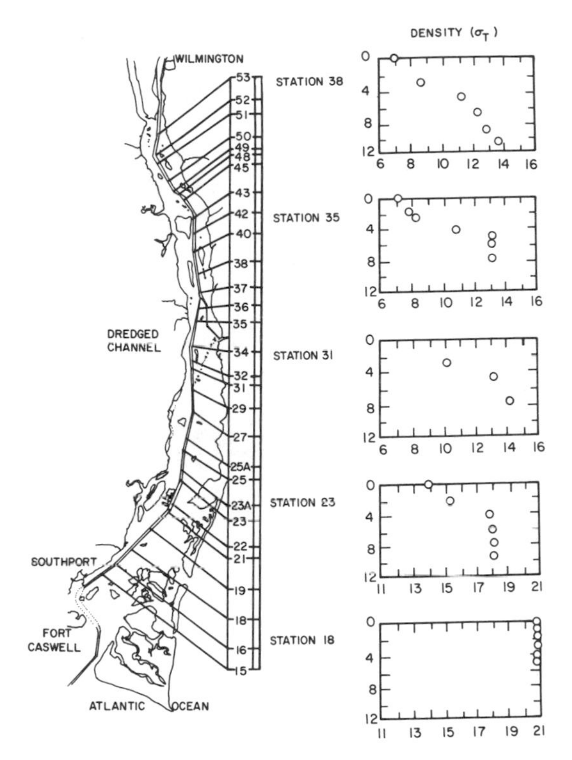

FIGURE 4.—Vertical density profiles in sigma-T units of ‰ along the axis of the Cape Fear River at five representative stations. Depths (vertical axes) are in meters.

tide issuing from the ocean by way of the inlets would be rapidly dampened away from the inlets. However, Singer and Knowles (1975) observed semidiurnal frequency currents in the Neuse River on the mainland side of the Sound.

The field data in Pamlico Sound indicate that it is well mixed in the vertical direction but that horizontal salinity changes can be appreciable over the sound's width (Roelofs and Bumpus 1953; Woods 1967). The horizontal density gradients integrated over depth may be an important local factor in water level variations, but data to verify this are lacking. Although temperature and salinity vary seasonally within Pamlico Sound, steric water level variations are relatively small due to the shallow depth. However, seasonal sea level changes along the coast may be transmitted to the sound water via the inlets.

Atmospheric pressure gradients over the sea surface result in an "inverse barometer effect" wherein a 100-Pa change in atmospheric pressure gives rise to a 0.01-m change in sea level. This is not the result of compressibility but rather an isostatic adjustment. Presumably, the effect breaks down when the atmospheric pressure scales are much larger than the local basins scales. This is the case for Pamlico Sound.

The relationship between wind stress and sea level has been the subject of recent investigations along the Atlantic coast between Cape Hatteras, North Carolina, and Charleston, South Carolina. Chao and Pietrafesa (1980) found that coastal sea level responded to local alongshore synoptic time scale winds within 8–10 h of the wind's onset. This is probably true over the entire North Atlantic seaboard.

Water level investigations in the Pamlico Sound have been limited to numerical simulations based upon simplifications of vertically integrated momentum and continuity equations (Jarret 1966; Smallwood and Amein 1967; Hammack 1969; Chu 1970; Amein 1971; Airan 1974; Amein and Airan 1976). They all support the observation of Roelofs and Bumpus (1953) that the wind is the major factor influencing circulation in Pamlico Sound. Nonetheless, these models all fail in their basic physics for a specific reason described below. We now introduce a more applicable model to define the physics of Pamlico Sound.

Wind-Driven Circulation Model

We consider the subtidal frequency time dependent circulation in Pamlico Sound (the circulation with time scales exceeding 12.4 h) to be due to the atmospheric wind field and take the water density field to be uniform. The ratio of the pressure gradient due to horizontal density variations to the pressure gradient due to water level slope for Pamlico Sound is 0.05, so a homogeneous model is acceptable. Also, because the ratio of water level variation η' to mean water depth H, is generally less than 0.01, and because nonlinearities in the dynamics are proportional to η'/H, the model is assumed to be linear. It is of note that spatial variations of salinity and temperature do exist within the sound; although they may be important as cues to the fish, the fronts are not important to the overall physical dynamics. We allow our model to be three-dimensional as opposed to vertically integrated. In the latter case, the bottom stress, which is the retarding force per unit area exerted by the bottom on the flow, is taken to oppose the vertically averaged motion but, in reality, the bottom stress opposes the near bottom motion, which is frequently in opposition to the vertically integrated motion. Hence, the bottom stress is incorrectly specified by a vertically integrated model for wind-driven cases. The time-dependent wind stress is assumed spatially uniform. Turbulent eddy stresses are modeled with a constant eddy viscosity coefficient. The Coriolis acceleration terms are retained in the model but the vertical Ekman number is of order unity, so the rotational effect is only slight.

The equations that govern the flow in the sound in a right-handed Cartesian coordinate frame, with x' positive cross-sound towards the coast (also the southeast), y' positive along-sound towards the northeast, and z' positive up, all rotating at $f_0/2$, are

$$\frac{\partial \boldsymbol{v}'}{\partial t'} + f_0 \mathbf{e}_z \times \boldsymbol{v}' = -g\nabla'\eta' + A\frac{\partial^2 \boldsymbol{v}'}{\partial z'^2} \quad (1)$$

and

$$\frac{\partial M'x}{\partial x'} + \frac{\partial M'y}{\partial y'} + \frac{\partial \eta'}{\partial t'} = 0; \quad (2)$$

$$M'x = \int_{-h'}^{0} u'dz', \; M'y = \int_{-h'}^{0} v'dz'. \quad (3)$$

Here, $\boldsymbol{v}'$ is the horizontal velocity vector, η' the free surface elevation, h' the local depth, f_0 the Coriolis parameter, A the constant eddy viscosity coefficient, $M'x$ and $M'y$ are the volume flux vector components in the x' and y' horizontal directions, u', and v' are the velocity components in the x' and y' directions, respectively, t' is time, g the acceleration due to gravity, ∇' the horizontal gradient operator, $\mathbf{e}_z$ a unit vertical vector, and $\times$ the vector cross product.

The boundary conditions which drive the model are as follows.

At the surface (depth $z = 0$),

$$\rho A \frac{\partial \boldsymbol{v}}{\partial z'} = \boldsymbol{\tau}'_s; \quad (4)$$

ρ is the mass density and $\boldsymbol{\tau}'_s$ is the effective surface wind stress.

At the bottom ($z' = -h'$),

$$\boldsymbol{v}' = 0. \quad (5)$$

Additionally, at the lateral boundaries, $\mathbf{M}' \cdot \mathbf{n} = 0$; $\mathbf{n}$ is a horizontal vector normal to the boundaries except at the inlets, where a normal flux proportional to $|\eta'|^{1/2}$ is taken, and at river junctions, where $\mathbf{M}' \cdot \mathbf{n}$ is specified.

These equations are solved numerically. The equations are nondimensionalized and then written in finite difference form. Central differences are used for spatial derivations and forward differences for the temporal derivations. In the horizontal momentum equations, the Coriolis and pressure gradient terms are evaluated at the present time step, the frictional terms at the next time step, and the temporal derivative contains the next as well as the present time step values, i.e., the equations are in implicit form. The new velocities at $N + 1$ levels in the vertical are then obtained by Gaussian elimination at each horizontal grid point. Updated values of $\mathbf{M}'$ are obtained by vertically integrating the updated velocities. The updated values of $\mathbf{M}'$ are used in the vertically integrated continuity equation to update the water level, and the entire process begins anew. All dependent variables are initialized to zero. At the periphery of the sound, the normal gradient of the sea level is adjusted to ensure that the normal volume flux vanishes. We now present results from the model.

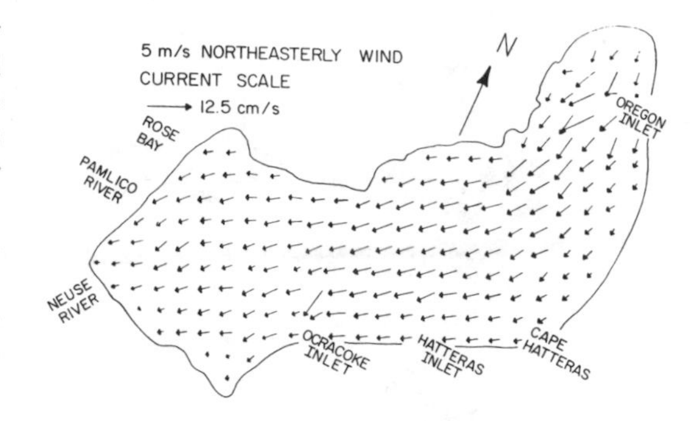

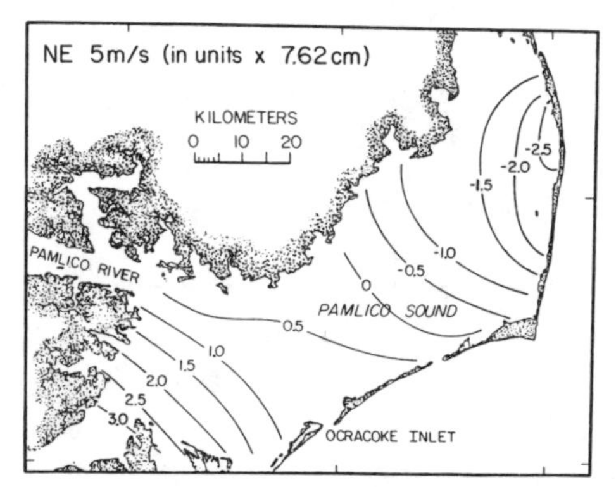

FIGURE 5.—Top. Wind-driven surface current vectors for Pamlico Sound, 10 h after the onset of a northeasterly wind of 5 m/s. Bottom. Water levels of Pamlico Sound after 10 h of northeasterly winds at 5 m/s; positive contour units indicate a rise in water level, negative units a drop. Results are from a numerical model.

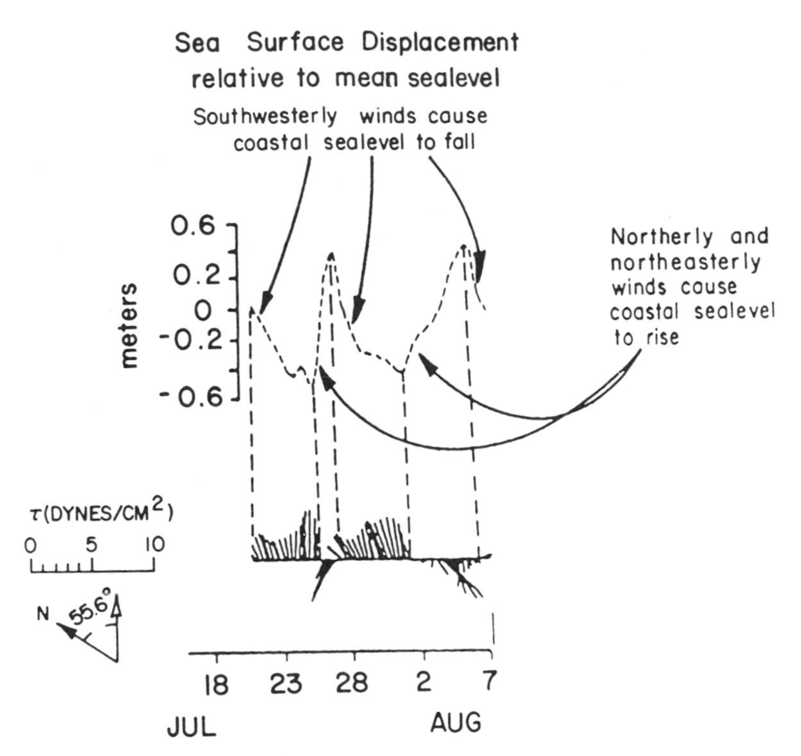

FIGURE 6.—Fluctuations of open ocean sea level at Cape Hatteras, North Carolina (see Figure 1), in relation to wind stress vectors (τ) at the same site.

Modeled surface velocities and water level variation 10 h after the onset of a 5 m/s wind blowing from the northeast, for an eddy viscosity of 10^{-2} m^2/s, are shown in Figure 5. The northeasterly wind forces surface currents towards the south to southwest and piles water up there. The model results suggest that Pamlico Sound "spins up," i.e., reaches a quasi-steady state condition, in a period of less than 10 h after the onset of a steady wind. Recent studies of the coastal meteorology in this region (Weisberg and Pietrafesa 1983) indicate that, although there are monthly and seasonal mean winds that generally repeat from year to year, the major portion of the wind variability occurs over time scales of 2 d to 2 weeks.

Basic Coastal Dynamics

For the coastal, seaward side of the barrier islands, the basic wind-driven dynamics have been described by Chao and Pietrafesa (1980) and Janowitz and Pietrafesa (1980). An example of sea level responses to local winds at Cape Hatteras, an open coastal station, is shown in Figure 6. In effect, southerly to southwesterly winds cause sea level to drop at the coast in concert with a surface Ekman transport offshore, and northerly to northeasterly winds cause sea level to rise at the coast. The response occurs within a period of 8–10 h. This response has been shown by Pietrafesa et al. (1981) to occur from Cape Hatteras south to Charleston, South Carolina. Northerly or southerly winds are important for coastal convergence or divergence, respectively. The question then is, does this phenomenon occur at the inlets, and, if so, what are its implications? We take Oregon Inlet as a case study.

Oregon Inlet Dynamics

Oregon Inlet is 71 km NNE of Hatteras Inlet (Figures 1, 7). The sound-side opening of the inlet is approximately 5.8 km south of Roanoke Island. Two channels link the inlet to the sound. One is a dredged channel that runs westward and joins the dredged channel network running along the east side of Roanoke Island and thence into Albemarle Sound. This channel is maintained to a depth of 3.7 m and a width of 30.5 m (U.S. Army Corps of Engineers 1973). The other is a natural channel that runs to the southwest and is known as the David Slough. It may range in depth from 2.7 m to 4.0 m and is subject to continual change (U.S. Coast and Geodetic Survey 1973). The inlet gorge itself has generally ranged in depth from 6.1 m to

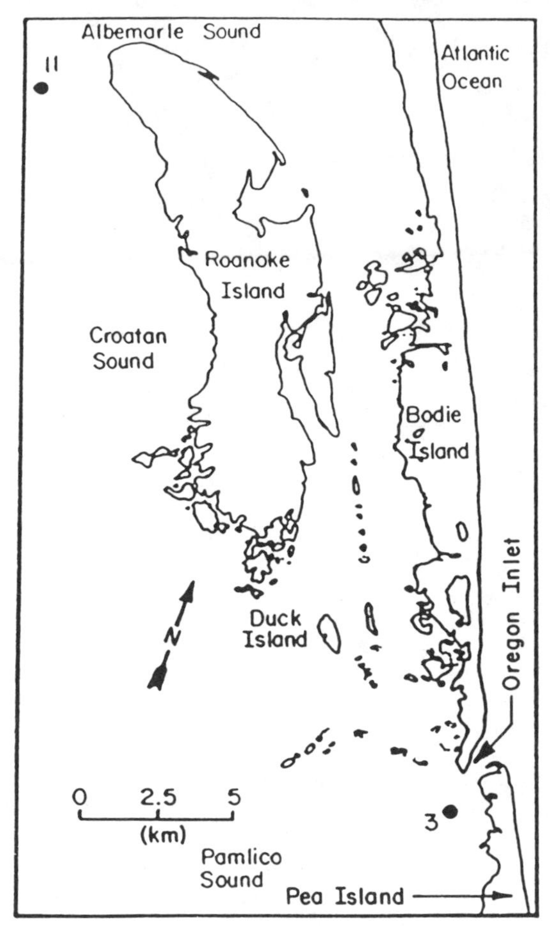

FIGURE 7.—Northern Pamlico Sound in the vicinity of Oregon Inlet, showing locations of stations 3 and 11 during a study of bottom currents (February 9–March 3, 1974) by Singer and Knowles (1975).

10.1 m and its width is nearly 0.8 km (U.S. Army Corps of Engineers 1968).

The mean tidal range at Oregon Inlet is 0.61 m and the spring tidal range is 0.73 m (U.S. National Climatic Center 1973). Tidal currents in the inlet may be as much as 2.5 m/s (U.S. Department of Commerce 1973). Temperatures reported at the inlet have ranged from 4°C in January to 21°C in August (Williams et al. 1973). Roelofs and Bumpus (1953) reported that the water flowing into the sounds through the inlets is isothermal. Because ocean and sound typically have different temperatures, however, temperature gradients may be found in the vicinity of the inlet within the area of tidal influence. Salinities at the inlet range from 8 to 33‰ and both extremes may be approached in the same month (Williams et al. 1973).

Singer and Knowles (1975) studied bottom currents and winds near Oregon Inlet (stations 3 and 11: Figure 7) during the period February 9–March 3, 1974. When winds were southerly, the subtidal frequency flow was out of the inlet at station 3 (an "ebb"); when the winds were northerly, the flow was into the sound (a "flood"; Figure 8). When winds persisted to the north, the inlet drained or "ebbed" even on flood tides. The opposite situation also held true, and it is this that enhances larval transport by currents through the inlets during winter. Currents from station 11 offered a point verification of the numerical model results (Figure 5).

Wind records at Charleston, South Carolina, represent those for the South Atlantic Bight (Cape Hatteras to south Florida) fairly closely. The October–January period is one of persistent northerlies (Weisberg and Pietrafesa 1983), which, according to rationale of Janowitz and Pietrafesa (1980) or Pietrafesa et al. (1981), would pile water up against the seaward side of the barrier islands (Figure 6); according to our Pamlico Sound model results, they would drive water away from the inside of Oregon Inlet (Figure 5).

Within the broad seasonal trends, the coastal marine climatology is dominated by synoptic wind events lasting 2–15 d (Figure 9). These create, within 8–14 h, drops or rises in sea level on either side of the barrier islands and related inlets. There may be a wind-forced rise in sea level on the ocean side and a wind-forced fall on the sound side, and the resulting pressure gradient force can drive a rapid current through the inlets (Figure 10). The opposite case is that a southerly wind will pile water up against the back side of the barrier islands while the coastal ocean will act like the plug was just pulled. If nontidal transport of fish larvae occurs through the inlets into Pamlico Sound, it will occur in pulses associated with northerly winds that may persist for 2–10 d with the passage of winter storms. We conclude that circulation through the barrier island inlets is forced directly by the 12.4-h (M_2) tide, and nonlocally by wind-induced pressure gradient forces due to the opposite responses of water level to the wind on the sea and sound sides of the inlets.

Cape Fear River Estuary

Data

Tide, salinity, and temperature data were collected in the Cape Fear River from July 22 to September 26, 1977, by North Carolina State

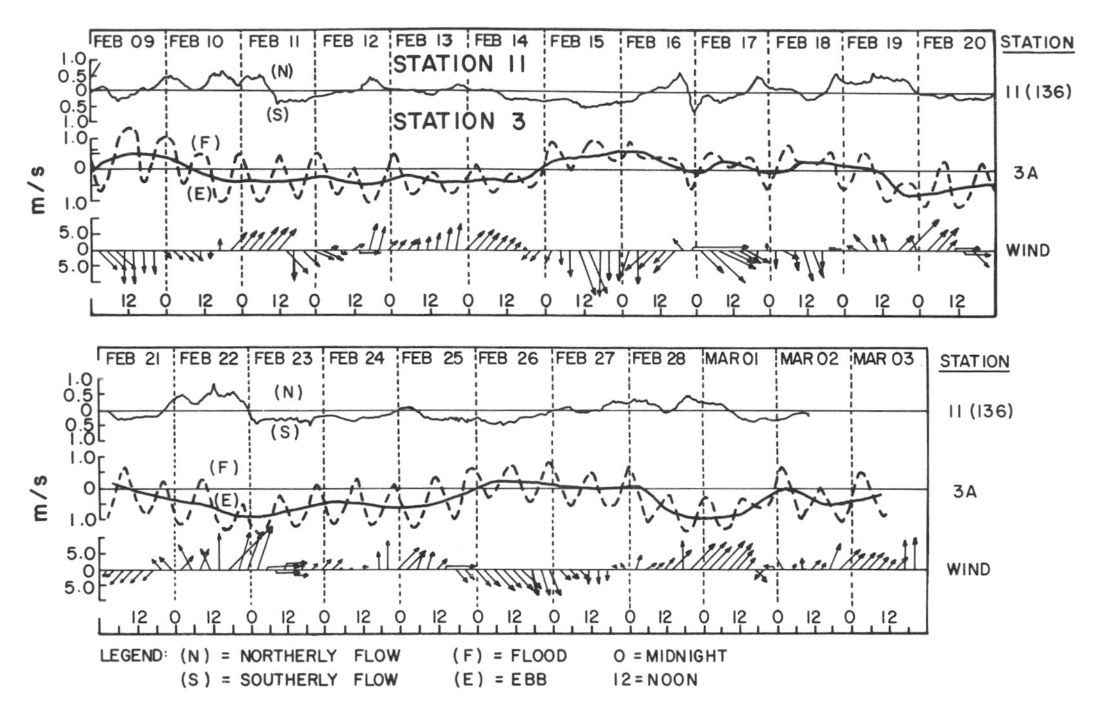

FIGURE 8.—Time series of current velocity and wind vectors for the 1974 study of Singer and Knowles (1975) at stations 3 and 11 in Pamlico Sound near Oregon Inlet (Figure 7). For station 3, the dashed line consists of 10-min series of observations and the solid line is a 40-h low-pass-filtered time series of the series of 10-min observations.

University (NCSU) personnel. Salinity and temperature data were analyzed and reported by D'Amato et al. (1980). The following are the main results drawn from the analyzed tide gage data. At the time of the NCSU field study, the Carolina Power and Light Company (CPL) was also conducting an intensive study of the Cape Fear estuary. The tide gage data collected by CPL are also presented here.

Tides and Seiches

Both the NCSU and CPL data reveal the presence of tidal constituents. The highest peaks of measured energy density are associated with the 12.41-h-period (M_2) semidiurnal constituent, as forced by the tide on the adjacent continental shelf (Redfield 1958; Pietrafesa et al. 1985). This shelf tide is a co-oscillation of the Atlantic Ocean tide and appears as a frictionally modified Poincare wave (Pietrafesa et al. 1985). Other tidal components are present; however, only the M_2 flood is sufficiently energetic to move larvae into the estuary.

If the estuary is considered as a bay closed at one end, the calculated period for the second seiche or free oscillation mode is about 2.3 h, which is near an observed 2.5 h peak in the unfiltered sea level data. Such phenomena are not important in larval recruitment.

The 40-h low-pass-filtered river level data (Figure 11) at stations CPL, OUR, and WIL display similar fluctuation patterns, suggesting high coherency between data pairs of location. The cross-spectra display high coherency at periods of 2.5–22.5 d and at the diurnal and semidiurnal periods. The water level autospectra are similar at the three stations with modest fluctuation peaks except at the diurnal and semidiurnal periods, when the peaks are prominent. The phase differences of the nontidal water level fluctuations show that the "up-estuary" stations lag the "down-estuary" station, which indicates a net up-estuary phase propagation. The time lag between CPL and WIL stations for periodicities of 2.5–22.5 d is about 9–10 h. This implies that these fluctuations propagate up-estuary with phase speed of about 0.9 m/s. The theoretical value, defined to be $(gh)^{1/2}$, is about 10 m/s, an order of magnitude larger than the observed value.

Subdiurnal Frequency Physics

Water levels at Wilmington tend to fall (rise) in concert with a southerly (northerly) wind (Figure

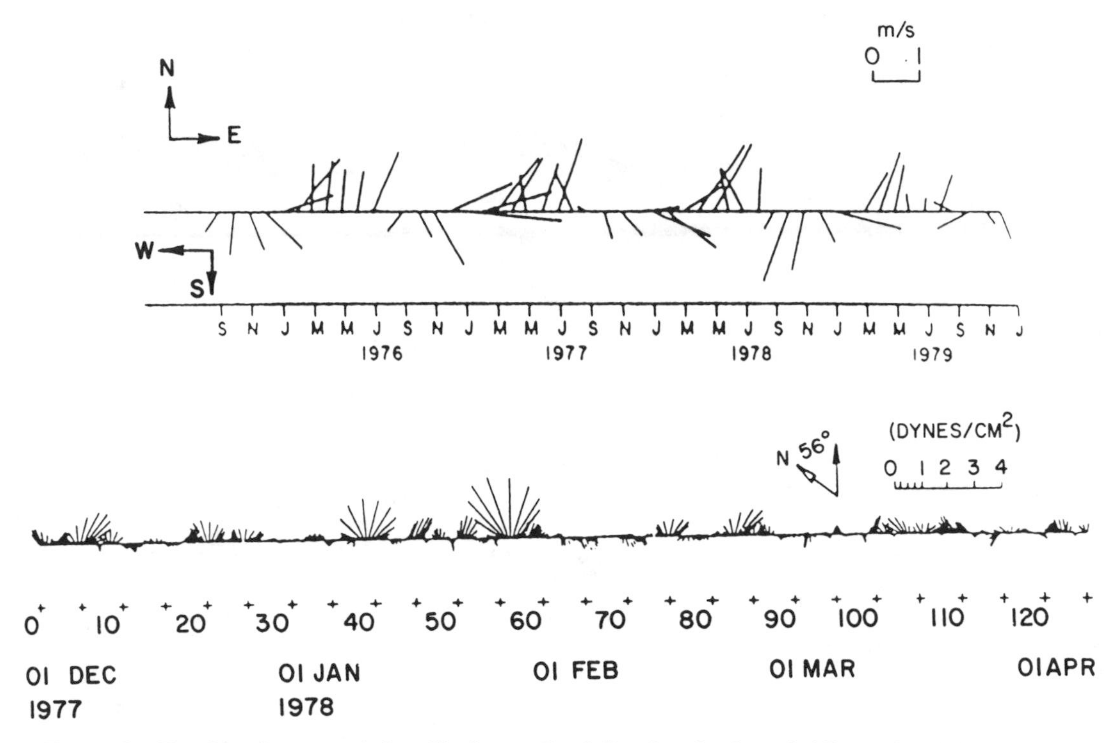

FIGURE 9.—Top. Monthy mean winds at Charleston, South Carolina, for the period September 1975–January 1980 (adapted from Weisberg and Pietrafesa 1983). Bottom. Forty-hour low-passed wind stress vector stick plots at Charleston, South Carolina, for the period December 1977–April 1978.

11), which is opposite to what would be expected. In a directly forced response situation, the onset of the wind should precede the change of water level along the axis of the estuary. We conclude that water level fluctuations are not induced directly by local wind stress but rather by nonlocal forcing mechanisms.

North of North Carolina, Wang and Elliot (1978) found that the dominant water level fluctuations in Chesapeake Bay and the Potomac River resulted from the up-bay propagation of coastal sea level fluctuations generated by the alongshore winds. Water was driven out of the bay by southerly (up-bay) winds and into the bay by northerly (down-bay) winds because of coastal Ekman flux. The river level fluctuations in the Potomac River (which empties into Chesapeake Bay) were induced nonlocally by the motion in the bay. South of North Carolina, Kjerfve et al. (1978) studied the low-frequency responses of estuarine water level to nonlocal wind forcing at North Inlet, South Carolina, and speculated that the water level fluctuations at the estuary were induced nonlocally. Coastal sea level fluctuations were attributed to atmospherically induced continental shelf waves passing by the mouth of the estuary.

In the Cape Fear River, plots of the low-pass-filtered atmospheric pressure data and water level data indicate a tendency for both to rise or fall contemporaneously, which is opposite to what would be expected from the inverse parameter effect. Their cross-spectrum shows poor coherence, which suggests a lack of sea level response to local atmospheric pressure. This phenomenon is not surprising because other investigators have noted the same phenomenon, which suggests that spatially extensive areas are required in order for sea level to respond directly to atmospheric pressure forcing.

The effects of freshwater inflow on water level fluctuations occasionally can be seen (Figure 11). When the Cape Fear River is heavily flooded by 566–850 m^3/s of freshwater inflow, water level at the head, middle, and mouth of its estuary show definitive rises, but these occasions of massive flooding are few, and, in general, coherence between sea level and river runoff in the Cape Fear River is poor.

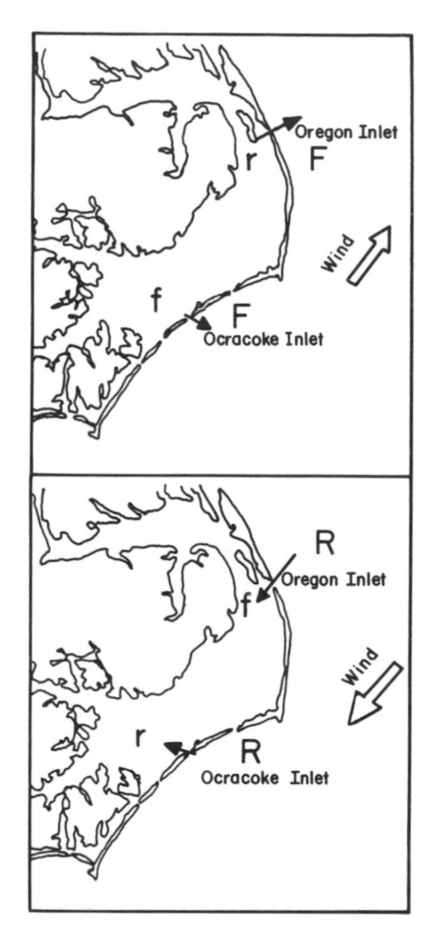

FIGURE 10.—Wind-driven pressure forces through the inlets to Pamlico Sound R (F) indicates a large rise (fall) in coastal sea level due to the wind; r (f) indicates a moderate rise (fall) in sound water level due to the wind. The arrows show the direction of the pressure force through the inlets due to the difference in coastal and sound water levels.

The cross-spectrum of atmospheric temperature and water level fluctuations show very poor coherence also, except at the diurnal period. It is likely that atmospheric temperature effects on water level fluctuations could be seen on much longer time scales, i.e., those the order of seasons.

The time domain plot of subdiurnal-frequency sea level fluctuations at Charleston, South Carolina (CHAR in Figure 11), and at the Cape Fear River stations indicate similar patterns. The autospectra of sea level at CHAR and at each of the estuary's stations display similar weak fluctuation patterns over the bandwidth period of 2.5–20 d, and show strong fluctuations at diurnal and semidiurnal periods. The cross-spectra between sea level at CHAR and each of the estuary's stations shown high coherence within the range of 2.9–10.5 d. We conclude that within this 2.9–10.5 d period, phenomena occur in the coastal ocean adjacent to the estuary and communicate their character through the mouth of the estuary. Water level fluctuations within the estuary are caused by these coastal events.

Subdiurnal-frequency sea level and the north–south wind stress time series at CHAR show mirror-like patterns, as discovered by Chao and Pietrafesa (1980). As the wind increases toward the south (north), sea level rises (falls). The two autospectra are similar, showing peaks at periodicities of 3–35 d, and their coherence is high over the same range of periods. This suggests that CHAR sealevel fluctuations are forced by the northeast–southwest wind stresses, in accordance with coastal Ekman flux concepts; that is, the southwesterly (northwesterly) winds that blow along the coast drive water seaward (coastward), causing a set-down (set-up) of sea level at the coast.

Forcing at the Mouth

Based on our evaluation of data presented in the previous section, we conclude that the nontidal water level fluctuations (over the period range of 2–12 d) in the Cape Fear River are generated in the open coastal ocean and communicated through its mouth. The coastal sea level fluctuations, which are induced by regional atmospheric forcing, can travel along the coast as free or forced events, or can occur via time-dependent coastal Ekman dynamics (Janowitz and Pietrafesa 1980). These fluctuations appear at the mouth and introduce water level changes within the estuary itself. This phenomenon suggests that nonlocal, atmospherically forced coastal events dominate the water level variations in this estuary. These are the events that carry larvae into the estuary.

Within the estuary, the water level fluctuations are propagated up-estuary from the mouth.

Cape Fear River Plume

We next address the question of what happens to the Cape Fear River plume as it enters the ocean. Does the river outwell at the surface while a current flows beneath it into the estuary? To address this question, we consider steady horizontal flows of two immiscible fluids. Layer 1, the estuarine plume, is of density ρ_1 and overlies layer 2 of density $\rho_1 + \Delta\rho$ ($\Delta\rho > 0$). The flow in the

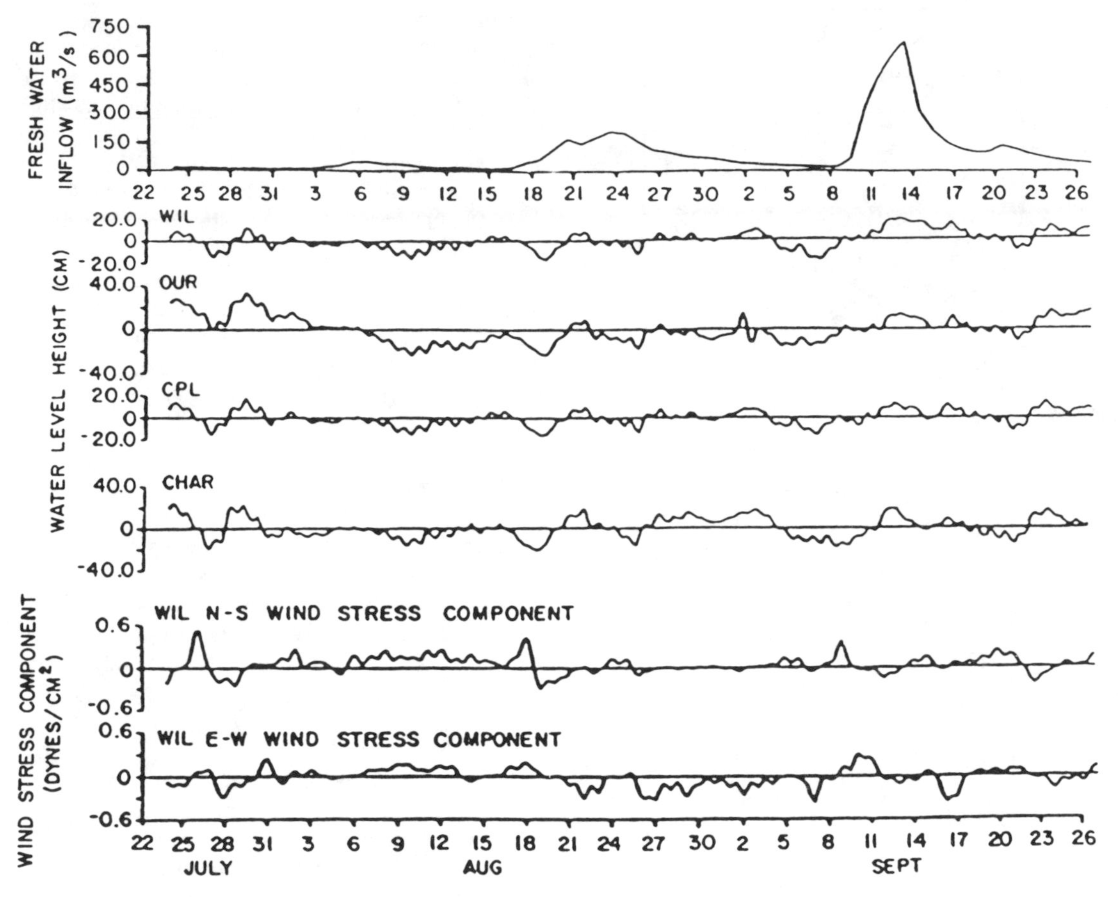

FIGURE 11.—Forty-hour low-pass-filtered time series of Cape Fear River discharges at station WIL (Wilmington); water levels at stations WIL, OUR, CPL (Figure 3), and CHAR (Charleston, South Carolina); and wind stress vector components N–S (positive is up-estuary or north and negative is down-estuary or south) and E–W (positive is east, negative is west). Data are for 1977.

upper layer originates at the mouth of the estuary and the denser coastal waters modeled by layer 2 are generally drawn into the mouth (Figure 12).

The interaction process takes place over a shelf that increases in depth linearly with offshore coordinate x and has slope ϕ. The alongshore coordinate is y and the vertical coordinate z is zero at sea level and decreases downward. The origin of the coordinate systems is the midpoint of the mouth. The interface between the two layers is located at $z = z_i = -D_1 + \eta$; D_1 is the depth of the interface at the origin and η is the vertical displacement of the interface above $z = -D_1$. The bottom is located at $z = z_0 = -D_1 - D_2 - \phi x$; $D_1 + D_2$ is the depth at the coast at $x = 0$.

The parametric constraints of the model are as follows. The thicknesses of the frictional Ekman layers, δ_e $(= \sqrt{2Av/f})$, which occur on both sides of the interface and at the bottom, are taken to be small compared to the depth of the two layers; here, Av is the eddy viscosity and f the Coriolis parameter. This assumption must break down at a front, if one occurs, because the thickness of the upper layer then vanishes. The cross-shelf components of the velocity in each layer (u_1 and u_2) are taken to be uniform at the mouth and equal to U_1 and U_2 respectively; U_1 generally is positive and U_2 generally negative. The width of the mouth, w, is sufficiently large that

$$\frac{U_1}{fw} < \frac{\delta_e}{D_1} < 1; \tag{6a}$$

$$w^2 > \rho g D_1/\rho_1 f^2; \tag{6b}$$

$$w^2 > U_1 D_1/f_\phi. \tag{6c}$$

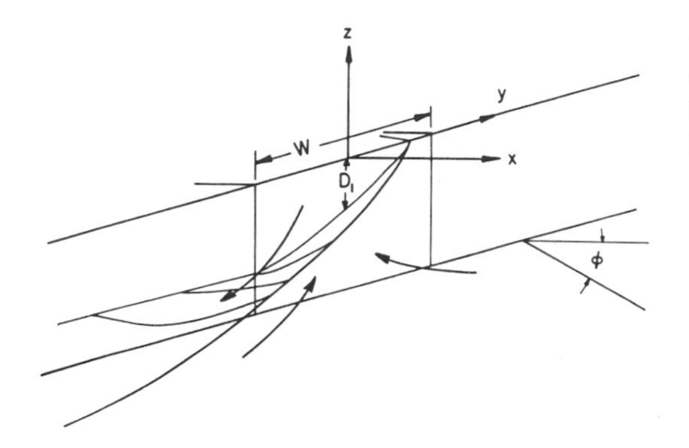

FIGURE 12.—Modeled geometry of the Cape Fear River mouth. Spatial coordinates are x (offshore, positive to the south), y (alongshore, positive to the east), and z (positive upward); w is the river width; D_i is depth to the bottom of the river plume at midmouth; ϕ is bottom slope. The arrow curving downward to the left denotes plume flow; arrows curving upwards denote ocean flow into the river.

Under assumption (6a), the flow can be taken as geostrophic and vertically uniform in each layer outside thin Ekman layers. The presence of friction at the bottom and the interface allows us to require that the total velocity vanish at the bottom and that the total velocity and the vertical shear at the interface be continuous across the interface. We can then show that the vertical velocity just above the frictional layer in layer 1 is equal to the vertical velocity just below the frictional layer at the interface in layer 2. The total pressure of the fluid layers is taken to be continuous across the interface, so the dynamically active part of the pressure P that geostrophically balances the Coriolis force then satisfies

$$P_2(x, y) - P_1(x, y) = \Delta\rho g \eta(x, y). \qquad (7)$$

We next consider the vorticity equation in each layer. Assumptions (6) allow us to neglect advection of vorticity, which implies that the vertical velocity is independent of depth in each layer. Because we assume there is no wind stress curl, the vertical velocity below the surface Ekman layer is zero; the vertical velocity just above the interface must vanish because the vertical velocity in layer 1 is uniform. Because the vertical velocity just above the interface in layer 1 vanishes, the vertical velocity just below the interface in layer 2 must also vanish, because it equals the former velocity. The vertical velocity in the lower layer is again depth-independent, so the vertical velocity just above the bottom must vanish. The vertical velocity outside the Ekman layer thus vanishes everywhere. The vanishing of the vertical velocity just above the bottom and just above the interface leads to the following two equations for P_1 and P_2.

$$\frac{\delta_e}{2} \nabla_H^2 P_2 + \phi \frac{\partial P_2}{\partial y} = 0; \qquad (8a)$$

$$\frac{\delta_e}{4} \nabla_H^2 (P_1 - P_2) - \frac{1}{\Delta\rho g}\left(\frac{\partial P_1}{\partial y}\frac{\partial P_2}{\partial x} - \frac{\partial P_1}{\partial x}\frac{\partial P_2}{\partial y}\right) = 0. \qquad (8b)$$

The coastal boundary conditions on P_1 and P_2 are as follows.

$$P_{1,2}(0,y) = \begin{cases} -\rho_1 f U_{1,2}\, w/2 & \text{for } y \geq w/2; \\ -\rho_1 f U_{1,2}\, y & \text{for } -w/2 \leq y \leq +w/z; \\ +\rho_1 f U_{1,2}\, w/2 & \text{for } y \leq -w/2. \end{cases} \qquad (9)$$

The solution of (8) subject to (9) is as follows, with η following from (7).

$$P_2 = \rho_1 f\, U_2\, w\, F(x,y) + \rho_1 f\, V_2\, x; \qquad (10a)$$

$$P_1 = \rho_1 f\, U_1\, wF(x,y) + \frac{\Delta\rho g \phi}{2U_2}(U_1 - U_2)x + \rho_1 f V_2\, U_1 x/U_2; \qquad (10b)$$

$$\eta = \rho_1 \frac{f(U_2 - U_1)w}{\Delta\rho g} F(x,y) + \frac{(U_2 - U_1)}{2U_2}(\phi + \frac{2\rho_1 f V_2}{\Delta\rho g})x; \qquad (10c)$$

$$F(x,y) = \frac{1}{2\pi w} \int_{-\infty}^{\infty} \frac{\exp(-\sqrt{t^2 + i\alpha t x} - it\eta)}{t^2}\, dt \Bigg|_{\eta = y - w/2}^{\eta = y + w/2}; \qquad (10d)$$

$$\alpha = 2\phi/\delta_e. \qquad (10e)$$

The parameter V_2 is an alongshore coastal current, which can be specified as a free parameter. We now discuss the flow in layer 2, the coastal waters. We consider the case first with no ambient longshore flow ($V_2 = 0$). The function $F(x, y)$, which totally describes the flow in the lower layer, has the following properties. For $y \leq -w/2$, $F(0, y) = +0.5$. For $y \geq w/2$, $F(0, y) = -0.5$; $F(0, y)$ decreases linearly with y in the range $-w/2 \leq y \leq +w/2$. To the right of the mouth when facing seaward, F ranges from $+0.5$ at the coast to -0.5

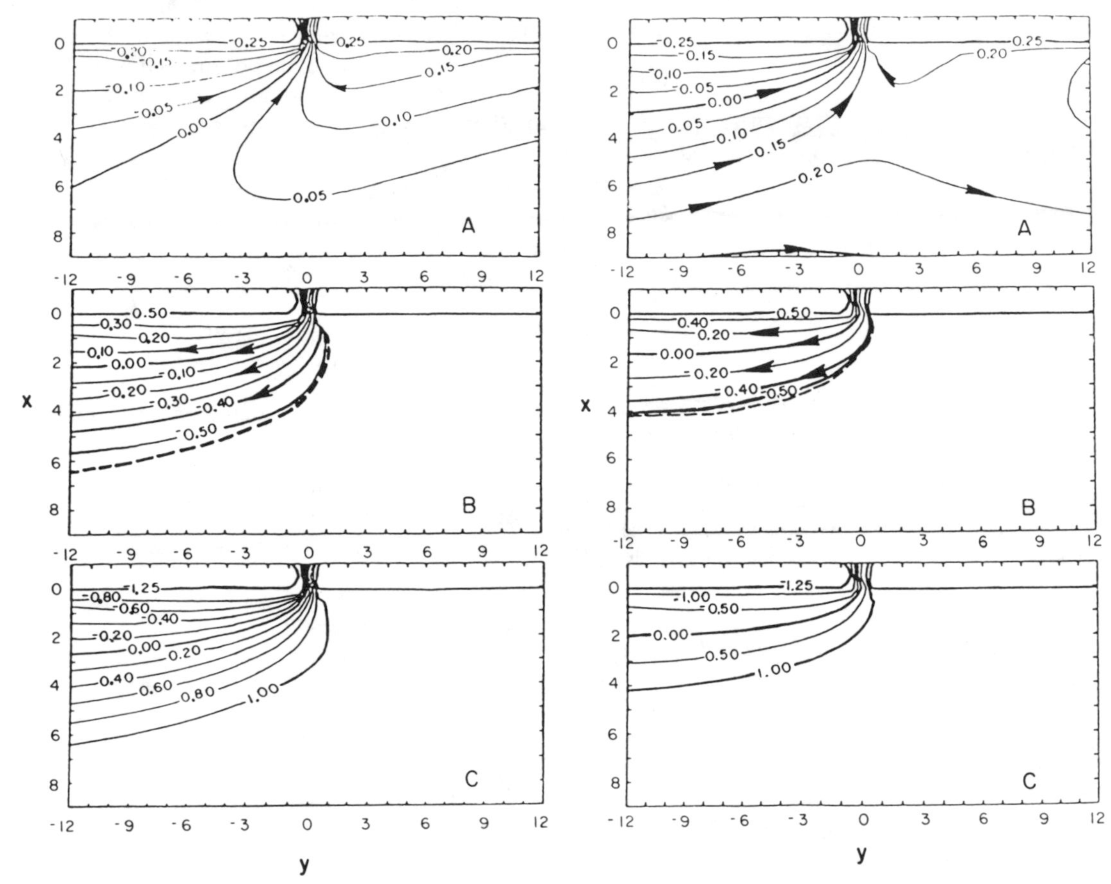

FIGURE 13.—Modeled circulation of the Cape Fear River plume for the case of no ambient coastal flow ($V_2 = 0$; plan view). (A) Lower layer of water. (B) Upper layer of water (plume). (C) Deflection of the interface between the two layers. There is inflow in the lower layer, outflow in the plume, and a front (dashed line) occurs. Coordinates and isopleths are dimensionless.

FIGURE 14.—Modeled circulation of the Cape Fear River plume for the case of ambient coastal flow to the east ($V_2 > 0$; plan view). (A) Lower layer of water. (B) Upper layer of water (plume). (C) Deflection of the interface between the two players. There is inflow in the lower layer, outflow in the plume, and a front (dashed line) occurs. Coordinates and isopleths are dimensionless.

far from the coast. Far to the left of the mouth, F is nearly a constant (-0.5). Thus, for $V_2 = 0$, most of the motion in the lower layer occurs to the right of the mouth. Fluid is either drawn in from the right ($U_2 < 0$) or outwelled to the right ($U_2 > 0$). For ambient coastal flow from right to left across the mouth ($V_2 > 0$), water is drawn into the estuary from the right ($U_2 < 0$) or expelled to the left ($U_2 > 0$). For ambient coastal flow from left to right across the mouth ($V_2 < 0$), water is drawn in from the left ($U_2 < 0$) or expelled to the right ($U_2 > 0$). The flow in the lower layer with $U_2 < 0$ is shown in Figures 13A, 14A, and 15A for $V_2 = 0$, $V_2 > 0$, and $V_2 < 0$, respectively. In these figures, $U_2 = -0.5U_1$ with $U_1 > 0$, $\phi w/D_1 = 0.1$, $\Delta\rho g (D_1/D_2) f U_1 w = 0.6$, and $\delta_e/D_1 = 0.2$.

We now turn to flow in the upper layer, the estuarine plume. Flow in the plume is somewhat more complex due to the term $\Delta\rho g\ \phi x(U_1 - U_2)/2U_2$ in equation (10b). Far from the mouth, $|\nabla F| \rightarrow 0$, so the alongshore flow in the plume far from the mouth is

$$V_{1_\infty} = \frac{1}{\rho_1 f}\frac{\partial P_1}{\partial x}\Big|_{|y| \rightarrow \infty} = \Delta\rho g\ \phi(U_1 - U_2)/2U_2\ \rho_1 f + \frac{V_2 U_1}{U_2}. \qquad (11)$$

In the case of an outwelling plume and inflowing coastal waters with no ambient coastal flow, $U_1 > 0$, $U_2 < 0$, $V_2 = 0$. In this case $V_{1_\infty} < 0$ and the

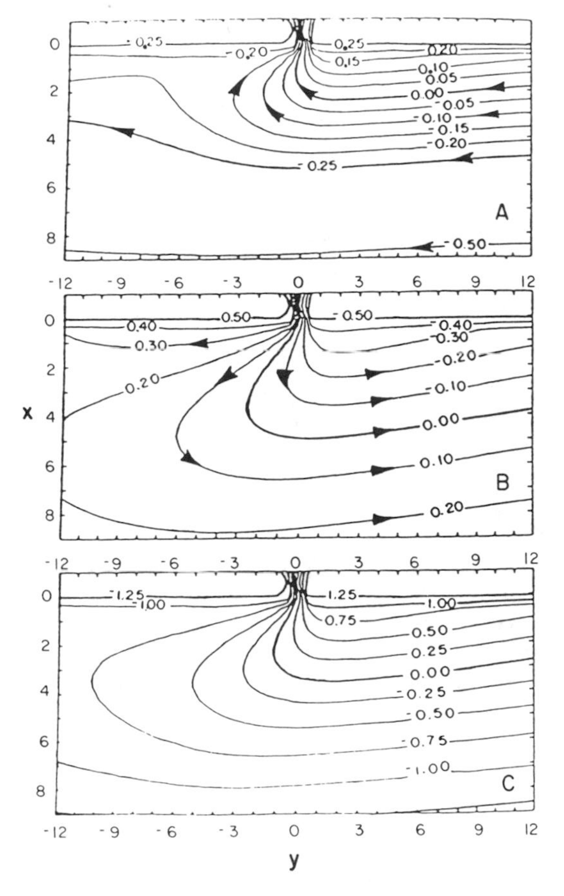

FIGURE 15.—Modeled circulation of the Cape Fear River plume for the case of ambient coastal flow to the west ($V_2 < 0$; plan view). (A) Lower layer of water. (B) Upper layer of water (plume). (C) Deflection of the interface between the two layers. There is inflow in the lower layer, outflow in the plume, and no front occurs. Coordinates and isopleths are dimensionless.

plume turns to the right. With outflow in both layers, $0 < U_2 < U_1$, the outflow is ultimately to the left. If $V_2 > 0$, then V_{1_∞} is increased in magnitude but not changed in sign, so the turning of the plume is simply enhanced. If $V_2 < 0$ but is weak, the turning of the plume is weakened. If

$$V_2 < -\left|\frac{\Delta\rho g\ \phi(U_1-U_2)}{2\rho_1 f U_2}\right|,$$

the motion of the plume is reversed compared to the cast $V_2 = 0$. In Figures 13B, 14B, and 15B, the motion in the plume is given for $V_2 = 0$, $V_2 > 0$, and $V_2 < 0$, respectively, with $U_1 > 0$ and $U_2 < 0$ in all three cases.

We now consider conditions under which a front will occur. This happens if $\eta = D_1$. In equation (10c), we see that this will occur if

$$\rho_1 f|(U_2-U_1)|\frac{w}{2}/\Delta\rho g D_1 > 1$$

or

$$(1 - U_1/U_2)(\rho_1 f V_2/\Delta\rho g + \phi/2) > 0. \quad (12)$$

If $U_2/U_1 < 0$, the normal case, fronts occur if $V_2 > -\Delta\rho g\ \phi/2\rho_1 f$, which includes the case $V_2 = 0$. Thus, under normal inflow–outflow conditions, a front will occur as long as the ambient coastal flow is not too strong from left to right across the estuary mouth. In Figures 13C, 14C, and 15C, the deflection of interface is again given for $V_2 = 0$, $V_2 > 0$, and $V_2 < 0$, respectively, with $V_1 > 0$ and $V_2 < 0$. In these figures, P_1 and P_2 have been made nondimensional by $\rho_1 f V w$, η by D_1, and x and y by w.

The model described above has been applied by Zhang et al. (1987) to the Changjiang River in China as well as to the Chesapeake Bay and Savannah River estuaries. Here, we apply the results to the Cape Fear River estuary system. An extensive series of measurements was made in the estuary proper by Carpenter and Younts (1978), who used marker dyes to show the motion of the plume out on the shelf. Due to its narrow width (1.7 km), shallow mean depth (6 m), and relatively high tidal speeds, the Cape Fear River is not the most appropriate system for model application. However, the 2-week mean currents shown in Figure 16 do suggest a two-layer inflow–outflow situation. These data are taken from the results of Carpenter and Younts (1978). The data suggest taking $U_1 = -12$ cm/s, $D_1 = 2$ m, $\Delta\rho g/\rho_1 = 1$ cm/s^2, and $\phi = 0.5 \times 10^{-3}$. No measurements were taken of currents off the shelf, so V_2 is unknown. However, based on the findings of Purba (1984), the south–southwesterly winds present during the period would drive an eastward alongshore current. Model predictions presented in Figures 13 and 15 are similar to observed currents at the mouth cross-section (Figure 16) and to the distribution of surface dye released across the mouth transect (Figure 17).

Conclusions

Local Forcing

In the case of the barrier island inlets of Pamlico Sound, the principal direct forcing mechanisms

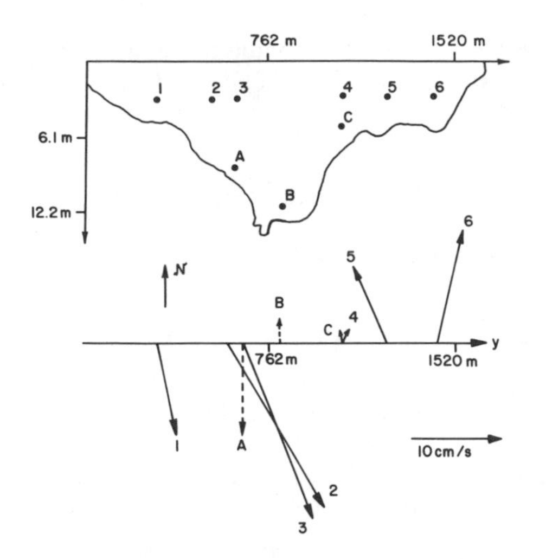

FIGURE 16.—Mean currents at the mouth of the Cape Fear River estuary (derived from Carpenter and Younts 1978).

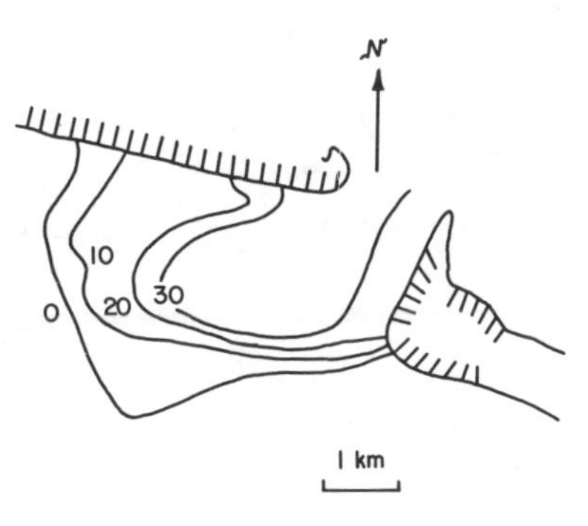

FIGURE 17.—Surface dye concentrations at the mouth of the Cape Fear River estuary in ng/L (derived from Carpenter and Younts 1978).

for flow across the inlet are due to the flood and ebb of the M_2 tide. Although this phenomenon is ultimately a flow driven by pressure gradients and could be considered "nonlocal" in origin, we take the liberty of decomposing the inlet dynamics into super- and subtidal frequency dynamics and claim that the former are "local" and the latter "nonlocal." Inlet jetting is dramatic, of the order of meters per second.

The Cape Fear River mouth is likewise heavily influenced by the M_2 tide, which floods and ebbs in phase with the offshore incursion of the co-oscillating Atlantic Ocean semidiurnal tide. Other tidal constituents are also in evidence, most prominently the diurnal species. Nonetheless, the M_2 is dominant. Atmospheric impulsive loading can create seiches at frequencies of 2.5 and 21.3 h in the Cape Fear basin, both fundamental modes.

The mouth of the Cape Fear is subject to subfortnightly frequency dynamics such that the surface plume turns southwest while there is an inflow from the southeastern side of the mouth. The flow in on the left (viewed seaward) is wider at the bottom than at the surface. The flow out at the mouth is wider in the upper part of the water column than at the bottom. There could well be a horizontal, counter-clockwise gyre located between the mouth of the estuary and the upper CPL station. Upstream of the CPL station, the estuarine circulation likely is more directly influenced by the wind but still retains elements of nonlocal forcing influence.

Nonlocal Forcing

The barrier island inlets experience the effect of nonlocal forcing, or one-sided divergences and convergences of water on either side. Given the basin–barrier island geometry, when the flux of water is away from the offshore side of the barrier islands, there is a flux towards the back, or sound, side of the islands. The opposite scenario also holds true. These inshore divergences (convergences) in synchrony with offshore convergences (divergencies) create pressure gradients; water level drops (rises) from ocean to sound through the inlets, which causes tremendous "flood" ("ebb") jets through the inlets.

The Cape Fear estuary mouth "floods" ("ebbs") in a subtidal frequency sense, when there is a buildup (drop) of sea level at the coast due to coastal Ekman dynamics induced by alongshore winds.

Cueing

Larvae may respond to hydrographic relationships to enter the inlets of Pamlico Sound or the Cape Fear mouth by tracking warm saline waters in the winter, which enter the estuaries on the flood.

Acknowledgments

The Office of Sea Grant, U.S. Department of Commerce (grant NA81AA-D-00026) and the U.S. Department of Energy (grants DOE-76-AS09-EY00902 and DE-FG09-85ER60376) provided support for this study and manuscript. C. Knowles provided data used in the study. J. Miller provided the biological impetus for the study.

References

Airan, D. S. 1974. Explicit modeling of circulation and water quality for two-dimensional unsteady flow. Doctoral dissertation. North Carolina State University, Raleigh.

Amein, N. 1971. Circulation in Pamlico Sound. Report to the North Carolina Board of Science and Technology, Raleigh.

Amein, N., and D. S. Airan. 1976. Mathematical modeling of circulation and hurricane surge in Pamlico Sound, North Carolina. University of North Carolina Sea Grant College Program, Publication 76-12, Raleigh.

Carpenter, J. H., and W. L. Younts. 1978. Dye tracer and current meter studies, Cape Fear Estuary, North Carolina. 1976, 1977, 1978, volume 1. Pages 159–187 *in* Brunswick steam electric plant Cape Fear studies. Carolina Power and Light Company, Raleigh, North Carolina.

Chao, S.-Y., and L. J. Pietrafesa. 1980. The subtidal response of sea level to atmospheric forcing in the Carolina capes. Journal of Physical Oceanography 10:1246–1255.

Chu, H. L. 1970. Numerical techniques for non-linear wave refraction and for propagation of long waves in a two-dimensional shallow water basin. Doctoral dissertation. North Carolina State University, Raleigh.

D'Amato, R., R. H. Weisberg, and L. J. Pietrafesa. 1980. Hydrographic observations in the Cape Fear Estuary: summer, 1977. North Carolina State University Department of Marine Science and Engineering, Report 80-4, Raleigh.

Dyer, K. R. 1973. Estuaries: a physical introduction. Wiley, New York.

Gilliam, J., J. Miller, L. Pietrafesa, and W. Skaggs. 1985. Water management and estuarine nurseries. University of North Carolina Sea Grant College Program, Publication 85-2, Raleigh.

Hammack, J. L. 1969. A mathematical investigation of freshwater flow through Pamlico Sound, North Carolina. Master's thesis. North Carolina State University, Raleigh.

Janowitz, G. S., and L. J. Pietrafesa. 1980. A model and observations of time-dependent upwelling over the mid-shelf and slope. Journal of Physical Oceanography 10:1574–1583.

Jarrett, T. J. 1966. A study of the hydrology and hydraulics of Pamlico Sound and their relation to the concentration of substances in the sound. Master's thesis. North Carolina State University, Raleigh.

Kjerfve, B., J. E. Greer, and R. L. Crout. 1978. Low-frequency response of estuarine sea level to nonlocal forcing. Pages 497–513 *in* M. L. Wiley, editor. Estuarine interaction. Academic Press, New York.

Marshall, N. 1951. Hydrography of North Carolina marine waters. Pages 1–76 *in* H. F. Taylor and Associates, editor. Survey of marine fisheries of North Carolina. University of North Carolina Press, Chapel Hill.

Miller, J. M., J. P. Reed, and L. J. Pietrafesa. 1984. Patterns, mechanisms and approaches to the study of estuarine dependent fish larvae and juveniles. Pages 209–225 *in* J. D. McCleave, G. P. Arnold, J. J. Dodson, and W. H. Neill, editors. Mechanisms of migration in fishes. Plenum, New York.

Pietrafesa, L. J., J. O. Blanton, J. D. Wang, V. Kourafalou, T. N. Lee, and K. A. Bush. 1985. The tidal regime in the South Atlantic Bight. American Geophysical Union Coastal and Estuarine Monograph Series 2:63–76.

Pietrafesa, L. J., S.-Y. Chao, and G. S. Janowitz. 1981. The variability of sea level in the Carolina capes. University of North Carolina Sea Grant College Program, Publication 81-11, Raleigh.

Pietrafesa, L., G. S. Janowitz, S.-Y. Chao, R. H. Weisberg, F. Askari, and E. Noble. 1986. The physical oceanography of Pamlico Sound. University of North Carolina Sea Grant College Program, Publication 86-5, Raleigh.

Posner, G. S. 1959. Preliminary oceanographic studies of the positive bar built estuaries of North Carolina, U.S.A. International Oceanographic Congress and American Association for the Advancement of Science, Washington, D.C.

Purba, M. 1984. A parametric evaluation of observations in the Georgia Bight March–May 19890. Doctoral dissertation. North Carolina State University, Raleigh.

Redfield, A. C. 1958. The influence of the continental shelf on the tides of the Atlantic coast of the U.S. Journal of Marine Research 17:432–448.

Roelofs, E. W., and A. W. Bumpus. 1953. The hydrography of Pamlico Sound. Bulletin of Marine Science of the Gulf and Caribbean 3:181–205.

Singer, J. J., and C. E. Knowles. 1975. Hydrology and circulation patterns in the vicinity of Oregon Inlet and Roanoke Island, North Carolina. University of North Carolina Sea Grant College Program, Publication 75-15, Raleigh.

Smallwood, C. A., and A. Amein. 1967. A mathematical model for the hydrology and hydraulics of Pamlico Sound. Proceedings of the symposium on the hydrology of coastal waters, North Carolina, report 5. University of North Carolina, Water Resources Research Institute, Raleigh.

U.S. Army Corps of Engineers. 1968. Manteo (Shallowbag) Bay, North Carolina. Review report, appendix A. Wilmington District, Wilmington, North Carolina.

U.S. Army Corps of Engineers. 1973. Location map for dredging between Manteo and Oregon Inlet, sheet 1 of 7. Wilmington District, Wilmington, North Carolina.

U.S. Coast and Geodetic Survey. 1973. Chart 1229, United States, east coast, North Carolina, Currituck beach light to Wimble Shoals. National Oceanic and Atmospheric Administration, National Ocean Survey, Washington, D.C.

U.S. Department of Commerce. 1973. United States coast pilot, Atlantic coast, Cape Henry to Key West. National Oceanic and Atmospheric Administration, National Ocean Survey, Washington, D.C.

U.S. National Climatic Center. 1973. Climatological data—North Carolina, 1973. Climatological Data 78(6). (Asheville, North Carolina.)

Wang, D.-P., and A. J. Elliot. 1978. Non-tidal variability in the Chesapeake Bay and Potomac River: evidence for non-local forcing. Journal of Physical Oceanography 8:225–232.

Weisberg, R. H., and L. J. Pietrafesa. 1983. Kinematics and correlation of the surface wind field in the South Atlantic Bight. Journal of Geophysical Research 88:4593–4610.

Welch, J. M. 1979. Aspects of the hydrodynamics and circulation of the lower Cape Fear River, North Carolina. Master's thesis. North Carolina State University, Raleigh.

Williams, A. B., G. S. Posner, W. J. Woods, and E. . Deubler, Jr. 1973. A hydrographic atlas of larger North Carolina sounds. University of North Carolina Sea Grant College Program, Publication 73-02, Chapel Hill.

Woods, W. J. 1967. Hydrographic studies in Pamlico Sound. Proceedings of the symposium on the hydrology of coastal waters North Carolina, report 5. University of North Carolina, Water Resources Research Institute, Raleigh.

Zhang, Q. H., G. S. Janowitz, and L. J. Pietrafesa. 1987. Estuarine and continental shelf water interaction: a model and applications. Journal of Physical Oceanography 17:455–469.

American Fisheries Society Symposium 3:51–67, 1988

Roles of Behavioral and Physical Factors in Larval and Juvenile Fish Recruitment to Estuarine Nursery Areas

George W. Boehlert and Bruce C. Mundy

Southwest Fisheries Center Honolulu Laboratory, National Marine Fisheries Service
2570 Dole Street, Honolulu, Hawaii 96822-2396, USA

Abstract.—Recruitment to and maintenance in estuaries are important parts of the early life history of many fish species. Field studies have documented patterns of estuarine recruitment for several species; although some studies have postulated passive mechanisms for recruitment, the majority suggest specific behavior patterns that clearly correlate with physical factors or other stimuli. We consider recruitment to the estuary of species spawned offshore as a two-stage process dependent first upon factors in the offshore planktonic environment and second upon estuarine factors related to tidal flux. Rather than a simple stimulus–response mechanism related to a single physical factor, we suggest that the suite of factors associated with tidal flux at particular locations may act as the zeitgeber for an endogenous rhythm with a tidal periodicity. In this manner, an animal may use tidal-stream transport both for movement into the estuary and for maintenance within the estuary. Further work in the laboratory is necessary to elucidate these behaviors in fishes, particularly those related to endogenous rhythms and the stimuli that serve as zeitgebers.

Use of estuarine nursery areas is an important phase of the life history of many marine organisms, including many commercially valuable shrimps and fishes (McHugh 1967; Staples 1980; Haedrich 1983). The net seaward movement of estuarine waters, combined with tidal flux, causes problems for taxa using estuaries; these problems are generally related to either export from or recruitment to the estuary. Different strategies of estuarine use are apparent among taxa. Several species are resident in estuaries throughout their life histories. Their primary recruitment problem is to prevent the export of early life history stages from the estuary (Johnson and Gonor 1982), and they often solve this by producing large, demersal eggs (Hempel 1979) and having brief larval stages. Species that visit periodically as adults for feeding or spawning face the problem of locating the estuary; if they spawn there, their larvae face the same export problems as the residents. The interactions between hydrographic features of an estuary and population maintenance have traditionally fascinated biologists. Early interest centered upon the ability of planktonic organisms, including larvae, to remain in estuarine systems despite tidal flushing (Rogers 1940; Carriker 1951; Bousfield 1955). Early hypotheses explaining retention in estuaries generally invoked the use by organisms of the net landward flow in the deeper waters of stratified estuaries. Rogers (1940), for example, suggested that larvae of rainbow smelt *Osmerus mordax* were retained in a stratified estuary through a mechanism of vertical migration, but were transported out of the system in another, intensively flushed, estuary. Thus, specific behavior patterns and estuarine hydrography may interact to retain larvae or plankton within estuaries. The hypothesis of vertical migration based upon a tidal rhythm (shallow at flood, deep at ebb) suggested by Carriker (1951) has been confirmed by Cronin and Forward (1979) for larvae of the crab *Rhithropanopeus harrisii*.

Among the least understood mechanisms of transport associated with estuaries is the one used by species that spawn offshore and subsequently enter estuarine systems as late larvae or early juveniles. These larvae must first move to nearshore areas, then locate an estuary mouth, and finally move into the estuary. This phenomenon is related to, but distinct from, retention in estuaries; it depends upon physical factors but is linked to active behavioral responses on the part of larvae. Movements to the nearshore environment from offshore are generally considered to be either passive or some form of modulated transport; in the latter case, behavioral attributes such as vertical migration or residence in convergences associated with internal waves play a role in movement (Shanks 1983; Kingsford and Choate 1986). Fishes spawning in the same offshore habitat may ultimately have different larval distributions, suggesting that small behavioral differences among species may alter susceptibility to passive transport (Powles 1981). The groups using estuaries typically are transported by drift to the nearshore environment (Rothlisberg et al. 1983; Miller et al. 1984; Boehlert and Mundy 1987). Nelson et al. (1977), for example, suggested that wind-driven Ekman transport is the mechanism of movement to the nearshore area, and that inter-annual variability in transport ultimately leads to variation in year-class strength.

Although movement to nearshore areas may be passive, the overwhelming evidence from both experimental and field work is that immigration of early life stages into the estuary is an active behavioral process. Such evidence exists for a variety of fish families, including anguillids (Creutzberg 1961; McCleave and Kleckner 1982), sparids (Pollock et al. 1983; Tanaka 1985), sciaenids (Weinstein et al. 1980), bothids (Weinstein et al. 1980), pleuronectids (Creutzberg et al. 1978; Tsuruta 1978; Rijnsdorp et al. 1985; Boehlert and Mundy 1987), and albulids (Pfeiler 1984). Still, the behavioral responses to physical factors influencing estuarine recruitment are poorly understood for the majority of taxa. In this paper, we consider the evidence for behaviorally mediated entry into estuarine nursery areas as a function of physical factors.

Physical Factors Influencing Estuarine Recruitment

We recognize two major phases of movement necessary for recruitment to estuaries by species spawned offshore. The first is accumulation of larvae in the nearshore or coastal zone as described for Atlantic coast fishes by Nelson et al. (1977) and Miller et al. (1984); the second is the process of accumulation near inlets and estuary mouths and eventual passage through them. Each process involves a distinct set of physical factors to which the larvae respond and thus a distinct set of behaviors needed by the larvae to reach their optimum environment.

Nearshore Accumulation

Accumulation in the nearshore zone is essentially passive, because larvae at this stage are typically planktonic; directed vertical movements, however, may modulate this transport to maximize shoreward movement. Accumulation can be prolonged, because many species are near ages of 60–90 d when immigration to estuaries occurs (Rosenberg 1982; Miller et al. 1984), and the youngest larvae are generally found most distant from the nearshore zone (Miller et al. 1984). Behaviors associated with shoreward movement are likely related to distribution in the water column and have evolved to take advantage of mean current conditions in the species' habitat. Behavioral differences between species can result in different distributions. Richardson and Pearcy (1977), for example, described coastal (0–28 km) and offshore (beyond 28 km) assemblages of ichthyoplankton that are persistent from year to year along the Oregon coast. Many species in the offshore assemblage are spawned along the coast, in some cases in the same season as species that remain in the nearshore assemblage. The species within these two groups show clear differences in their use of estuaries (Table 1). Thus, passive movements by larvae must be modulated by a behavioral component that, under mean conditions, results in the observed distributions.

Surface drift has been implicated in many studies of planktonic distribution, but the evidence from larval distributions is not always clear. The physical environment of the Atlantic menhaden *Brevoortia tyrannus,* including the surface and bottom currents, was investigated in the Chesapeake Bight by Harrison et al. (1967), who concluded that benthic orientation of larvae and their occurrence in deep water could result in passive transport to the mouth of Chesapeake Bay. Nelson et al. (1977), however, suggested that surface drift from Ekman transport was the operating transport mechanism. Later studies suggested an even more complex picture; Miller et al. (1984) proposed that both surface and bottom waters move offshore in winter months but that midlevel waters move onshore. Abundance of Atlantic menhaden larvae does not differ between deep and shallow water (Kendall and Reintjes 1975); thus a midwater distribution and avoidance of the surface and bottom offshore transport layers may result in onshore transport.

Other evidence exists for the importance of surface water transport. Early life stages of the English sole *Parophrys vetulus,* one of the few fish species on the west coast of North America that spawns offshore and uses estuaries as nursery grounds (Krygier and Pearcy 1986), are members of the nearshore ichthyoplankton assemblage (Richardson and Pearcy 1977; Mundy 1984). Recruitment to estuaries occurs on night flood tides for fish already beginning morphological transformation. A negative correlation exists between recruitment pulses of early stage larvae and the upwelling index (Figure 1). Larvae in early stages of transformation typically enter the estuary throughout the water column, whereas later-stage larvae arrive deeper in the water column. The correlation with the upwelling index is strong for

TABLE 1.—Selected species of fish larvae collected in abundance from Oregon marine waters (Richardson and Pearcy 1977), classified by distribution, by abundance in Yaquina Bay, Oregon, and by stage of development when found in the bay. Asterisks indicate species using the bay as a nursery. Distribution categories were modified from Richardson and Pearcy (1977); taxa categorized as "bay" include those found to be abundant by Pearcy and Myers (1974) and Mundy and Boehlert (unpublished data).

Distribution	Abundance in Yaquina Bay	Species	Developmental stage in Yaquina Bay
Offshore	Absent	*Tarletonbeania crenularis*	
		Microstomus pacificus	
	Rare	*Engraulis mordax*	Postflexion
		Stenobrachius leucopsarus	Preflexion
		Glyptocephalus zachirus	Preflexion
	Common	*Sebastes* spp.	Preflexion
Offshore and coastal	Common	*Hemilepidotus hemilepidotus*	Preflexion
Coastal	Absent	*Radulinus asprellus*	Preflexion
	Rare	*Microgadus proximus*	Preflexion, flexion
		Isopsetta isolepis	Preflexion
		Platichthys stellatus	Pre- to postflexion
	Common	*Clinocottus acuticeps*	Pre- to postflexion
		Psettichthys melanostictus	Postflexion
	Abundant	*Clupea harengus pallasi**	Pre- to postflexion
		Osmeridae	Pre- to postflexion
		Anoplarchus spp.*	Pre- to postflexion
		Pholis spp.*	Pre- to postflexion
		*Enophrys bison**	Pre- to postflexion
		*Leptocottus armatus**	Pre- to posflexion
		*Parophrys vetulus**	Preflexion (few), postflexion
		*Cottus asper**(?)	Pre- to postflexion
Bay only	Rare	*Apodichthys flavidus**(?)	Pre- to postflexion
	Common	*Gobiesox maeandricus**	Pre- to postflexion
		*Ascelichthys rhodorus**	Preflexion
	Abundant	*Lepidogobius lepidus**	Preflexion to ?

the abundance of surface-captured early larvae, but weak for numbers of deep-captured late larvae (Figure 1). Boehlert and Mundy (1987) suggested that early larvae were moving inshore along the entire coast during periods of onshore transport. Transforming larvae of English sole are present in the neuston and depend upon surface Ekman transport to arrive at the nearshore zone; sampling with large neuston nets has shown relatively high densities of transforming larvae of this species in the upper meter during months of peak recruitment (Shenker 1985). In contrast, late larvae enter the estuary along the bottom; they apparently recruit first to nearshore habitats, then migrate to the estuary mouth (Krygier and Pearcy 1986; Boehlert and Mundy 1987). The mean current patterns on the west coast may have resulted in the evolution of reproductive patterns such that most fish species spawn in the winter and early spring months when onshore surface transport dominates (Parrish et al. 1981). Deviations from the mean pattern of surface water transport may result in poor year-class strengths for species such as the Pacific hake *Merluccius productus* (Bailey 1981).

Several other mechanisms have also been proposed for movement to the nearshore zone, including population maintenance in eddies (Sale 1970). Eddy mechanisms have typically been used to explain the maintenance of populations associated with islands (Boden 1952; Emery 1972), but coastal studies off California suggest that eddy mechanisms might be important there as well (Hewitt 1981; Owen 1981). Population maintenance in the nearshore zone in an upwelling area may also be accomplished by use of countercurrent systems, whereby drift is modulated by changes in vertical distribution (Peterson et al. 1979; Wroblewski 1982; Figure 2). Such a mechanism, termed "larval navigation" by Crisp (1974), could maintain the nearshore ichthyoplankton assemblage noted off Oregon. Finally,

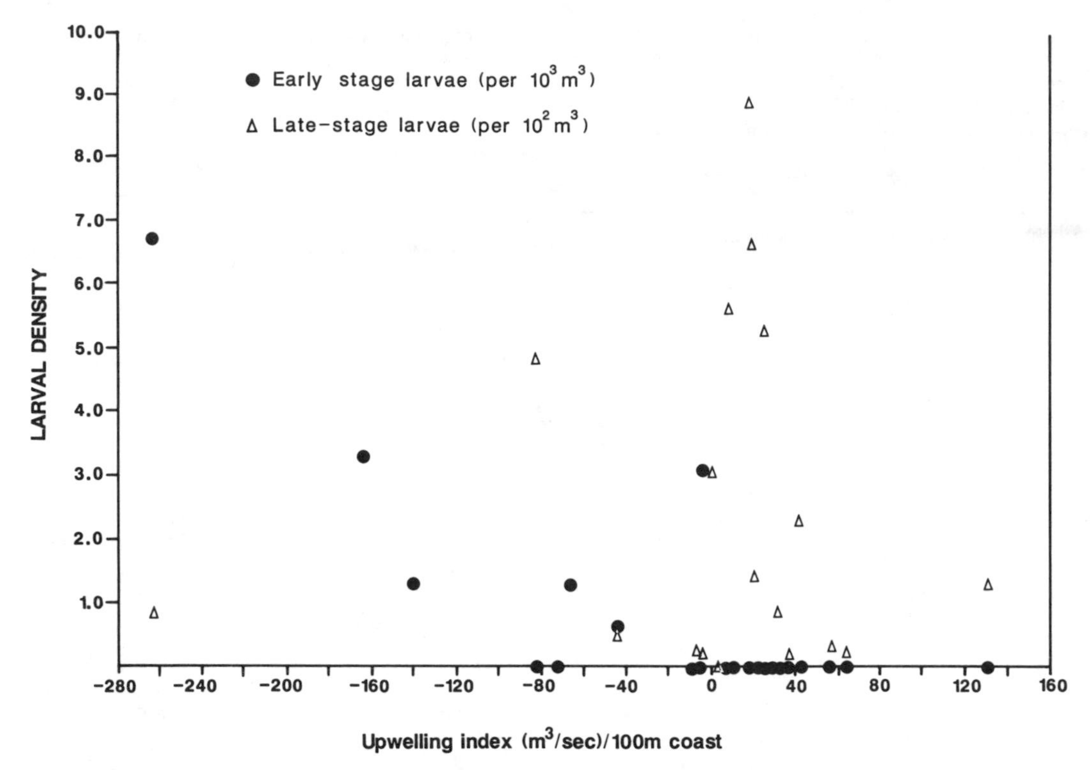

FIGURE 1.—Density of transforming larvae of English sole recruiting to Yaquina Bay, Oregon, as a function of the upwelling index. Early-stage larvae were taken by shallow nets; late-stage larvae were caught in nets set near the bottom, where these larvae are in higher abundance. Early larvae are abundant only during those times when the upwelling index is negative, suggesting onshore Ekman transport.

actual inshore movement may be modulated by accumulation in surface slicks, which in turn may be associated with internal waves (Shanks 1983, 1985; Kingsford and Choate 1986). Norris (1963) suggested a behavioral mechanism of orientation to these slicks; avoidance of the lower temperature at internal wave fronts by young opaleyes *Girella nigricans* would facilitate shoreward movement.

Estuarine Recruitment

Once larvae have recruited to the nearshore environment, a new set of physical factors influence their accumulation at inlets or estuarine mouths and their movements upstream into the estuary. Alongshore drift must play a key role in the movement of larvae to areas under estuarine influence. This influence may be more important on the west coast of North America, where only 10–20% of the coast is estuarine, than along the Atlantic and Gulf of Mexico coasts, which are 80% estuarine (Emery 1967). Krygier and Pearcy (1986) observed increases in the density of larval English sole in a nearshore nursery area distant from any estuarine influence. As densities at that location decreased, however, densities in Yaquina Bay, some 10 km to the south, increased, demonstrating a linkage between the nearshore and estuarine nursery areas. A southern alongshore drift may have brought the larvae near the estuarine mouth. The transition from northward to southward alongshore drift off Oregon typically occurs during March (Huyer et al. 1975), a period when the metamorphosing larvae enter Yaquina Bay (Krygier and Pearcy 1986; Boehlert and Mundy 1987).

Several physical factors near inlets or estuaries may serve as "point source" stimuli that could elicit short-term behavioral responses by larvae. Tidal flux may ultimately lead to accumulation, but tidal flux is a complex of related factors that cannot be isolated in field studies. This complex includes current speed, salinity (as affected by both river discharge and magnitude of tidal mixing and exchange), temperature, olfactory cues, turbidity, bottom composition (grain size), and lunar

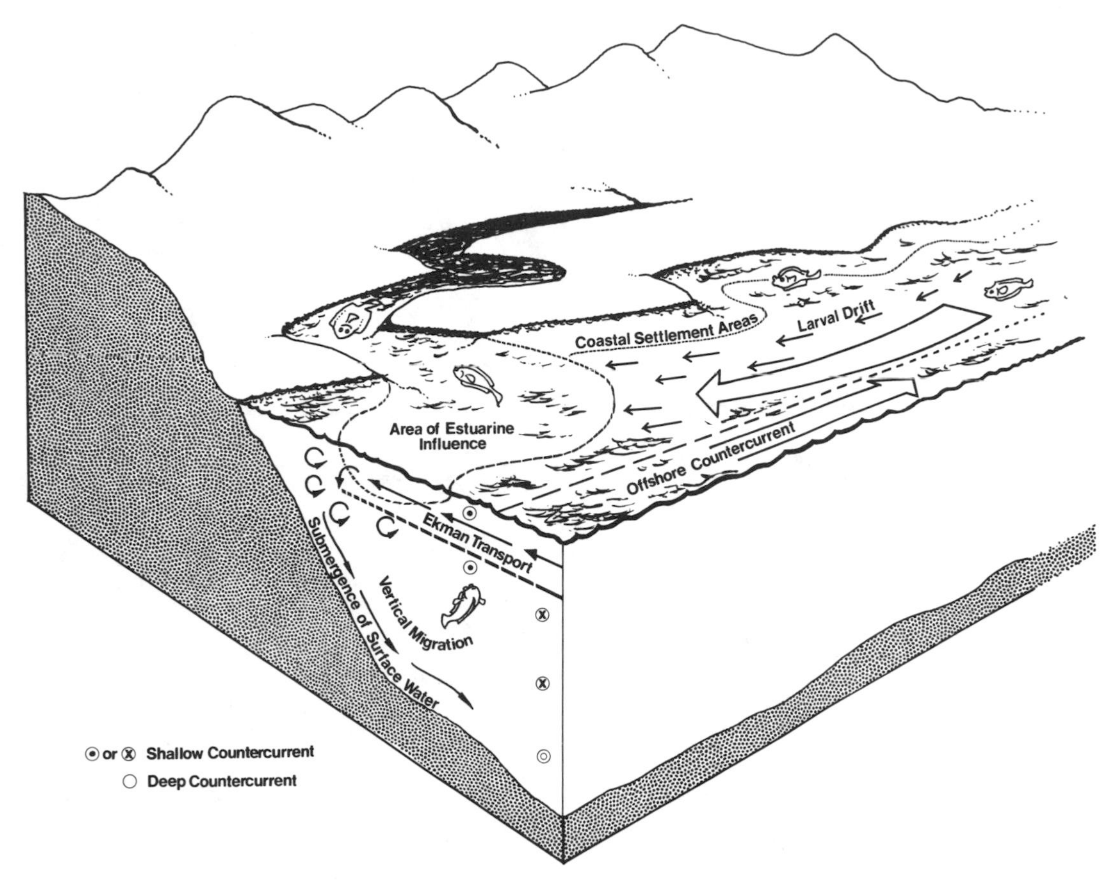

FIGURE 2.—Diagramatic representation of movements of larval fishes in the Oregon upwelling zone. Pelagic larvae may control onshore–offshore transport by their vertical movements, using surface Ekman transport to move shoreward (this condition pertains during the winter season off the Oregon coast). After larvae move near shore and settle, alongshore transport may result in movement to areas under estuarine influence.

phase. Individually and in concert, each factor may play a role in altering larval behavior patterns that facilitate recruitment. The complexity was pointed out by Hoar (1953), who suggested that "more rapid progress might be made if migrations which are not complicated by changes in osmotic medium are studied."

Many authors, including those cited previously, have monitored both recruitment to and maintenance in estuaries for a variety of taxa; results of these studies often are complex. Weinstein et al. (1980), who conducted one of the most detailed sampling studies, monitored movement of some commercially important species in an inlet in North Carolina. Catches of flounder *Paralichthys* spp. were greater on flood tides, particularly at night, which led the authors to suggest that the flounder postlarvae moved to the bottom during ebb tides. At night flood tides, flounders and spot *Leiostomus xanthurus* occurred high in the water column. Weinstein et al. (1980) suggested that this mechanism transported these two taxa to the tidal flats and creeks, which they used as nursery areas (Shenker and Dean 1979; Weinstein 1979, 1983). In contrast, postlarvae of Atlantic croaker *Micropogonias undulatus,* which use the head of the estuary, remained deeper in the water column, even at night. The model devised by Weinstein et al. (1980) may be used to explain both recruitment to and maintenance in estuaries (Figure 3); their results clearly show that species-specific behavioral responses to physical factors may result in different distributions within the estuary.

A suite of physical factors may serve as cues for such behavior. Several authors have investigated the relationship of animal distribution in the field to salinity and also the behavior of fish in laboratory tanks as a function of salinity. Many studies

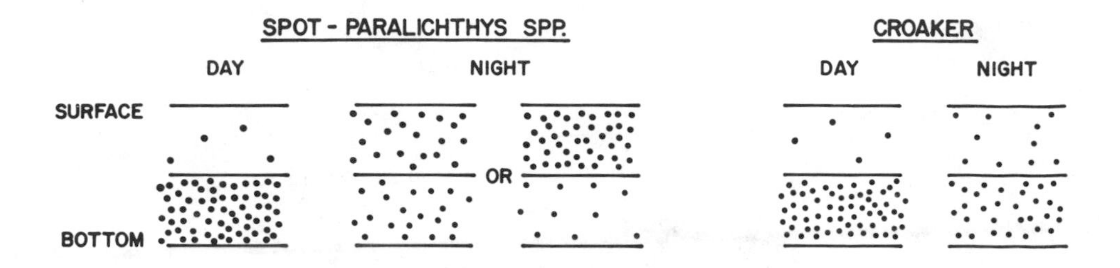

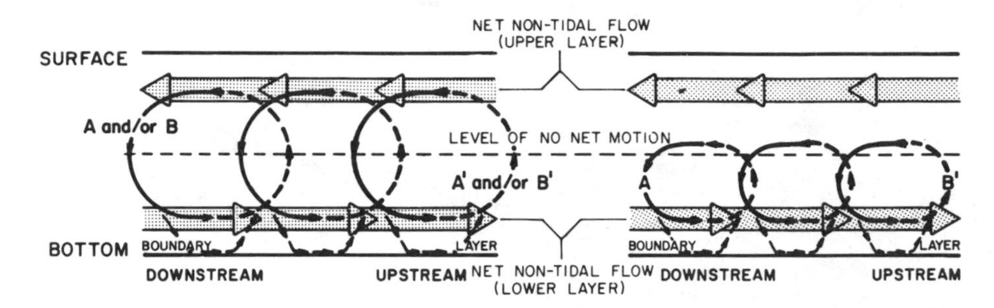

FIGURE 3.—A model of movements with the tide for three taxa of postlarval fishes in a North Carolina estuary, showing different strategies of tidal-stream transport. Spot and flounders use shallow tidal areas, whereas Atlantic croakers move to the upper reaches of the estuary, presumably by remaining deep in the water column at all times. (From Weinstein et al. 1980.)

have been based upon the classic work of Creutzberg (1961), who studied the immigration of European eel elvers *Anguilla anguilla* to tidal areas from the open sea. Field sampling demonstrated that the elvers swam actively in the water column on flood tides, but they were either near bottom or clinging to the bottom during ebb tides; this behavior pattern resulted in inshore movement. Creutzberg investigated the behavioral responses to increasing and decreasing salinity, but with "disappointing" results. More recent studies, particularly those with crustaceans, show clear responses to changes in salinity. Postlarval penaeid shrimp *Penaeus duorarum* move into estuarine nursery areas, whereas juveniles move out. Direct salinity changes, rather than some other factor, caused changes in behavior that resulted in the appropriate movements. Decreasing salinity caused benthic orientation by postlarvae and negative rheotaxis by juveniles; increasing salinity resulted in swimming behavior by the postlarvae but positive rheotaxis by the juveniles (Hughes 1969). The importance of salinity in determining the use of estuarine nursery areas by young shrimp is dramatically demonstrated by the strong correlation of offshore commercial shrimp catch with rainfall in the preceding year. Years with low rainfall have low salinity change near the estuaries and recruitment of postlarvae to estuaries is less successful. Salinity was also suggested as the controlling factor for emigration of the swimming crab *Macropipus holsatus* from estuarine areas (Venema and Creutzberg 1973); as with the juvenile shrimp, the crabs swim during decreasing salinity and settle to the bottom during times of increasing salinity, resulting in seaward movement. Physical variables are often highly intercorrelated, however, and speculations of causal relationships in such cases may be spurious. In our work with English sole, for example, there was a strong negative correlation between bottom salinity at the start of a flood tide and density of larvae recruiting to an estuary (Boehlert and Mundy 1987). Bottom salinity was highly correlated with other physical factors, however, including freshwater input, sea level, surface salinity, surface temperature, and other factors that may be related to recruitment. When larvae migrate onto nonestuarine tidal flats, they may encounter only very slight salinity differences, as in Elkhorn Slough, California (Yoklavich 1982), in the Gulf of Carpenteria, Australia (Staples 1980), or in some tropical areas (Pfeiler 1984). In the tropics, increased salinity caused by evaporation in hypersaline lagoons may allow salinity gradients to develop, but the salinity change is opposite that in estuarine systems. One must thus use caution in interpreting cause and effect in relationships between recruitment and single physical factors.

Temperature gradients between estuarine and offshore waters may be strong in temperate regions, particularly in winter months. Temperature is a candidate physical orientation factor for larvae because fish show temperature tolerances and preferences (Brett 1970). The majority of studies of temperature and migration, however, describe temperature-initiated migration out of an area. Hoar (1951), for example, showed that increasing temperature causes a change from positive to negative rheotaxis by young salmonids *Oncorhynchus* spp., facilitating downstream migration. Olla et al. (1980) showed that the tautog *Tautoga onitis* migrates to deeper water as temperature increases. Animals moving into estuaries might be seeking a preferred temperature, but more research is needed before any role of temperature in recruitment can be understood.

Other cues associated with water chemistry may play important orientation roles. Creutzberg's "disappointing" results with European eel elvers occurred when he altered salinity with tap water and observed no response, but when he diluted seawater with natural inland waters the elvers showed swimming behavior during increasing salinity and benthic orientation during decreasing salinity. Creutzberg (1961) suggested that some "attractive substance" in inland waters was the important factor in the immigration, rather than salinity alone. These results were later confirmed by Miles (1968) with elvers of the American eel *Anguilla rostrata*. He observed that some biodegradable component of either dissolved or particulate organic matter played a role. Thus some unknown "inland water odor," sensed through olfaction and similar to home-stream odors detected by adult salmonids (Hasler et al. 1978), may be involved. An alternative, but related, "odor" hypothesis is that a species- or population-specific pheromone may be involved (Nordeng 1977), but little evidence exists for this hypothesis.

Another olfactory cue is food odor, which may trigger activity if predatory larvae are unsatiated. Research on the movements of larval plaice *Pleuronectes platessa* in tidal areas showed that larvae were three times as abundant in the water column on flood than on ebb tides, resulting in net inshore transport (Creutzberg et al. 1978). Preliminary experimental data, however, suggested a mechanism of immigration different from that of other species studied. In the laboratory, neither salinity nor nursery "odors" stimulated behavior appropriate for transport, but starved animals exhibited greater swimming activity in the water column than fed animals. Although Creutzberg et al. (1978) proposed that tidal areas act as "traps" due to the abundance of food (as compared to offshore areas), they could not explain the stimulus for settlement. Gradients of food abundance may play a role in accumulation of young fish; Tanaka (1985) suggested that a gradient of copepod abundance, increasing from offshore to inshore, could lead immigrating red sea bream *Pagrus major* to inshore areas where the preferred prey, gammaridean amphipods, are present. Larval Atlantic herring *Clupea harengus harengus* may show vertical distributions within estuaries that are adaptive for population maintenance, but Fortier and Leggett (1983) suggested that these movements are simply a behavioral response to vertical movements of their prey organisms.

A variety of other candidate physical factors exist near entrances to estuaries that may elicit behavior appropriate for recruitment. Currents, a dominant feature of tidal flux, may be used, but only by fishes able to orient to a surface. Rheotaxis by young salmonids, for example, is lost in darkness, resulting in downstream movements (Hoar 1958). Transforming plaice, on the other hand, show different rheotactic responses depending upon their distance from a solid surface; within 3 cm of the surface, they are able to show normal rheotaxis even in darkness, but at greater distances they lose the ability to orient (Arnold 1969).

The use of electric or magnetic fields for orientation by fish larvae has not yet been demonstrated, but the elvers of American eels can detect relatively weak electric fields (McCleave and Power 1978). The thresholds for detection may be such that the relatively strong water currents (which cause electric fields) in regions of high tidal flux could be used for orientation. If this were the case, there would be no need for visual or tactile contact with the bottom, as discussed above with respect to rheotaxis.

Turbulence, turbidity, and light may interact to affect distributions. Fore and Baxter (1972) suggested that larval Atlantic menhaden seek less turbulent water and thereby accumulate in shallows and tidal creeks. In contrast, Blaber and Blaber (1980) observed the highest densities of juvenile fishes in the most turbid reaches of estuaries; negative phototropism may be involved in accumulation in such areas. Residence in turbid areas may result in decreased predation (Moore and Moore 1976; Gardner 1981). Some larvae, such as those of Pacific herring *Clupea harengus pallasi,* may indeed be adapted for feeding and survival in turbid estuaries (Boehlert and Morgan 1985). Rijnsdorp et al. (1985) suggested another relationship with turbulence; they observed a high correlation of suspended matter with the abundance of immigrating plaice larvae, suggesting that larvae may be swept up by stronger tides along with suspended materials. Flood tides at their study sites were typically stronger than the ebb tides, so their observations suggest that passive transport alone, with no behavioral change, could be important for immigration. Working with the young larvae of Atlantic herring, however, Fortier and Leggett (1983) noted no overall vertical dispersion during times of greatest vertical mixing of an estuary, contrary to the expectation that larvae would be passively mixed by turbulence (Fortier and Leggett 1982).

Lunar phase may affect transport of larvae into estuaries, but the mechanism is unknown. From a long-term study, Williams and Deubler (1968) suggested that lunar phase affects the timing of entry into estuaries for *Paralichthys* postlarvae and also for postlarval penaeid shrimps *Penaeus duorarum;* both groups were captured in greater numbers near the times of new moons than of full moons. This was later confirmed for penaeid shrimps by Roessler and Rehrer (1971). Similarly, catches of *Anguilla* elvers may show lunar periodicities (Jellyman 1979; Tzeng 1985). Tzeng (1985) obtained peak catches of riverine elvers at both new and full moons, correlated with spring tides. In coastal waters, however, peak catches occurred only during the new moon, suggesting that light inhibited activity there (Tzeng 1985). In all these studies, net avoidance is a confounding factor that may decrease catches during full moon. We observed no lunar periodicity in the immigration of larval English soles to Yaquina Bay, Oregon (Boehlert and Mundy 1987). Lunar phase is also coupled to changing tidal magnitude and, therefore, to salinity and temperature changes, current speed, and turbulence. Field or laboratory studies designed specifically to test hypotheses about the interrelationships among physical factors affecting immigration rates would be necessary to determine the roles of these physical factors.

Behavioral Responses to Estuarine Stimuli

The ability of organisms to orient to environmental stimuli underlies their movement to and retention in habitats that are optimal for survival. Aquatic organisms that inhabit different environments during ontogenetic development must respond to changing stimuli to orient to each new habitat. Recruitment to estuaries, we believe, is a behaviorally mediated rather than passive migration. To investigate migration, Hoar (1953) suggested that research " . . . should start with the behavior patterns of the fish; it should involve an analysis of the internal physiological states responsible for this restless appetitive behavior, and eventually, an understanding of the way in which this activity is guided and directed by the variables in the external environment." Behavior has generally been inferred from distribution in the field (changes in abundance), and correlation or statistical approaches have been used to relate results to physical variables. The appetitive behaviors necessary for recruitment may begin as either random movements (kineses) or directed movements (taxes). Organisms may accumulate during kinesis by changing rates of activity or rates of turning (Hunter and Thomas 1974) and during taxis by slowing movement when they reach the source of a directional stimulus. There are relatively few physical factors with vector properties; therefore, directed movements may be limited to phototaxis, geotaxis, and rheotaxis in response to light direction, gravity, and currents, respectively (Crisp 1974). Each of these factors might guide estuarine immigration, but they must be modulated by scalar factors such as salinity, temperature, turbidity, and olfactory cues; sequential sampling of a scalar factor through time or space may allow development of a directional sense through a learned response. These more advanced behavioral sequences, however, may be beyond the physical and sensory capability of larval or juvenile fishes, because current speeds in tidal inlets may prevent young fishes from controlling their horizontal position.

Tidal-stream transport is commonly cited as an example of vertical movement affecting horizontal displacement (Tsuruta 1978; Weinstein et al.

1980). Generally, the control of larval vertical distribution may arise either through taxis or kinesis. Fish larvae are frequently concentrated near thermoclines (Ahlstrom 1959; Loeb 1979), and kineses may explain accumulation over time within specific thermal regimes. Kinetic motion may be the most efficient way for larvae to find food under some circumstances (Hunter and Thomas 1974). Random kinetic motions cannot accomplish directed vertical movements, which must, therefore, be associated with a taxis or (as suggested by Fortier and Leggett 1983) with feeding excursions. No single mechanism has yet emerged as a general control of vertical placement for tidal-stream transport. Phototaxis of fish larvae has been documented extensively in the field and laboratory (Ahlstrom 1959; Blaxter and Staines 1970; Blaxter 1974; Hunter and Sanchez 1976; Kendall and Naplin 1981; Yamashita et al. 1985), but it typically operates on diel cycles. Reaction to both direction and intensity of light may explain the presence of immigrating larvae in night rather than day flood tides (Weinstein et al. 1980; Boehlert and Mundy 1987), but it cannot provide the temporal cue for activity on flood and not ebb tides. Similarly, geotaxis cannot account for observed periodicities in activity by larvae.

Rheotaxis of fishes plays an important part in tidal-stream transport; the subject has been reviewed by Arnold (1974, 1981). Fish can apparently detect current speeds of 1–10 cm/s (Arnold 1981), which are within or below the range of tidal currents, and the lateral line system, which detects water movements, develops relatively early in fish larvae (Iwai 1967). Thus, young fish may use currents and rheotaxis for recruitment to an estuary. This use is constrained by the need for either visual or tactile contact (Arnold 1969), as described earlier. Tidally generated turbulent fields within estuaries, however, may provide tactile cues under certain conditions (McCleave and Kleckner 1982).

Development and Behavioral Ontogeny in Relation to Movements

The stage of development at which fishes are present in inlets or estuaries may in large part determine their ability to behaviorally alter their distribution. For species spawned within an estuary, early larvae may be advected out, particularly from intensively flushed estuaries (Rogers 1940; Johnson and Gonor 1982). For those species using the estuary for a nursery, development may continue in offshore waters; Pacific herring, for example, are abundant in estuaries at certain times of year (Pearcy and Myers 1974), but they are also found in the nearshore assemblage of ichthyoplankton (Richardson and Pearcy 1977; Table 1). Resident species may be temporarily exported only to return to the estuary at a later stage (Johnson and Gonor 1982). In general, species that recruit to estuaries from offshore do so at advanced developmental stages, usually near or after metamorphosis; this is true for several orders of fishes, including Anguilliformes (Creutzberg 1961), Elopiformes (Pfeiler 1984), Clupeiformes (Miller et al. 1984), Perciformes (Weinstein et al. 1980; Tanaka 1985), and Pleuronectiformes (Tsuruta 1978; Boehlert and Mundy 1987).

Important changes occur at metamorphosis in sensory systems, behavior, and morphology that affect immigration to estuaries. Metamorphosis entails full development of fins and associated structural calcification, which result in better swimming capabilities (Blaxter and Staines 1971). Fukuhara (1985), for example, noted an increase in swimming speed of red sea bream from 1 body length/s for larvae to 4 body lengths/s for metamorphosed juveniles. Greater swimming ability may decrease a larva's chance of being flushed from the estuary, even during times of high freshwater input (Rogers et al. 1984). Many species undergo rapid changes in color and morphology in association with the change from pelagic to benthic existence (Hubbs 1958; Hunter 1967; Kendall et al. 1984). Major changes in structure and function of the retina occur (Blaxter 1974; Boehlert 1978; Kawamura and Hara 1980), including development of rods in the retina (Blaxter 1974) that are necessary for vision in dim light. Flatfish larvae undergo great changes in behavior at the time of metamorphosis (Fluchter 1965; Finger 1976; Gibson et al. 1978); they become benthic, and their rheotactic behavior effectively ceases (Arnold 1969). Kawamura and Ishida (1985) gave a detailed account of the changes in sensory systems and behavior from hatching to 72 d posthatch for the flounder *Paralichthys olivaceus*. In particular, they noted development of retinal rods at 25 d, disappearance of positive phototaxis at 30 d, and nocturnal activity at 33 d posthatch. It is during this 9-d period that the transition from the pelagic to benthic habitat occurs. Pfeiler (1984) described changes during metamorphosis of bonefish *Albula* sp., which shrink 60–65% as they enter hypersaline lagoons.

Generalizations about the methods of immigration for different fishes are difficult, but we wish to distinguish between perciform, anguilliform, and pleuronectiform fishes. Tanaka (1985) suggested that tidal-stream transport, which is important to flatfish and eels (Creutzberg 1961; Tsuruta 1978; McCleave and Kleckner 1982), is of secondary importance to perciform juveniles, which have more advanced swimming abilities but, perhaps more importantly, no ability to settle on the bottom. As they near metamorphosis, larval plaice and flatfish generally spend increasing amounts of time near the bottom, often alternating pelagic and benthic behavior (Fluchter 1965; Blaxter and Staines 1971). Given the proper behavior patterns with respect to tidal movements, these larvae could immigrate inshore using tidal-stream transport during flood tides and settling on the bottom during ebb tides; the energetic savings of this mode of migration approach 90% for juvenile fish (Weihs 1978). Blaxter and Staines (1971) observed that when larval plaice metamorphosed, their movement rates dropped from about 70 to 4 cm/min, the time they spent actively swimming changed from 95 to 10%, and the volume of water they searched (for food) dropped from 2,500 to 50 mL/h. This is a clear distinction from the changes noted above for red sea bream, a perciform (Fukuhara 1985). Although perciform fish use tidal stream transport (Weinstein et al. 1980), the benefits may be relatively low due to their inability to settle out on the bottom.

Relatively little work has been done on the influence of habitat type on larval settlement. Certain species may be able to delay transformation until they encounter suitable juvenile habitat (Pearcy et al. 1977; Moser 1981); this has recently been demonstrated, based on daily growth increments, for a coral reef fish (Victor 1986). Marliave (1977) found significant substrate preferences by settling larvae in five of six benthic species tested. The preferences seemed to be for structural features of the substrate that gave tactile cues to the settling larvae. Metamorphosing flatfish alternate between pelagic swimming and benthic resting behavior (Fluchter 1965), a behavior that may aid settling larvae in locating preferred habitats.

Role of Endogenous Rhythms

Another behavioral consideration important in recruitment to estuarine nursery areas is endogenous activity rhythms. As already noted, many of the physical factors associated with tidal flux have clear temporal signals but lack directionality. Temporal cues, or "zeitgebers," exist for setting internal clocks for individuals in nearshore areas under the influence of an estuarine system. A wide variety of evidence supports the presence of endogenous rhythms in fishes (see Thorpe 1978), including studies of activity patterns (Gibson 1965), retinomotor movements (Olla and Marchoni 1968), photosensitivity (Davis 1962), and biochemical patterns (de Bras 1979). The actual timekeeping mechanism may be a chemical phenomenon mediated by the pineal gland (Binkley et al. 1978; Matty 1978).

Endogenous rhythms may be daily, tidal, or lunar in periodicity. They can be observed experimentally, in the absence of temporal stimuli, for varying periods before they revert to a free-running rhythm. They are synchronized by normal environmental cycles to the appropriate periodicity (Enright 1975). Within species, endogenous rhythms may change during ontogeny. Planktonic plaice larvae are most active near the surface at night, but the rhythmicity is a function of light, rather than of any endogenous rhythm, because it disappears in constant light or darkness (Gibson et al. 1978). After planktonic larvae undergo a short period of relatively little activity (Blaxter and Staines 1971) and settle, newly benthic larvae maintain diel activity peaks. Populations remaining offshore retain this diel pattern, but nearshore or estuarine populations exhibit marked tidal activity patterns that are often of an endogenous nature (Gibson 1978; Gibson et al. 1978). That these rhythms change with changing habitat suggests that they may play an important role in habitat selection.

Endogenous rhythms in general were reviewed by Enright (1975); tidal rhythms have been reviewed by Palmer (1973) and Gibson (1978). Tidal rhythms generally relate to activities important for foraging, respiration, and movement and to other activities affected by tidal cycles. Tidal rhythms have been demonstrated for many species of fishes (Gibson 1978). The entrainment of the circatidal rhythm must occur after young fish come under the influence of estuaries in nearshore waters. Appropriate zeitgebers for synchronizing circatidal rhythms include pressure (Gibson 1971) and turbulence (Enright 1965) and, possibly, inundation and temperature (Palmer 1973). Cyclic changes in salinity in estuarine areas could also act as a zeitgeber. It is important that rhythms are set by local conditions, because tides differ in magnitude and phase over the range of most species. Some species such as the English sole may use different cues in different areas, because some estuaries they occupy have freshwater inputs and others have nearly none (Yoklavich 1982; Boehlert and Mundy 1987). If such species develop endogenous rhythms as they immigrate to estuaries, it is likely that the suite of physical factors associated with tidal flux, rather than a single factor, may serve as the zeitgeber. This is supported by the lack of a conclusive experimental identification of the zeitgeber for a tidal rhythm. Also, laboratory-reared animals have circadian rhythms instead of the circatidal rhythms found in field-collected animals of the same species (Gibson et al. 1978; Cronin and Forward 1979, 1983).

Hughes (1972) demonstrated endogenous, circatidal rhythms of swimming by the shrimp *Penaeus duorarum* over 3 d under constant conditions; he suggested that the postlarval shrimp were pelagic, drifting with the current during flood tide but exhibiting benthic behavior during ebb, resulting in movement into and up the estuary, much in the way described for fish by Weinstein et al. (1980; Figure 3). The significance of such a circatidal rhythm is obvious; if a single cue such as salinity or odor is used as a stimulus for pelagic or benthic behavior, there is no mechanism for the animal to detect the end of flood tide (McCleave and Kleckner 1982). Because the drifting animal will remain in the same water mass until it mixes with ebbing water, considerable downstream transport may occur. Linking the behavioral pattern to a circatidal rhythm would allow the animal to settle from the water column at the end of flood tide in the absence of stimuli associated with water quality, resulting in more efficient migration. Such rhythms must play a key role in tidal stream transport.

Behavioral Model of Recruitment to Estuaries

In studies of fish recruitment to estuaries, researchers typically consider spatial and temporal patterns of distribution and correlate them with a wide range of physical variables to infer the behavioral responses by the immigrating fish (Tsuruta 1978; Boehlert and Mundy 1987). In the laboratory, however, effects of physical factors generally are considered in isolation (Creutzberg 1961), often with somewhat surprising or enigmatic results (Creutzberg et al. 1978). Problems may arise due to artifacts associated with the laboratory environment, as shown by different behaviors of laboratory-reared and field-captured animals (Gibson et al. 1978; Cronin and Forward 1979, 1983). In this section, we consider a conceptual model for the behavior of fish and their responses to physical variables or other stimuli as they affect recruitment to estuaries. Such a model may be useful in the development of testable hypotheses.

Certain field studies have suggested mechanisms of estuarine recruitment that are entirely passive and require no behavioral response on the part of the larvae (Rijnsdorp et al. 1985). The majority of information on fishes, however, suggests otherwise. Most movements can be classified as a form of tidal-stream transport, and thus behavioral considerations associated with migration are pertinent, but we frame our model in terms of the process of habitat selection. Sale (1969) considered habitat selection as a negative feedback mechanism; stimuli from the environment, monitored by sense organs, feed into a "selection mechanism" which then triggers "appetitive exploration." In fishes, appetitive exploration may take many forms, from simple activity to changes in vertical distribution to rheotaxis or phototaxis. As a kinetic movement, such searching behavior may ultimately result in accumulation in more favorable environments where random activity will be reduced (Hunter and Thomas 1974). Sale suggested that in an adequate or optimal environment, the intensity of feedback is relatively low, which may eliminate or reduce appetitive exploration. In an inadequate environment, the feedback intensity increases, resulting in a greater percentage of the animal's time being spent in searching behaviors. Sale (1969) provided support for this hypothesis for habitat selection by showing that young manini *Acanthurus triostegus* reduce their exploratory behavior in an appropriate environment. Because fishes change optimal habitats with growth, the stimuli important for the "selection mechanism" most likely change throughout ontogeny as habitat requirements change. This was observed by Norris (1963) in a study of the selection of thermal habitats by opaleyes.

In the context of an estuarine system, Sale's model is more applicable than are simple concepts of taxes or kineses. The suite of physical factors is complex and, during ontogeny, larvae must select among several habitats. Our conceptual model (Figure 4) is generalized to allow application to a variety of species. We consider four habitats sequentially occupied by migrants; the dominant physical process in the habitat, the stimuli inducing the fishes' behavior, and the dominant activity pattern of the fish all determine the transition from

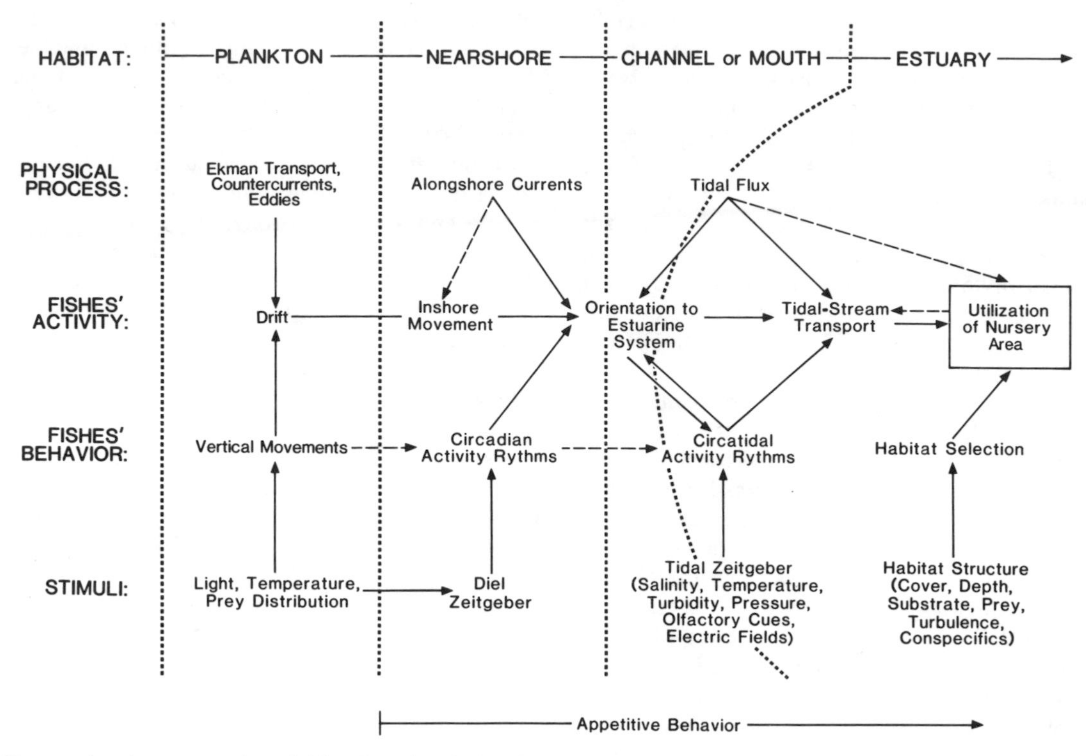

FIGURE 4.—A conceptual model for the role of stimuli and behavior in fish movement to estuarine nursery areas.

one habitat to the next. We first consider the offshore planktonic habitat, where the eggs and larvae are distributed. Here, the dominant physical factors are Ekman transport, countercurrents, and eddies (Parrish et al. 1981; Miller et al. 1984; Figure 2). Light, temperature, and prey distribution may be the key stimuli affecting vertical distribution, which may control the passive pattern of drift, both onshore–offshore and alongshore. This process may be one of critical importance to survival and one that may vary interannually, resulting in variations in year-class strength for both estuarine- and nonestuarine-dependent species (Nelson et al. 1977; Bailey 1981; Parrish et al. 1981).

Later larvae enter the phase of appetitive behavior as they begin searching for benthic habitat. A successful transition from the planktonic to the nearshore habitat involves "settlement," which may be taken literally in the case of a pleuronectiform but which may be referred to as "benthic orientation" for perciforms. In this initial settlement, substrate type and depth may be important factors (Marliave 1977). This is a major change usually involved with the process of metamorphosis. For species utilizing estuaries, the nearshore zone is a transitory habitat. Although we generally suggest that only late larvae recruit to the benthic nearshore habitat, more recent studies are documenting surprisingly large concentrations of various stages of fish larvae near bottom or rocky reefs (Brewer et al. 1981; Marliave 1986). Additional data may change our thinking about the process of recruitment to the nearshore zone. Later-stage larvae have increased powers of locomotion. At this stage, rhythmic activity may begin to play an important role in behavior and a circadian rhythm may develop (Gibson et al. 1978). Since the majority of fishes immigrate to estuaries at night (Pfeiler 1984; Tzeng 1985; Boehlert and Mundy 1987), it is probable that nocturnal activity is the rule; swimming activity, along with the movement induced by alongshore currents, may result in movement to the next transitory habitat, the channel or mouth of the estuary (the area of "estuarine influence" in Figure 2).

In Sale's (1969) hypothesis for habitat selection, the "intensity" of stimuli near the channel mouth would vary with a tidal periodicity. During ebb tides, settlement or bottom orientation might be induced by the factors associated with tidal flux (Figure 4), whereas the intensity of these stimuli would change during flood tides, as offshore water moves into the estuary. This condition would occur if a simple stimulus–response behavior exists. Such behavior patterns have been suggested for shrimp with salinity as the stimulus (Hughes 1969), for *Anguilla* elvers with "inland water odor" as the stimulus (Creutzberg 1961), and for plaice larvae with olfactory cues from prey as the stimulus (Creutzberg et al. 1978). Although such behavior would result in a form of tidal-stream transport, response to a single stimulus is a relatively simplistic mechanism and would not occur until a tidal change was well underway (McCleave and Kleckner 1982). Further, flow patterns around the mouths of estuaries are sufficiently complex to decrease the probability that a single factor changes consistently with a tidal periodicity (Ozsoy and Unluata 1982). Instead, we propose that a suite of stimuli associated with tidal flux acts as a zeitgeber to superimpose a circatidal rhythm upon the circadian rhythm which already may be present (O'Connor 1972; Gibson 1978). In this case, nocturnal swimming activity on flood tides and benthic orientation during other times could result in tidal-stream transport and movement up the estuary.

The process of estuarine recruitment may be a continuing one for some species that depend upon the process of habitat selection once inside the estuary (Figure 4). Weinstein et al. (1980) suggested that different vertical distribution patterns during tidal flux result in species-specific distributions within the estuary (Figure 3). The wide range of habitats present, and the new set of stimuli (vegetation, turbulence, prey distribution, substrate type) they provide, can lead to selection of shallow areas where flushing from the estuary is not a problem (Weinstein 1983; Rogers et al. 1984) or, conversely, to selection of areas near inlets where tidal-stream transport may be a continuing process necessary to maintain the fish within the estuary.

Sampling studies near the mouths of estuaries have provided a wealth of information on the movements of larval and juvenile fishes through inlets. Generally, these have led authors to generate hypothetical schemes of behavior involved in recruitment to and maintenance within estuaries. To date, laboratory studies (Creutzberg 1961; Miles 1968; Arnold 1969; Creutzberg et al. 1978; McCleave and Power 1978) on behavior of fishes relative to estuarine recruitment have tested some of the field-generated hypotheses on behavior, but they typically have used simplified designs that may not be meaningful to the field situation. More comprehensive studies on endogenous rhythms in fishes, including the array of zeitgebers as they relate to tidal-stream transport, are necessary; however, the behavioral complexity and activity patterns of fish larvae during transformation require that experimental facilities be larger than they have been so they can mimic natural environments.

Acknowledgments

Support for a portion of this work was provided by National Science Foundation grant OCE-80-25214. We thank Michael Weinstein for his work in the organization of the symposium where this paper was presented. Earlier drafts were improved by comments from reviewers and Robert Kendall.

References

Ahlstrom, E. H. 1959. Vertical distribution of pelagic fish eggs and larvae off California and Baja California. U.S. Fish and Wildlife Service Fishery Bulletin 60:106–146.

Arnold, G. P. 1969. The orientation of plaice larvae *Pleuronectes platessa* in water currents. Journal of Experimental Biology 50:785–801.

Arnold, G. P. 1974. Rheotropism in fishes. Biological Reviews of the Cambridge Philosophical Society 49:515–576.

Arnold, G. P. 1981. Movements of fish in relation to water currents. Pages 55–80 *in* D. J. Aidley, editor. Animal migration. Cambridge University Press, London.

Bailey, K. M. 1981. Larval transport and recruitment of Pacific hake, *Merluccius productus*. Marine Ecology Progress Series 6:1–9.

Binkley, S. A., J. B. Riebman, and K. M. Reilly. 1978. The pineal gland; a biological clock in vitro. Science (Washington, D.C.) 202:1198–1200.

Blaber, S. J. M., and T. G. Blaber. 1980. Factors affecting the distribution of juvenile estuarine and inshore fish. Journal of Fish Biology 17:143–162.

Blaxter, J. H. S. 1974. The eyes of larval fish. Pages 427–443 *in* M. A. Ali, editor. Vision in fishes. Plenum, New York.

Blaxter, J. H. S., and M. E. Staines. 1970. Pure cone retinae and retinomotor responses in larval teleosts. Journal of the Marine Biological Association of the United Kingdom. 50:449–460.

Blaxter, J. H. S., and M. E. Staines. 1971. Food searching potential in marine fish larvae. Pages 467–485 *in* D. J. Crisp, editor. Fourth European marine biological symposium. Cambridge University Press, Cambridge, England.

Boden, B. P. 1952. Natural conservation of insular

plankton. Nature (London) 169:697–699.

Boehlert, G. W. 1978. Intraspecific evidence for the function of single and double cones in the teleost retina. Science (Washington, D.C.) 202:309–311.

Boehlert, G. W., and J. B. Morgan. 1985. Turbidity enhances feeding success of larval Pacific herring, *Clupea harengus pallasi*. Hydrobiologia 123: 161–170.

Boehlert, G. W., and B. C. Mundy. 1987. Recruitment dynamics of metamorphosing English sole, *Parophrys vetulus,* to Yaquina Bay, Oregon. Estuarine, Coastal and Shelf Science 25:261–281.

Bousfield, E. L. 1955. Ecological control of the occurrence of barnacles in the Miramichi estuary. National Museum of Canada Bulletin 137.

Brett, J. R. 1970. Temperature—Pisces. Pages 515–560 *in* O. Kinne, editor. Marine ecology, volume 1. Wiley, New York.

Brewer, G. D., R. J. Lavenberg, and G. E. McGowan. 1981. Abundance and vertical distribution of fish eggs and larvae in the southern California bight: June and October 1978. Rapports et Procès-Verbaux des Réunions Commission Internationale pour l'Exploration Scientifique de la Mer 178:165–168.

Carriker, M. R. 1951. Ecological observation on the distribution of oyster larvae in New Jersey estuaries. Ecological Monographs 21:19–38.

Creutzberg, F. 1961. The orientation of migrating elvers (*Anguilla anguilla* Turt.) in a tidal area. Netherlands Journal of Sea Research 1:257–338.

Creutzberg, F., A. T. G. W. Eltink, and G. J. Van Noort. 1978. The migration of plaice larvae, *Pleuronectes platessa*, into the western Wadden Sea. Pages 243–251 *in* D. S. McLusky and A. J. Berry, editors. Physiology and behavior of marine organisms. Pergamon, New York.

Crisp, D. J. 1974. Factors influencing the settlement of marine invertebrate larvae. Pages 177–265 *in* P. T. Grant and A. M. Mackie, editors. Chemoreception in marine organisms. Academic Press, New York.

Cronin, T. W., and R. B. Forward, Jr. 1979. Tidal vertical migration; and endogenous rhythm in estuarine crab larvae. Science (Washington, D.C.) 205:1020–1022.

Cronin, T. W., and R. B. Forward, Jr. 1983. Vertical migration rhythms of newly hatched larvae of the estuarine crab, *Rhithropanopeus harrisii*. Biological Bulletin (Woods Hole) 165:139–153.

Davis, R. 1962. Daily rhythm in the reaction of fish to light. Science (Washington, D.C.) 137:430–432.

de Bras, Y. M. 1979. Circadian rhythm in brain catecholamine concentrations in the teleost *Anguilla anguilla* L. Comparative Biochemistry and Physiology C, Comparative Pharmacology 62:115–117.

Emery, A. R. 1972. Eddy formation from an oceanic island: ecological effects. Caribbean Journal of Science 12:121–128.

Emery, K. O. 1967. Estuaries and lagoons in relation to continental shelves. American Association for the Advancement of Science Publication 83:9–11.

Enright, J. T. 1965. Entrainment of tidal rhythm. Science (Washington, D.C.) 147:864–867.

Enright, J. T. 1975. Orientation in time: endogenous clocks. Pages 917–944 *in* O. Kinne, editor. Marine ecology, volume 2, part 2. Wiley, New York.

Finger, T. E. 1976. An asymmetric optomotor response in developing flounder larvae (*Pseudopleuronectes americanus*). Vision Research 16:395–403.

Fluchter, J. 1965. Versuche zur Brutaufzucht der Seczunge *Solea solea* in kleinen Aquarien. Helgolaender Wissenschaftliche Meeresuntersuchungen 12:395–403.

Fore, P. L., and K. N. Baxter. 1972. Diel fluctuations in the catch of larval menhaden *Brevoortia patronus* at Galveston entrance, Texas. Transactions of the American Fisheries Society 101:729–732.

Fortier, L., and W. C. Leggett. 1982. Fickian transport and the dispersal of fish larvae in estuaries. Canadian Journal of Fisheries and Aquatic Sciences 39:1150–1163.

Fortier, L., and W. C. Leggett. 1983. Vertical migrations and transport of larval fish in a partially mixed estuary. Canadian Journal of Fisheries and Aquatic Sciences 40:1543–1555.

Fukuhara, O. 1985. Functional morphology and behavior of early life stages of red sea bream. Bulletin of the Japanese Society of Scientific Fisheries 51:731–743.

Gardner, M. B. 1981. Effects of turbidity on feeding rates and selectivity of bluegills. Transactions of the American Fisheries Society 110:446–450.

Gibson, R. N. 1965. Tidal rhythmic activity in littoral fish. Nature (London) 207:544–545.

Gibson, R. N. 1971. Factors affecting the rhythmic activity of *Blennius pholis* (Teleostei). Animal Behaviour 19:336–343.

Gibson, R. N. 1978. Lunar and tidal rhythms in fish. Pages 201–214 *in* J. E. Thorpe, editor. The rhythmic activity of fishes. Academic Press, New York.

Gibson, R. N., J. H. S. Blaxter, and S. J. deGroot. 1978. Developmental changes in the activity rhythms of the plaice *Pleuronectes platessa,* L. Pages 169–186 *in* J. E. Thorpe, editor. The rhythmic activity of fishes. Academic Press, New York.

Haedrich, R. L. 1983. Estuarine fishes. Ecosystems of the World 26:183–207.

Harrison, W., J. J. Norcross, N. A. Pore, and E. M. Stanley. 1967. Circulation of shelf waters off the Chesapeake Bight. U. S. Department of Commerce, Environmental Science Services Administration Professional Paper 3, Washington, D.C.

Hasler, A. D., A. T. Scholz, and R. M. Horrall. 1978. Olfactory imprinting and homing in salmon. American Scientist 66:347–355.

Hempel, G. 1979. Early life history of marine fish. The egg stage. University of Washington, Washington Sea Grant Program, Seattle.

Hewitt, R. 1981. Eddies and speciation in the California Current. California Cooperative Oceanic Fisheries Investigations Report 22:96–98.

Hoar, W. S. 1951. The behavior of chum, pink, and coho salmon in relation to their seaward migration. Journal of the Fisheries Research Board of Canada 8:241–263.

Hoar, W. S. 1953. Control and timing of fish migration. Biological Reviews of the Cambridge Philosophical Society 28:437–452.

Hoar, W. S. 1958. Evolution of migratory behavior. Journal of the Fisheries Research Board of Canada 15:391–428.

Hubbs, C. L. 1958. *Dikellorhynchus* and *Kanazawaichthys:* nominal fish genera interpreted as based on prejuveniles of *Malacanthus* and *Antennarius*, respectively. Copeia 1958:282–285.

Hughes, D. A. 1969. Responses to salinity change as a tidal transport mechanism of pink shrimp *Penaeus duorarum*. Biological Bulletin (Woods Hole) 136: 43–53.

Hughes, D. A. 1972. On the endogenous control of tide associated displacements of pink shrimp, *Penaeus duorarum* Burkenroad. Biological Bulletin (Woods Hole) 142:271–280.

Hunter, J. R. 1967. Color changes of pelagic prejuvenile goatfish, *Pseudupeneus grandisquamis,* after confinement in a shipboard aquarium. Copeia 1967:850–852.

Hunter, J. R., and C. Sanchez. 1976. Diel changes in swimbladder inflation of the larvae of the northern anchovy, (*Engraulis mordax*). U.S. National Marine Fisheries Service Fishery Bulletin 74:847–855.

Hunter, J. R., and G. L. Thomas. 1974. Effect of prey distribution and density on the searching and feeding behavior of larval anchovy (*Engraulis mordax*) Girard. Pages 559–574 *in* J. H. S. Blaxter, editor. The early life history of fish. Springer-Verlag, New York.

Huyer, A., R. D. Pillsbury, and R. L. Smith. 1975. Seasonal variation of the alongshore velocity field over the continental shelf off Oregon. Limnology and Oceanography 20:90–95.

Iwai, T. 1967. Structure and development of lateral line cupulae in teleost larvae. Pages 27–44 *in* P. H. Cahn, editor. Lateral line detectors. Indiana University Press, Bloomington.

Jellyman, D. J. 1979. Upstream migration of glass-eels (*Anguilla* spp.) in the Waikato River. New Zealand Journal of Marine and Freshwater Research 13:13–22.

Johnson, G. E., and J. J. Gonor. 1982. The tidal exchange of *Callianassa californiensis* (Crustacea, Decapoda) larvae between the ocean and the Salmon River estuary, Oregon. Estuarine, Coastal and Shelf Science 14:501–516.

Kawamura, G., and S. Hara. 1980. The optomotor reaction of milkfish larvae and juveniles. Bulletin of the Japanese Society of Scientific Fisheries 46:929–932.

Kawamura, G., and K. Ishida. 1985. Changes in the sense organ morphology and behavior with growth in the flounder *Paralichthys olivaceus*. Bulletin of the Japanese Society of Scientific Fisheries 51:155–165.

Kendall, A. W., Jr., E. H. Ahlstrom, and H. G. Moser. 1984. Early life history stages of fishes and their characters. American Society of Ichthyologists and Herpetologists Special Publication 1:11–22.

Kendall, A. W., Jr., and N. A. Naplin. 1981. Diel-depth distribution of summer ichthyoplankton in the Middle Atlantic Bight. U.S. National Marine Fisheries Service Fishery Bulletin 79:705–727.

Kendall, A. W., Jr., and J. W. Reintjes. 1975. Geographic and hydrographic distribution of Atlantic menhaden eggs and larvae along the middle Atlantic coast from RV *Dolphin* cruises, 1965–66. U.S. National Marine Fisheries Service Fishery Bulletin 73:317–335.

Kingsford, M. J., and J. H. Choate. 1986. Influence of surface slicks on the distribution and onshore movement of small fish. Marine Biology (Berlin) 91:161–172.

Krygier, E. E., and W. G. Pearcy. 1986. The role of estuarine and offshore nursery areas for young English sole, *Parophrys vetulus* Girard, of Oregon. U.S. National Marine Fisheries Service Fishery Bulletin 84:119–132.

Loeb, V. J. 1979. Vertical distribution and development of larval fishes in the north Pacific central gyre during summer. U.S. National Marine Fisheries Service Fishery Bulletin 77:777–793.

Marliave, J. B. 1977. Substratum preferences of settling larvae of marine fishes reared in the laboratory. Journal of Experimental Marine Biology and Ecology 27:47–60.

Marliave, J. B. 1986. Lack of planktonic dispersal of rocky intertidal fish larvae. Transactions of the American Fisheries Society 115:149–154.

Matty, A. J. 1978. Pineal and some pituitary hormone rhythms in fish. Pages 21–30 *in* J. E. Thorpe, editor. The rhythmic activity of fishes. Academic Press, New York.

McCleave, J. D., and R. C. Kleckner. 1982. Selective tidal stream transport in the estuarine migration of glass eels of the American eel (*Anguilla rostrata*). Journal du Conseil, Conseil International pour l'Exploration de la Mer 40:262–271.

McCleave, J. D., and J. H. Power. 1978. Influence of weak electric and magnetic fields on turning behavior in elvers of American eel *Anguilla rostrata*. Marine Biology (Berlin) 46:29–34.

McHugh, J. L. 1967. Estuarine nekton. American Association for the Advancement of Science Publication 83:581–620.

Miles, S. G. 1968. Rheotaxis of elvers of the American eel *Anguilla rostrata* in the laboratory to water from different streams in Nova Scotia. Journal of the Fisheries Research Board of Canada 25:1591–1602.

Miller, J. M., J. P. Reed, and L. J. Pietrafesa. 1984. Patterns, mechanisms, and approaches to the study of migrations of estuarine-dependent fish larvae and juveniles. Pages 209–225 *in* J. D. McCleave, G. P. Arnold, J. J. Dodson, and W. H. Neill, editors. Mechanisms of migration in fishes. Plenum, New York.

Moore, J. W., and I. A. Moore, 1976. The basis of food selection in flounders, *Platichthys flesus* (L.) in the Severn estuary. Journal of Fish Biology 9:139–156.

Moser, H. G. 1981. Morphological and functional aspects of marine fish larvae. Pages 89–131 *in* R. Lasker, editor. Marine fish larvae: morphology, ecology, and relation to fisheries. University of Washington Press, Seattle.

Mundy, B. C. 1984. Yearly variation in the abundance and distribution of fish larvae in the coastal upwelling zone off Yaquina Head, Oregon, from June 1969 to August 1972. Master's thesis. Oregon State University, Corvallis.

Nelson, W. R., M. C. Ingham, and W. E. Schaff. 1977. Larval transport and year-class strength of Atlantic menhaden, *Brevoortia tyrannus*. U.S. National Marine Fisheries Service Fishery Bulletin 75:23–41.

Nordeng, H. 1977. A pheromone hypothesis for homeward migration in anadromous salmonids. Oikos 28:155–159.

Norris, K. S. 1963. The functions of temperature in the ecology of the percoid fish *Girella nigricans* (Ayres). Ecological Monographs 33:23–62.

O'Connor, J. M. 1972. Tidal activity rhythm in the hogchoker, *Trinectes maculatus* (Bloch and Schneider). Journal of Experimental Marine Biology and Ecology 9:173–177.

Olla, B. L., and W. Marchoni. 1968. Rhythmic movements of cones in the retina of bluefish, *Pomatomus saltatrix*, held in constant darkness. Biological Bulletin (Woods Hole) 135:530–536.

Olla, B. L., A. L. Studholme, A. J. Bejda, and C. Samet. 1980. Role of temperature in triggering migratory behavior of the adult tautog *Tautoga onitis* under laboratory conditions. Marine Biology (Berlin) 59:23–30.

Owen, R. W. 1981. Fronts and eddies in the sea: mechanisms, interactions, and biological effects. Pages 197–233 *in* A. R. Longhurst, editor. Analysis of marine ecosystems. Academic Press, New York.

Ozsoy, E., and U. Unluata. 1982. Ebb-tidal flow characteristics near inlets. Estuarine, Coastal and Shelf Science 14:251–263.

Palmer, J. D. 1973. Tidal rhythms: the clock control of the rhythmic physiology of marine organisms. Biological Reviews of the Cambridge Philosophical Society 48:377–418.

Parrish, R. H., C. S. Nelson, and A. Bakun. 1981. Transport mechanisms and reproductive success of fishes in the California Current. Biological Oceanography 1:175–203.

Pearcy, W. G., M. J. Hosie, and S. L. Richardson. 1977. Distribution and duration of pelagic life of larvae of Dover sole, *Microstomus pacificus;* rex sole, *Glyptocephalus zachirus;* and petrale sole, *Eopsetta jordani,* in waters off Oregon. U.S. National Marine Fisheries Service Fishery Bulletin 75:173–184.

Pearcy, W. G., and S. S. Myers. 1974. Larval fishes of Yaquina Bay, Oregon; a nursery ground for marine fishes? U.S. National Marine Fisheries Service Fishery Bulletin 72:201–213.

Peterson, W. T., C. B. Miller, and A. Hutchinson. 1979. Zonation and maintenance of copepod populations in the Oregon upwelling zone. Deep-Sea Research, Part A, Oceanographic Research Papers 26: 467–494.

Pfeiler, E. 1984. Inshore migration, seasonal distribution and sizes of larval bonefish, *Albula,* in the Gulf of California. Environmental Biology of Fishes 10:117–122.

Pollock, B. R., H. Weng, and R. M. Morton. 1983. The seasonal occurrence of postlarval stages of yellow fin bream, *Acanthopagrus australis* (Gunther), and some factors affecting their movement into an estuary. Journal of Fish Biology 22:409–415.

Powles, H. 1981. Distribution and movements of neustonic young estuarine dependent (*Mugil* spp,. *Pomatomus saltatrix*) and estuarine independent (*Coryphaena* spp.) fishes off the southeastern United States. Rapports et Procès-Verbaux des Réunions Commission Internationale pour l'Exploration Scientifique de la Mer 178:207–210.

Richardson, S. L., and W. G. Pearcy. 1977. Coastal and oceanic fish larvae in an area of upwelling off Yaquina Bay, Oregon. U. S. National Marine Fisheries Service Fishery Bulletin 75:125–145.

Rijnsdorp, A. D., M. van Stralen, and H. W. van der Veer. 1985. Selective tidal transport of North Sea plaice larvae *Pleuronectes platessa* in coastal nursery areas. Transactions of the American Fisheries Society 114:461–470.

Roessler, M. A., and R. G. Rehrer. 1971. Relation of catches of postlarval pink shrimp in Everglades National Park, Florida, to commercial catches on the Tortugas grounds. Bulletin of Marine Science 21:791–805.

Rogers, H. M. 1940. Occurrence and retention of plankton within the estuary. Journal of the Fisheries Research Board of Canada 5:164–171.

Rogers, S. G., T. E. Targett, and S. B. Van Sant. 1984. Fish-nursery use in Georgia salt-marsh estuaries: the influence of springtime freshwater conditions. Transactions of the American Fisheries Society 113:595–606.

Rosenberg, A. A. 1982. Growth of juvenile English sole, *Parophrys vetulus,* in estuarine and open coastal nursery areas. U.S. National Marine Fisheries Service Fishery Bulletin 80:245–252.

Rothlisberg, P. C., J. A. Church, and A. M. G. Forbes. 1983. Modelling the advection of vertically migrating shrimp larvae. Journal of Marine Research 41:511–538.

Sale, P. F. 1969. A suggested mechanism for habitat selection by the juvenile manini, *Acanthurus triostegus sandvicensis* Streets. Behaviour 35: 27–44.

Sale, P. F. 1970. Distribution of larval Acanthuridae off Hawaii. Copeia 1970:765–766.

Shanks, A. L. 1983. Surface slicks associated with tidally forced internal waves may transport pelagic larvae of benthic invertebrates and fishes shoreward. Marine Ecology Progress Series 13:311–315.

Shanks, A. L. 1985. Behavioral basis of internal-wave-induced shoreward transport of megalopae of the crab *Pachygrapsus crassipes*. Marine Ecology Progress Series 24:289–295.

Shenker, J. M. 1985. Biology of neustonic larval and juvenile fishes and crabs off Oregon, 1984. Doctoral dissertation. Oregon State University, Corvallis.

Shenker, J. M., and J. M. Dean. 1979. The utilization of

an intertidal salt marsh creek by larval and juvenile fishes: abundance, diversity, and temporal variation. Estuaries 2:154–163.

Staples, D. J. 1980. Ecology of juvenile and adolescent banana prawns, *Penaeus merguiensis*, in a mangrove estuary and adjacent off-shore area of the Gulf of Carpenteria. Australian Journal of Marine Freshwater Research 31:635–652.

Tanaka, M. 1985. Factors affecting the inshore migration of pelagic larval and demersal juvenile red sea bream *Pagrus major* to a nursery ground. Transactions of the American Fisheries Society 114: 471–477.

Thorpe, J. E., editor. 1978. The rhythmic activity of fishes. Academic Press, New York.

Tsuruta, Y. 1978. Field observations on the immigration of larval stone flounder into the nursery ground. Tohoku Journal of Agricultural Research 29(3–4): 136–145.

Tzeng, W. N. 1985. Immigration timing and activity rhythms of the eel, *Anguilla japonica*, elvers in the estuary of northern Taiwan, with emphasis on environmental influences. Bulletin of the Japanese Society of Fisheries Oceanography 47–48:11–28.

Venema, S. C., and F. Creutzberg. 1973. Seasonal migration of the swimming crab *Macropipus holsatus* in an estuarine area controlled by tidal streams. Netherlands Journal of Sea Research 7:94–102.

Victor, B. C. 1986. Delayed metamorphosis with reduced larval growth in a coral reef fish (*Thalassoma bifasciatum*). Canadian Journal of Fisheries and Aquatic Sciences 43:1208–1213.

Weihs, D. 1978. Tidal stream transport as an efficient method for migration. Journal du Conseil, Conseil International pour l'Exploration de la Mer 38: 92–99.

Weinstein, M. P. 1979. Shallow marsh habitats as primary nurseries for fishes and shellfish, Cape Fear River, North Carolina. U.S. National Marine Fisheries Service Fishery Bulletin 77:339–358.

Weinstein, M. P. 1983. Population dynamics of an estuarine-dependent fish, the spot (*Leiostomus xanthurus*), along a tidal creek–seagrass meadow coenocline. Canadian Journal of Fisheries and Aquatic Sciences 40:1633–1638.

Weinstein, M. P., S. L. Weiss, R. G. Hodson, and L. R. Gerry. 1980. Retention of three taxa of postlarval fishes in an intensively flushed tidal estuary, Cape Fear River, North Carolina. U.S. National Marine Fisheries Service Fishery Bulletin 78:419–436.

Williams, A. B., and E. E. Deubler. 1968. A ten year study of meroplankton in North Carolina estuaries: assessment of environmental factors and sampling success among bothid flounders and penaeid shrimps. Chesapeake Science 9:27–41.

Wroblewski, J. S. 1982. Interaction of currents and vertical migration in maintaining *Calanus marshallae* in the Oregon upwelling zone—a simulation. Deep-Sea Research 29:665–686.

Yamashita, Y., D. Kitagawa, and T. Aoyama. 1985. Diel vertical migration and feeding rhythm of the larvae of the Japanese sand-eel, *Ammodytes personatus*. Bulletin of the Japanese Society of Scientific Fisheries 51:1–5.

Yoklavich, M. 1982. Growth, food consumption, and conversion efficiency of juvenile English sole (*Parophrys vetulus*). Pages 97–105 *in* G. M. Cailliet and C. A. Simenstad, editors. Gutshop 81, fish food habits studies. University of Washington, Washington Sea Grant Program, Seattle.

American Fisheries Society Symposium 3:68–76, 1988

Physical Processes and the Mechanisms of Coastal Migrations of Immature Marine Fishes

JOHN M. MILLER

Zoology Department, North Carolina State University, Raleigh, North Carolina 27695, USA

Abstract.—Migrations of immature marine fishes begin as passive drift, then develop into active swimming as the fish grow. The migration mechanisms of larval plaice *Pleuronectes platessa*, anguillid eels *Anguilla* spp., capelin *Mallotus villosus*, and Atlantic herring *Clupea harengus harengus* illustrate various degrees of active responses to hydrographic regimes. Passively and actively migrating larvae must have opposite orientations to coastal gradients of odors, temperature, salinity, or turbidity. To be transported shoreward, passively migrating larvae should be attracted to water with the opposite characteristics of their destination. Although an active mechanism of selective tidal stream transport has been suggested in several cases, a mechanism based on a passive buoyancy response is suggested as an alternative. The hydrographic data requirements to test any mechanism are discussed.

Despite the wide variations in abundance exhibited by most fish stocks, fishery biologists have pursued the concept of density-dependent regulation of stock size. The vast reproductive potential of most fishes is itself an argument for the concept. But, despite vast amounts of effort expended to establish its validity, the proof remains elusive. One difficulty is separating what may be a small fraction of density-dependent regulation from a large component of density-independent variation. It has been generally established, however, that the process(es) that controls year-class strength occurs early in the life history of most species. Sharp (1987) suggested that the average fish dies in less than a week. Evidence is beginning to mount that certain meteorological variables are related to year-class strength. In the case of species that use coastal systems as nurseries, it seems likely that meteorological variability may affect year-class strength by altering the physical environment through, or to, which immature stages must migrate. Once biologists achieve a sufficient level of understanding of such density-independent variation in abundance, a density-dependent component may surface.

Understanding the effects of abiotic variables is severely limited by the lack of data on physical forces that operate on many scales relevant to biology. Most physical oceanographers are not working on micro- or mesoscale processes. Thus, we need information on the dynamics and distribution of physical variables as well as the responses of fish to these in order to determine their effects upon fish populations. The purpose of this paper is to review some of the literature and ideas on larval and juvenile fish migration and to point out where we biologists need to interface with physical oceanographers in order to better understand the role of physical processes in the distribution and abundance of immature fish. I will focus on the mechanism(s) that the more or less planktonic stages of immature fishes use to migrate to estuarine nursery areas and on the physical and biological information we must have to predict the success (or failure) of these migrations.

Migration has been defined, by Hasler (1966), Harden Jones (1968), McKeown (1984), and others, as any displacement (passive or active) that has a return component. Leggett (1984) has argued that we have been too concerned with the behavioral facets of migration and too little with ecological and model-centered approaches, and that this has led us to an expectation of more precision in migration than is probably necessary. Saila and Shappy (1963) and Patten (1964) felt that migration need not be very precise, though their ideas have not been widely accepted (but see Quinn and Groot 1984). Recently, DeAngelis and Yeh (1984) showed that models of fish migration based on advection–diffusion and biased random walk predict the arrival of about the same number of fish at the pass of a hypothetical estuary. Although different mathematically, the advection and bias in the two models could represent either orientation or directional transport by currents. Diffusion and random walk both represent unoriented movement; because they are both "loss" terms with respect to migration to a particular area, they could also represent mortality. Thus, mathematical techniques exist to simulate migrations of fish, but we need better physical descriptions of the migratory environments before such models can be tested in the field. And, because we do not have good estimates of the numbers of fish surviving in the ocean, or thus of the number of potential migrants, we do not have a basis for judging the actual or necessary precision of their migrations. However, if random walk models with a slight bias are adequate predictors of the percentage of returning adult salmon, as Saila and Shappy (1963) suggested, it seems likely they would suffice for the orders-of-magnitude greater numbers of immature fish.

An equally fundamental problem is the lack of information on the responses of immature fishes to environmental scalars and currents. Even if we had the necessary physical description of the waters through which fish migrate and knew the number of fish expected to migrate, we still could not predict migration success. Do the fish behave as passive particles? Do they accumulate at sharp clines? Are they "attracted" by favorable scalars? Do they learn as they encounter new conditions? These are but a few of the questions we must answer before we can build quantitative models of migrations by immature fish. A first step is to consider the alternative hypothesis that young fish migrate by passive (advective) transport rather than by swimming or by active selection of water masses moving in the required direction.

Mechanisms of Migration

Aggregation of Animals

The simplest kind of response of an animal to a scalar gradient is to slow down when it encounters relatively favorable conditions. Such responses are termed kineses, because speed (usually modeled as step length) is the important variable. Where no directional bias exists, such movements are termed *orthokinetic*; where the probability of the animal changing direction is the variable, the movements are *klinokinetic*. In the latter case, favorable conditions increase the probability that an animal will change direction (thereby remaining in that vicinity). Simple klinokineses do not effectively produce aggregations or displacements; simple orthokineses may result in local aggregations, but not effective displacements (DeAngelis and Yeh 1984).

Topotaxes are movements whereby an animal detects gradients and orients its own movements accordingly. The key is the ability to perceive the gradient direction, which means the animal must be able to compare the stimulus strength at two different locations on its body. Most scalar gradients in the ocean are usually much less than 1°C or 1‰ salinity/m which suggests it is unlikely that a 10- or 20-mm-long fish could distinguish a difference in stimulus intensity. This makes taxis as a method of aggregation along most scalar gradients in the ocean seem unlikely for immature fishes. Exceptions may be found at ocean fronts or in vertically stratified estuaries and unscattered light near the surface. But, unlike kineses, taxes are effective at producing aggregations.

Vagility

One or two body lengths per second is generally accepted as the sustainable swimming speed of fish (Miller et al. 1985). For a 10-mm fish larva, this translates into a ground speed of about 1 km/d or about 500 m/tidal cycle. Thus, even a 20-mm fish could hardly oppose the typical currents in estuaries, much less the tidal currents in passes. Furthermore, larvae and juveniles are known to migrate faster than 1 or 2 body lengths/s (Miller et al. 1984). These considerations, plus the energy required for such migrations, have led to an hypothesis of selective tidal stream transport (Weihs 1978; Arnold 1981), whereby fish ride favorable currents and, alternately, escape unfavorable ones. This hypothesis requires that fish be able to detect and move into different water masses. The hypothesized movements are most often vertical, that is, between horizontal strata, so the distances are small and the gradients are relatively steep, making such taxes more feasible than similar responses to typical horizontal gradients. Two basically similar tactics occurring in different water strata have been suggested. First is a vertical migration between downstream-flowing surface waters and upstream-flowing deeper layers. This presumably functions to retain fishes in a particular region of an estuary (Graham 1972; Leggett 1976; Weinstein et al. 1980). Second is the behavior whereby fish rest on the bottom, then ascend into flooding waters near the bottom, and then escape flow reversal by returning to the bottom. It is this latter pattern that has been termed selective tidal stream transport or modulated drift (Arnold 1981). This could also include escape from ebbing waters by migration to the edges of passes (Beckley 1985).

It is not understood, however, how fish distinguish favorable current regimes from unfavorable ones—if indeed they do. Nevertheless, these and other kinds of movements have been modeled in the context of fish migrations (Balchen 1976; Neill 1979, 1984; DeAngelis and Yeh 1984). The salient conclusion from these exercises and the above considerations is that low vagility and energy stores, plus the requirement of orienting long-distance movements, probably preclude larval or juvenile fish from migrating to, or remaining in, estuarine nursery areas by any other means than advection by currents. Furthermore, selection of the appropriate water mass by small fish can only be accomplished over small distances. Thus, the ability to select favorable currents in the vertical axis, perhaps by responses to gradients of certain scalars, may be the major determinant of the precision of migrations by small fishes. Estuaries, systems where scalar gradients are relatively strong, seem most likely to yield insight into mechanisms of migration by immature fishes.

Migratory Environments of Passes and Estuaries

Bar-built estuaries and drowned river valleys represent extremes of estuarine environments to and through which immature fishes of many species that are spawned offshore must migrate to reach their juvenile nursery areas. Bar-built estuaries are separated from coastal waters by relatively narrow passes and include lagoons like Pamlico Sound in North Carolina and the Dutch Waddensee. Depending on the amplitude of the ocean tides, currents through their passes are strong and reversing. Such passes present narrow windows to locate and are complicated mediums through which immature fishes must migrate. Inside lagoons, the migratory environment is no less complicated. The physics of shallow lagoons are tightly coupled to meteorological forcing and river flow. Although lunar tides are of relatively small amplitude in most lagoons, small vertical perturbations of the surface effect large horizontal movements of water masses and their attendant physical and chemical properties. Small fishes that migrate upstream may be unpredictably carried long distances in any of many "wrong" directions at the whim of a storm.

In contrast to lagoons, drowned river valley estuaries are more of a continuum between coastal waters and nurseries. Those with wide mouths (passes) may not experience tidal flow reversals, but simply changes in current velocity—especially near the bottom. There may be a persistent downstream flow of diluting river water at the surface and a persistent upstream flow of diluting sea water near the bottom. Horizontal flow in response to lunar tidal forces is stronger than in lagoons, but more predictable. What scalar quantities might be tightly linked to the appropriate vectors of upstream flow in estuaries?

Clues to Water Masses

Among the many possible scalars that might be used by a migrating fish to orient its movements upstream in an estuary are odor, temperature, salinity, turbidity, and pH. Some of these increase toward the nursery and should be "attractive" to fishes. The same direction of movement could be accomplished by "avoiding" those that increase toward the ocean.

Scalars That Increase toward the Ocean

Salinity and associated ions.—Salinity increases toward the ocean as its dilution by freshwater inflow decreases. From upstream nursery areas to the pass, a relatively persistent horizontal gradient may exist, although its magnitude and position respond to changes in inflow. In the vicinity of narrow passes, however, salinity is more variable. Tidal jets of salty water regularly penetrate the less saline estuarine waters, creating horizontal gradients perpendicular to the general direction of flow. The other region of an estuary experiencing relatively high variation in salinity is at the margin of a salt wedge. Other, more stable, horizontal gradients perpendicular to the flow exist in estuaries wide enough to respond to Coriolis forces.

Temperature in winter.—At temperate latitudes during winter, inshore waters are generally cooler than the ocean (Miller et al. 1985). Fresh water flowing into estuaries can create a horizontal thermal gradient wherein temperature increases from the nursery area to the sea. But temperature is less conservative than salinity. Being tightly linked to the atmosphere, temperature may change without mass transfer. Because heat may be lost to the atmosphere during the passage of a cold front, vertical temperature gradients may be rapidly destroyed, altering the temperature of bottom waters as well. Especially in the surface waters of shallow bar-built estuaries, thermal isopleths may be distorted along the principal gradient, particularly near points of lateral inflow. The temperature of waters of deeper estuaries may be more tightly linked to horizontal vectors but it is still a less reliable predictor of upstream flow than salinity.

pH.—In most cases, fresh water flowing into estuaries has a lower pH than the ocean. Therefore, pH generally mirrors salinity. However, pH responds to biological processes. Where the retention time of estuarine waters is long, the distribution of pH may be complex, especially in surface waters where phytoplankters consume CO_2 and in stratified bottom waters where community respiration increases CO_2. Such variations due to biological activity would be progressively dampened with salinity increase, creating a horizontal gradient in pH *variability* as well.

Scalars That Increase toward the Nurseries

Odor.—Chemical signatures probably exist in most waters flowing into estuaries. What they are and how labile they may be are unknown, but they should generally increase horizontally upstream and vertically towards the surface.

Turbidity.—Because river water is generally more turbid than ocean water, a gradient of turbidity would be expected in estuaries. The inorganic turbidity of a river is highly variable, being a function of erosion and discharge. In shallow estuaries, turbidity is also a function of wind-generated currents that scour the sediments. The organic fraction of turbidity caused by phytoplankton may be substantial but variable in estuaries. Thus, the distribution of turbidity in most estuaries would be expected to be complex, and hence a relatively unreliable predictor of current direction. Notwithstanding this complexity, Blaber and Blaber (1980) considered turbidity to be important in attracting juvenile fish in estuaries.

Temperature in summer.—In temperate estuaries, the summer temperature of inflowing fresh water is frequently warmer than that of the ocean. This reverses the horizontal thermal gradient of winter. The vertical gradient may be amplified, however, because atmospheric warming tends to stabilize the density gradient. Summer thermal gradients are subject to the same atmospheric forcing as winter ones, but, in general, thermal gradients are more stable and more tightly linked to water mass vectors in summer than in winter.

In order to be transported upstream to estuarine nursery areas, fish could select water masses with relatively high temperature in winter (low in summer), low turbidity, low odor, high salinity or high pH. Such selection would generally displace them toward the bottom and, thus, in the water mass most likely to transport them toward the nurseries. That is, they should select water with the characteristics of the ocean water from which they originated, *not* water with the same characteristics as their upstream nurseries (destination). However, oriented horizontal swimming in the same gradient direction would tend to displace fish toward the ocean, not toward the nurseries. But, compared to the advective displacement, active swimming by immature fishes, even in the wrong direction, is likely to be insignificant in most cases. Transport toward the nurseries could also be accomplished by avoiding water with the opposite characteristics—i.e., the (usually surface) water flowing downstream from the nurseries. Either mechanism would be effective as long (or as far upstream in an estuary) as there is upstream-flowing seawater, i.e., to the limit of salt penetration. The bottom circulation in any particular estuary could be complex; for example, each tributary could have its own upstream-flowing "sub-system."

If, however, the fish are migrating actively, their movements would be oriented correctly if they swam against the current and in the direction of decreasing salinity, increasing odor, increasing turbidity or decreasing temperature in winter (increasing temperature in summer)—i.e., actively migrating fish should select water with characteristics *opposite* those selected by passively migrating fish. Furthermore, actively migrating fish should tend to be in the surface layers of an estuary, because selection of low-salinity water (for example) would tend to direct them toward the surface as well as toward the nurseries. Of the possible scalars, salinity and odor would seem to be the most reliable predictors in either case.

Immature Estuarine Fishes

The juvenile fishes of estuaries exhibit an enormous variety of life history strategies, preferred habitats, and migratory patterns (McHugh 1967; Day 1981; Haedrich 1983; Beckley 1984; Wallace et al. 1984; Miller et al. 1985; Miller and Merriner 1985). I will consider only anadromous or catadromous fishes whose larvae or juveniles are transitory residents of estuaries. These species exhibit the clearest migration patterns. Among these, I will contrast pelagic and demersal species, which seem to prefer the upper and lower strata of estuarine waters, respectively.

The pelagic species that have been studied with respect to their migratory activities as larvae or juveniles are capelin *Mallotus villosus* and Atlantic herring *Clupea harengus harengus*. Both capelin and Atlantic herring spawn in estuaries, so the young face the problem of being flushed from their esturarine nursery areas.

Among demersal species are plaice *Pleuronectes platessa*, American eel *Anguilla rostrata*, and European eel *A. anguilla*, the larvae of which migrate into estuaries after having hatched offshore.

Examples of Migrations by Juvenile Fish

Plaice

The migrations of larval and juvenile plaice into their Waddensee nursery areas have received more scientific attention than those of any other nonsalmonid. Eggs are spawned in the southern North Sea and the English Channel in winter (November–January) after the adults migrate some 200–300 km south (Cushing 1972; Talbot 1977; Arnold 1981). The larvae develop as they drift northward. A fraction (estimated at 15% by Arnold and Cook 1984) of these drift to the east along the Dutch coast and migrate as late larvae through the channels into the Waddensee during March and April. In the nursery, the juveniles migrate with the tides between channels and flats to feed (Kuipers 1973). After spending their first summer in the Waddensee nursery, the juveniles return to the North Sea to overwinter. Age-I and -II juveniles return to the Waddensee in the following two summers (de Veen 1978).

The fraction of the population that enters the Waddensee through the Marsdiep channel between the island of Texel and Den Helder is the most studied.

Creutzberg et al. (1978) sampled migrating larvae in the channels between the North Sea and Waddensee. Owing to high turbulence, water is not stratified vertically during flood or ebb. Oblique hauls made on ebb and flood tides showed a net inward migration of larvae. About a third of the larvae entering on a particular flood tide exited on the next ebb. The percentage retained increased with developmental stage (Rijnsdorp et al. 1985). Once inside the Waddensee, the larvae dispersed. The time lapse between peak entry numbers and peak abundance in one area (Balgzand) 15 km from the pass was 35–45 d, so displacement inside the nursery was apparently very slow. To elucidate the mechanism of selective tidal transport, Creutzberg et al. (1978) subjected plaice larvae to laboratory changes in salinity, temperature, and odor (estuarine water) to see which elicited swimming behavior. None of the experiments suggested larvae were able to respond to flooding waters. On the other hand, the smell of food elicited a strong swimming response from hungry (unfed) larvae, leading to the conclusion that their net movement into the Waddensee was a response to food. Specifically, the authors suggested that the food-rich flats acted as a "trap" for hungry larvae.

Gibson (1973a, 1973b), working with juvenile plaice caught off the Scottish coast, found a bimodal daily activity pattern in the laboratory, which was replaced in darkness by a unimodal circadian rhythm. This bimodal activity pattern correlated well with movements of juveniles in shallow coastal waters (not estuaries)—shoreward on flooding tides and seaward on ebbs (Gibson 1973b). How (or if) plaice detected flood currents was uncertain, but Gibson (1973b) suggested they were cued by tidal changes in water depth. Their feeding movements perpendicular to the beach may be analogous to the tidal flat movements of the Waddensee plaice.

European and American Eels

Atlantic anguillid eels spawn at sea and migrate as leptocephali to the vicinity of estuaries and then as glass eels (unpigmented juveniles) and elvers into rivers where they spend several years. The migration of American glass eels into the Penobscot estuary in Maine is accomplished by selective tidal stream transport (McCleave and Kleckner 1982). Glass eels rise into flooding waters and are transported as far as 4.2 km/tidal cycle. They return to the bottom waters on the ebb, thus avoiding transport back downstream. On the average, about 20 times as many glass eels were caught moving upstream as down, suggesting a more active (and efficient) transport process than the plaice's migration into the Waddensee. The authors considered odor, turbulence, electric field detection, and circatidal clocks to be potential clues to water masses, but had no data.

Gandolfi et al. (1984) came to quite a different conclusion about the mechanism by which European elvers migrate into the Arno River in Italy. They caught most elvers during the first phase of ebbing tides, but the difference in numbers caught on falling and rising tides was insignificant. It is possible that their catches at the edge of the river reflected some lateral movements by the elvers. The month-to-month variation in migratory activity was attributed to the difference between river and ocean temperatures. They suggested that, in April, when catches were lower than in March and May and the river temperature (<8°C) was lower than that of the ocean owing to snow melt, elvers slowed their migration. In March there was little difference in temperature; in May the river was warmer than the ocean.

The migration of European elvers into the Waddensee occurs at about the same time as that of plaice (Tesch 1977), and its timing depends on temperature. European eels seem to be attracted to the odor of fresh water (Creutzberg 1961). This means they should prefer ebbing water and thus must swim against the current, a pattern opposite that observed by McCleave and Kleckner (1982) for American eels.

Capelin and Atlantic Herring

In the St. Lawrence River, Canada, capelin larvae hatch from eggs deposited in the estuary, and the juveniles eventually move downstream into the ocean. Atlantic herring larvae move upstream after hatching in the estuary. Fortier and Leggett (1982, 1983, 1984) and Frank and Leggett (1983) related the distribution and movements of these fish to the hydrography of the estuary. The area of the estuary studied was persistently stratified and the surface layer moved rapidly seaward at about 30 cm/s, while the deeper (about 22–50m) layer moved upstream at about 10 cm/s. Capelin larvae occupied the upper layer and always moved downstream at a speed roughly predictable by the current speed. Small Atlantic herring larvae, by staying closer to the bottom, effected a net upstream movement. The distribution of larger Atlantic herring larvae was centered at about the depth of null velocity. By migrating twice daily between the two layers they remained more or less in the same vicinity of the estuary.

The movements of both species were somewhat size-dependent; early larvae of both species showed less tendency to migrate and their movements were passive. Neither vertical or horizontal shear forces were especially important, but a response to zooplanktonic food was suggested. Fortier and Leggett (1982) concluded that the changes in distribution of at least the smaller larvae could be predicted by the distribution of salinity; i.e., dispersal was passive.

In summary, the above species illustrate a range of mechanisms of migration from passive to active. Plaice, once in the vicinity of a pass, are carried passively into the estuary on flood tides. Once inside the estuary, they avoid being washed back out by remaining near the bottom. Thus, the migration of plaice seems to be mostly passive; active selection of depth or food may occur once the fish are inside the estuary.

European eels, on the other hand, apparently migrate actively, presumably being attracted by some property of fresher water. They must actively oppose ebbing currents; in the Arno River, more migrate upstream on ebb tides than flood.

Between these extremes, other species migrate by selecting appropriate currents (water masses). Early larvae of Atlantic herring are transported upstream by remaining in the bottom water masses of the St. Lawrence River, whereas larval capelin effect a seaward migration in the same system by selecting the surface water masses. Later stages of the Atlantic herring vertically migrate diurnally into the vicinity of the horizontal shear layer. There, their migrations through the shear layer serve to retain them in the same region of the estuary. Early stages of the American eel migrate by selecting flooding waters; they avoid ebbing waters by migrating to the bottom, thus effecting a net movement upstream.

Discussion

Migrations of young fish appear to be size-(stage-) dependent (Boehlert and Mundy 1987), beginning with completely passive transport, followed by selection of particular water masses and, finally, by active swimming. The stages in the transition from passive to active migration depend upon increased vagility, and would be expected to vary among species and the particular estuarine current regime. These migratory behaviors result in movements that increasingly differ from the trajectories of passive particles with hydrodynamic properties similar to fish. The active component of migration could be quantified as the difference between the trajectories of such passive particles and the actual trajectories of fish. The active component would be expected to be inversely related to current speed and directly related to fish size. The trajectories of fish and particles would also vary with their respective specific gravities and with water density.

While it appears from the above that selective tidal stream transport may involve oriented movements of fish into and out of particular water masses, certain of these movements could also be passive. In estuaries, ebb currents are likely to be less saline than flood currents. Because fish are more buoyant in high-salinity water than in low salinities, a simple buoyancy response could produce the appearance of selecting water masses. Fish near the bottom would tend to rise into and sink out of flooding and ebbing waters, respectively. Likewise, feeding increases the density, hence sinking rate, of larval fish (Blaxter and Ehrlich 1974), and would tend to cause accumulations of fish where feeding was most efficient. This, also, could easily be interpreted as an active selection of a particular habitat. Accumulation of fish in a lagoon, as observed for Waddensee plaice, may be a simple function of increased density caused by increased feeding, rather than a cued response to estuarine conditions. Further, the increased efficiency with which larger plaice larvae are retained in the Waddensee (Rijnsdorp et al. 1985) may result simply because plaice larvae sink faster as they grow (Blaxter and Ehrlich 1974). In any case, the retention of fish in a lagoon is likely to be a function of the tendency for residual circulation of the estuary to advect flood water (and the fish) away from the pass, thus reducing the probability that the same water mass will be entrained into the ebb flow.

Whether or not immature fish orient their (limited) swimming abilities by environmental gradients, or migrate actively, it is clear that advection is likely to play an important part in the migration process and in the resulting distribution of immature fish (Miller 1974; Leis and Miller 1976; Nelson et al. 1977; Bailey 1981; Hamann et al. 1981; Melville-Smith et al. 1981; Powles 1981; Pietrafesa et al. 1986). To detect an active component of migration, a three-dimensional hydrodynamic description of the migratory environment is necessary. In most cases, this means that synoptic physical studies must accompany biological investigations of immature fish migration, because adequate three-dimensional hydrodynamic models do not exist for most estuarine systems.

We also need more precise descriptions of the vertical distributions of fish to predict their trajectories in different water strata (Power 1984). Relatively small differences in vertical distribution can result in very large differences in horizontal transport. Whether or not an active component (if any) in the migrations of immature fishes in such systems can be separated from the inevitable stochastic variability in the behaviors of both fish and currents remains to be seen (John 1984). In any case, the relationships between fish movements and physical features of environments, if quantified, could be used to predict the distribution of fish during or after migration, if the initial distribution were known. Likewise, the effects of engineering activities, such as jetty construction or channelization, on fish migration could be estimated. At least, better hydrodynamic resolution would help reconcile apparent differences among systems through which fish migrate.

Acknowledgment

This work was supported by the U.S. Office of Sea Grant, National Oceanic and Atmospheric Administration, under grant NA85AA-D-00012 to the University of North Carolina.

References

Arnold, G. P. 1981. Movements of fishes in relation to water currents. Pages 55–79 *in* D. J. Aidley, editor. Animal migration. Cambridge University Press, Cambridge, England.

Arnold, G. P., and P. H. Cook. 1984. Fish migration by selective tidal stream transport: first results with a computer simulation model for the European continental shelf. Pages 227–261 *in* McCleave et al. 1984.

Bailey, K. M. 1981. Larval transport and recruitment of Pacific hake *Merluccius productus*. Marine Ecology Progress Series 6:1–9.

Balchen, J. G. 1976. Modeling, prediction, and control of fish behavior. Pages 99–146 *in* C. T. Leondes, editor. Control and Dynamic Systems 15. Academic Press, New York.

Beckley, L. E. 1984. The ichthyofauna of the Sundays Estuary, South Africa, with particular reference to the juvenile marine component. Estuaries 7: 248–258.

Beckley, L. E. 1985. Tidal exchange of ichthyoplankton in the Swartkops Estuary mouth, South Africa. South African Journal of Zoology 20:15–20.

Blaber, S. J. M., and T. G. Blaber. 1980. Factors affecting the distribution of juvenile estuarine and inshore fish. Journal of Fish Biology 17:143–162.

Blaxter, J. H. S., and K. F. Ehrlich. 1974. Changes in behavior during starvation of herring and plaice larvae. Pages 575–588 *in* J. H. S. Blaxter, editor. The early life history of fish. Springer-Verlag, Berlin.

Boehlert, G. W., and B. C. Mundy. 1987. Recruitment dynamics of metamorphosing English sole, *Parophrys vetulus*, to Yaquina Bay, Oregon. Estuarine, Coastal and Shelf Science 25:261–281.

Creutzberg, F. 1961. On the orientation of migrating elvers (*Anguilla vulgaris* Turt.) in a tidal area. Netherlands Journal of Sea Research 1:257–338.

Creutzberg, F., A. T. G. W. Eltink, and G. J. van Noort. 1978. The migration of plaice larvae *Pleuronectes platessa* into the western Wadden Sea. Pages 243–251 *in* D. S. McLusky and A. J. Berry, editors. Physiology and behavior of marine organisms. Pergamon Press, Oxford, England.

Cushing, D. H. 1972. The production cycle and the numbers of marine fish. Symposia of the Zoological Society of London 29:213–232.

Day, J. H., editor. 1981. Estuarine ecology. Balkema, Cape Town, South Africa.

DeAngelis, D. L., and G. T. Yeh. 1984. An introduction to modeling migratory behavior of fishes. Pages 445–469 *in* McCleave et al. 1984.

de Veen, J. F. 1978. On selective tidal transport in the migration of North Sea plaice (*Pleuronectes platessa*) and other flatfish species. Netherlands Journal of Sea Research 12:115–147.

Fortier, L., and W. C. Leggett. 1982. Fickian transport and the dispersal of fish larvae in estuaries. Canadian Journal of Fisheries and Aquatic Sciences 39:1150–1163.

Fortier, L., and W. C. Leggett. 1983. Vertical migrations and transport of larval fish in partially mixed estuary. Canadian Journal of Fisheries and Aquatic Sciences 40:1543–1555.

Fortier, L., and W. C. Leggett. 1984. Small-scale covariability in the abundance of fish larvae and their prey. Canadian Journal of Fisheries and Aquatic Sciences 41:502–512.

Frank, K. T., and W. C. Leggett. 1983. Multispecies larval fish associations: accident or adaptation? Canadian Journal of Fisheries and Aquatic Sciences 40:754–762.

Gandolfi, G., M. Pesaro, and P. Tongiorgi. 1984. Environmental factors affecting the ascent of elvers, *Anguilla anguilla* (L.), into the Arno River. Oebalia 10 (N.S.):17–35.

Gibson, R. N. 1973a. The intertidal movements and distribution of young fish on a sandy beach with special reference to the plaice (*Pleuronectes platessa* L.). Journal of Experimental Marine Biology and Ecology 12:79–102.

Gibson, R. N. 1973b. Tidal and circadian activity rhythms in juvenile plaice, *Pleuronectes platessa*. Marine Biology (Berlin) 22:379–386.

Graham, J. J. 1972. Retention of larval herring in the Sheepscot estuary of Maine. U.S. National Marine Fisheries Service Fishery Bulletin 70:229–305.

Haedrich, R. L. 1983. Estuarine fishes. Ecosystems of the World 26:183–207.

Hamann, I., H.-C. John, and E. Mittelstaedt. 1981. Hydrography and its effect on fish larvae in the Mauritanian upwelling area. Deep Sea Research, Part A, Oceanographic Research Papers 28: 561–575.

Harden Jones, F. R. 1968. Fish migration. Edward Arnold, London.

Hasler, A. D. 1966. Underwater guideposts. University of Wisconsin Press, Madison.

John, H.-C. 1984. Drift of larval fishes in the ocean: results and problems from previous studies and a proposed field experiment. Pages 39–59 *in* McCleave et al. 1984.

Kuipers, B. 1973. On the tidal transport of young plaice (*Pleuronectes platessa*) in the Wadden Sea. Netherlands Journal of Sea Research 6:376–388.

Leggett, W. C. 1976. The American shad (*Alosa sapidissima*), with special reference to its migration and population dynamics in the Connecticut River. American Fisheries Society Monograph 1:169–225.

Leggett, W. C. 1984. Fish migrations in coastal and estuarine environments: a call for new approaches to the study of an old problem. Pages 159–178 *in* McCleave et al. 1984.

Leis, J. M., and J. M. Miller. 1976. Offshore distributional patterns of Hawaiian fish larvae. Marine Biology (Berlin) 36:359–367.

McCleave, J. D., and R. C. Kleckner. 1982. Selective tidal stream transport in the estuarine migration of glass eels of the American eel (*Anguilla rostrata*). Journal du Conseil, Conseil International pour l'Exploration de l'Mer 40:262–271.

McCleave, J. D., G. P. Arnold, J. J. Dodson, and W. H. Neill, editors. 1984. Mechanisms of migration in fishes. Plenum, New York.

McKeown, B. A. 1984. Fish migration. Timber Press, Portland, Oregon.

McHugh, J. L. 1967. Estuarine nekton. American Association for the Advancement of Science Publication 83:581–620.

Melville-Smith, R., D. Baird, and P. Woolridge. 1981. The utilization of tidal currents by the larvae of an estuarine fish. South African Journal of Zoology 16:10–13.

Miller, J. M. 1974. Nearshore distribution of Hawaiian marine fish larvae: effects of water quality, turbidity and currents. Pages 217–231 *in* J. H. S. Blaxter, editor. The early life history of fish. Springer-Verlag, Berlin.

Miller, J. M., L. B. Crowder, and M. L. Moser. 1985. Migration and utilization of estuarine nurseries by juvenile fishes: an evolutionary perspective. Contributions in Marine Science 27 (Supplement):338–352.

Miller, J. M., and J. V. Merriner. 1985. Determinants of habitat dependency in marine sport fishes. Pages 119–130 *in* R. L. Stroud, editor. Proceedings of 1st world angling conference, Cap d'Agde, France. International Game Fish Association, Fort Lauderdale, Florida.

Miller, J. M., J. P. Reed, and L. J. Pietrafesa. 1984. Patterns, mechanisms and approaches to the study of migration of estuarine-dependent fish larvae and juveniles. Pages 209–225 *in* McCleave et al. 1984.

Neill, W. H. 1979. Mechanisms of fish distribution in heterothermal environments. American Zoologist 19:305–317.

Neill, W. H. 1984. Behavioral enviroregulation's role in fish migration. Pages 61–66 *in* McCleave et al. 1984.

Nelson, W. R., M. C. Ingham, and W. E. Schaaf. 1977. Larval transport and year-class strength of Atlantic menhaden, *Brevoortia tyrannus*. U.S. National Marine Fisheries Service Fishery Bulletin 75:23–41.

Patten, B. C. 1964. The rational decision process in salmon migration. Journal du Conseil, Conseil International pour l'Exploration de la Mer 28:410–471.

Pietrafesa, L. P., G. S. Janowitz, J. M. Miller, E. B. Noble, S. W. Ross, and S. P. Epperly. 1986. Abiotic factors influencing the spatial and temporal variability of juvenile fish in Pamlico Sound, North Carolina. Pages 341–353 *in* D. A. Wolfe, editor. Estuarine variability. Academic Press, New York.

Power, J. H. 1984. Advection, diffusion, and drift mechanisms. Pages 27–37 *in* McCleave et al. 1984.

Powles, H. 1981. Distribution and movements of neustonic young estuarine dependent (*Mugil* spp., *Pomatomus saltatrix*) and estuarine independent (*Coryphaena* spp.) fishes off the southeastern United States. Rapports et Procès-Verbeaux des Réunions, Conseil International pour l'Exploration de la Mer 178:207–209.

Quinn, T. P., and C. Groot. 1984. Pacific salmon (*Oncorhynchus*) migrations: orientation versus random movement. Canadian Journal of Fisheries and Aquatic Sciences 41:1319–1324.

Rijnsdorp, A. D., M. van Stralen, and H. W. van der Veer. 1985. Selective tidal transport of North Sea plaice larvae *Pleuronectes platessa* in coastal nursery areas. Transactions of American Fisheries Society 114:461–470.

Saila, S. B., and R. A. Shappy. 1963. Random movement and orientation in salmon migration. Journal du Conseil, Conseil International pour l'Exploration de la Mer 28:153–166.

Sharp, G. D. 1987. Averaging the way to inadequate information in a varying world. American Institute of Fishery Research Biological Briefs 16:3–4.

Talbot, J. W. 1977. The dispersal of plaice eggs and larvae in the Southern Bight of the North Sea. Journal du Conseil, Conseil International pour l'Exploration de la Mer 37:221–248.

Tesch, F.-W. 1977. The eel. Chapman and Hall, London.

Wallace, J. H., H. M. Kok, L. E. Beckley, B. Bennett, S. J. M. Blaber, and A. K. Whitfield. 1984. South African estuaries and their importance to fishes. South African Journal of Science 80:203–207.

Weihs, D. 1978. Tidal stream transport as an efficient method of migration. Journal du Conseil, Conseil International pour l'Exploration de la Mer 38: 92–99.

Weinstein, M. P., S. L. Weiss, R. G. Hodson, and L. R. Gerry. 1980. Retention of three taxa of postlarval fishes in an intensively flushed tidal estuary, Cape Fear River, North Carolina. U.S. National Marine Fisheries Service Fishery Bulletin 78: 419–436.

American Fisheries Society Symposium 3:77–89, 1988

Ocean–Estuary Coupling of Ichthyoplankton and Nekton in the Northern Gulf of Mexico

RICHARD F. SHAW

Coastal Fisheries Institute, Center for Wetland Resources
Louisiana State University
Baton Rouge, Louisiana 70803, USA

BARTON D. ROGERS

U.S. Fish and Wildlife Service
Louisiana Cooperative Fish and Wildlife Research Unit
Louisiana State University Agricultural Center
Baton Rouge, Louisiana 70803, USA

JAMES H. COWAN, JR.[1]

Coastal Ecology Institute, Center for Wetland Resources

WILLIAM H. HERKE

U.S. Fish and Wildlife Service
Louisiana Cooperative Fish and Wildlife Research Unit
Louisiana State University Agricultural Center

Abstract.—Both gulf menhaden *Brevoortia patronus* and sand seatrout *Cynoscion arenarius* spawn offshore in the Gulf of Mexico during winter and spring. After a cross-shelf transit, their larvae later enter estuarine nursery areas. We have integrated new and existing early life history data on the occurrence of these species in the continental shelf ichthyoplankton and as late-stage larvae and juveniles in Louisiana estuaries to document this recruitment process. The sequential appearances and length-frequency data from larvae in the offshore plankton and larvae and juveniles in the estuarine marsh, as well as growth rates inferred from daily otolith increments, imply shelf-to-estuary transit times of 40–73 d for gulf menhaden and an upper range of 30–94 d for sand seatrout. During these times, water mass movements near the coast displace larvae to the west-northwest for distances that may measure several hundreds of kilometers, i.e., larvae do not necessarily recruit to estuaries nearest their offshore spawning areas. Additional information on hydrodynamics, e.g., coastal ocean–estuarine exchange mechanisms, and on the behavior of fish larvae with respect to vertical stratification and water movements will be needed to better define the controls on estuarine recruitment of these and other species in the northern Gulf of Mexico.

Little is known of the ecological coupling or linkage between the estuaries and the coastal ocean (Turner et al. 1979). Some progress has been made recently toward understanding the transport and recruitment mechanisms that link offshore spawning grounds to estuarine nursery areas (for reviews see Miller et al. 1984; Norcross and Shaw 1984; Sherman et al. 1984). Fisheries ecologists are beginning to recognize that the study of finfish transport and immigration to estuarine ecosystems as it is coupled with estuarine outwelling of nutrients and biomass, including finfish emigration, is crucial to a better understanding of recruitment processes (Ad Hoc Group of the Ocean Sciences Board 1980; Bakun et al. 1982; Rothschild and Rooth 1982; NSF Advisory Committee on Ocean Sciences 1985).

Most marine research in the northern Gulf of Mexico has focused on the estuarine and marsh environment, and limited effort has been directed towards the coastal ocean and continental shelf. This coastal ocean receives the discharge from the Mississippi River and this nutrient influx helps to support the most productive fisheries (i.e., weight of fish and shellfish landed in Louisiana) in the USA (USNMFS 1983–1986). Several of these fisheries depend upon species whose eggs are spawned offshore and whose larvae must be transported into coastal and estuarine nursery grounds.

[1]Present address: Center for Environmental and Estuarine Studies, University of Maryland, Chesapeake Biological Laboratory, Solomons, Maryland 20688, USA.

In this paper, we demonstrate the coupling between the coastal ocean and an adjacent estuarine marshland for gulf menhaden *Brevoortia patronus* and sand seatrout *Cynoscion arenarius*. We have amassed and integrated new data or data from final contract reports by the authors. We have also drawn upon data from unpublished theses (i.e., Marotz 1984; Cowan 1985; Deegan 1985) and have reviewed and synthesized published literature on early life history. We present catch and length-frequency data and analyze age data from continental shelf and estuarine regions of Louisiana. Finally, we calculate the transit times from spawning ground to nursery area.

Life Histories

Gulf Menhaden

Gulf menhaden eggs are spawned along the continental shelf of the northern Gulf of Mexico at depths ranging from 11 to 128 m, but most spawning occurs within the 10- to 60-m isobaths (Roithmayr and Waller 1963; Fore 1970b; Shaw et al. 1985a). Spawning may begin in late August or September and extend to April or May, but peak egg densities occur in December (Fore 1970b; Christmas and Waller 1975; Shaw et al. 1985a). Larvae hatch out at lengths of 2.6–3.2 mm (Hettler 1968, 1984) and later enter estuarine nursery areas when they are about 15–25 mm long (Suttkus 1956; Hoese 1965; Norden 1966; Perret et al. 1971; Gallaway and Strawn 1974; Simoneaux 1979; Deegan 1985). Larval gulf menhaden have been collected in Louisiana and Texas estuaries as early as September and as late as May and June (Reid 1955a, 1955b, 1956; Simmons 1957; Copeland 1965; Perret et al. 1971; Dunham 1972; Gallaway and Strawn 1974; Simoneaux 1979). Transformation to juvenile morphology is completed by 30 to 35 mm standard length (Tagatz and Wilkens 1973; Deegan 1985). Juveniles as small as 45 mm total length (TL) have been collected in October (Dugas 1970).

Hettler (1984) reported that the average growth rate for larval gulf menhaden during their first 90 d in the laboratory was 0.3 mm/d at 20°C and 30‰ salinity and also verified that growth increments are deposited daily on the otoliths. In the estuarine environment, juvenile growth is believed to be rapid. Small fish in cooler waters grow about 0.2 mm/d, whereas larger fish in warmer waters grow approximately 0.8–1 mm/d (Loesch 1976; Deegan 1985). Juveniles usually spend between 4 and 10 months in the estuary, and emigrate during summer and fall, although some may remain over winter (Gunter 1945; Christmas 1960; Springer and Woodburn 1960; Christmas and Waller 1975; Christmas et al. 1981; Deegan 1985).

Sand Seatrout

Early life history information on sand seatrout is limited. Sand seatrout are endemic to the Gulf of Mexico and spawn at depths ranging from 5 to 70 m, but most spawning occurs between the 5- and 25-m isobaths (Moffett et al. 1979; Shlossman and Chittenden 1981; Cowan 1985). Larvae have been collected in all months of the year in shelf waters of the northern Gulf (Finucane et al. 1977; Cowan 1985). Spawning is strongly bimodal; a spring spawn extends from January to May and peaks in April, and a late-summer spawn occurs from August to September; little spawning occurs in June or July (Daniels 1977; Shlossman and Chittenden 1981).

Larvae hatch out at 1.8–2.0 mm (Daniels 1977). Young sand seatrout (10–17 mm) first appear in estuarine nursery areas in March or April and again in October through November (Hoese et al. 1968; Dunham 1972; Gallaway and Strawn 1974; Shlossman and Chittenden 1981; Rogers and Herke 1985b). There is some evidence that sand seatrout may also use shallow parts of the Gulf of Mexico (<18 m deep) as a nursery, at least off Texas (Shlossman and Chittenden 1981).

Growth rates of juvenile sand seatrout, both near shore and in the estuaries, are seasonally variable. Growth is slowest during winter (0.17–0.33 mm/d), and fastest (1.17 mm/d) during May–October (Shlossman and Chittenden 1981). Juvenile sand seatrout are present in northern Gulf estuaries year-round (Gunter 1945; Kilby 1955; Roessler 1970; Dunham 1972; Galloway and Strawn 1974; Thompson and Verret 1980; Rogers and Herke 1985b). However, most juveniles emigrate from estuaries in summer to late fall after a residency of 4–8 months (Shlossman and Chittenden 1981).

Methods

Offshore study.—A sampling grid consisting of 37 stations, spaced on five transects normal to the coast, was established off western Louisiana (Figure 1). Samples were collected on each of five cruises: 15–16 December 1981 and 18–21 January, 11–15 February, 7–10 March, and 16–20 April 1982. In December, only the westernmost transect (A) was sampled due to adverse weather conditions. All plankton collections were made with an opening and closing 60-cm, paired-net frame, plankton sampler. Net meshes on this "bongo-type" sampler were 500 μm and 335 μm.

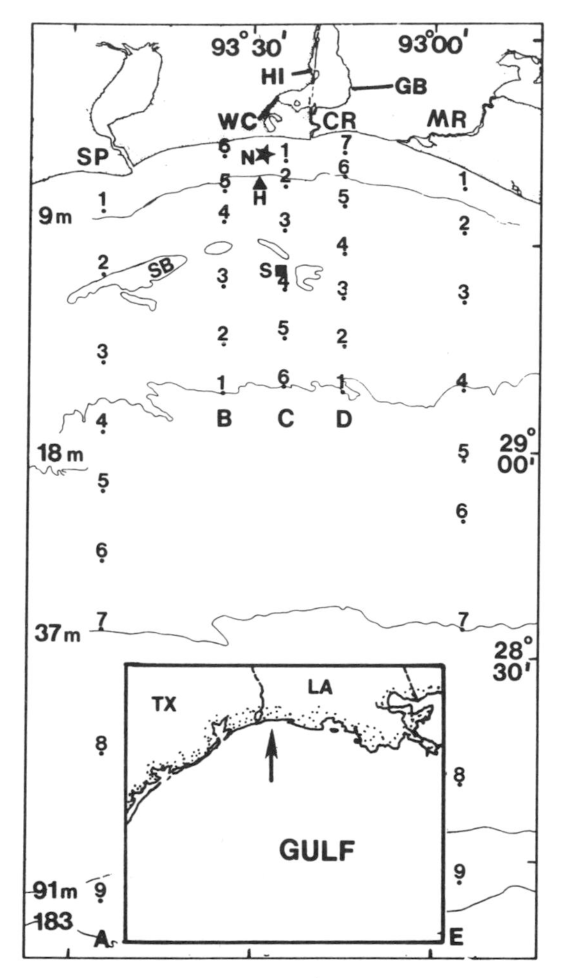

FIGURE 1.— Sampling grid showing offshore ichthyoplankton cruise transects (A–E) in the vicinity of the Calcasieu River (CR) estuary, Louisiana (LA) and the sites of larval and juvenile fish collections at three marsh stations (HI, Hog Island Gully; WC, West Cove Canal; GB, Grand Bayou). Depth contours (in meters) are labeled at the left margin. Sites N, H, and S represent moored current meter locations. SP = Sabine Pass, Texas (TX); MR = Mermentau River, Louisiana.

A flowmeter (General Oceanics model 2030) was secured in each net to determine volume of water filtered. Towing speed was about 1 m/s. Most plankton collections consisted of 10-min stepped-oblique tows from near-bottom to surface (Shaw et al. 1985a).

In order to delineate possible vertical stratification of the ichthyoplankton, 10-min simultaneous surface and near-bottom horizontal tows were taken at transect stations A-3, A-6, A-9, B-1, C-6, D-1, E-3, E-6, and E-9 (Figure 1). These nets were set closed, opened at the appropriate depth, and then closed for retrieval. Plankton tows were made during day and night, depending on the time of day the ship occupied a particular station. The majority of samples were preserved in 10% buffered formalin at sea and later transferred into 4% buffered formalin. Gulf menhaden and sand seatrout larvae were counted from the 335-μm-mesh collections and catches are reported as standardized densities (larvae/100 m^3). Ichthyoplankton data from both stepped-oblique and horizontal tows were combined for the seasonality analysis.

Sand seatrout larvae used in otolith analysis were taken from 500-μm-mesh collections at stations A-1, A-8, B-2, B-6, C-1, C-5, D-2, D-7, E-1, and E-8. Samples were stored in 100% isopropanol at sea and changed again when returned to the laboratory. Age determination of sand seatrout larvae was based on the methods of Taubert and Coble (1977). Saccular otoliths were removed and mounted in Permount on a glass microscope slide. Otoliths sufficiently thick to prevent a direct increment count were ground in the saggital plane with number 600 ultra-fine silicon carbide sand paper until the core was reached. Ground otoliths were then etched in 5% EDTA for 15–45 s, placed under a cover slip, and observed with a polarized oil-immersion light microscope at 400–1,000 × magnification. Regularly spaced increments along a radius from the core area were enumerated. Each increment consisted of one light (translucent) and one dark (opaque) concentric band. Three independent increment counts were made on each otolith without knowledge of date of capture or length of specimen. Replicate counts of an otolith never differed by more than 7.5%.

Water current data were collected with Endeco 174 current meters moored at site N, 6.2 km offshore in 6.4 m of water; at site H, 11.4 km offshore in 9.8 m of water; and at site S, 35 km offshore in 12.5 m of water (Figure 1). At each of these sites a meter was moored 1.8 m off the bottom. Sites H and S had a second current meter 3.7 m below the surface. Progressive vector diagrams (PVDs) were constructed by successive graphical addition of daily-averaged unfiltered current data computed from half-hour velocity observations taken between 22 October 1981 and 1 June 1982. Every 24-h displacement is represented by the curve between two graphed points (Figure 2). The PVDs show eulerian data (velocities measured at a fixed point) and are based on an assumption that the observed motion is representative of the larger water mass. They do not

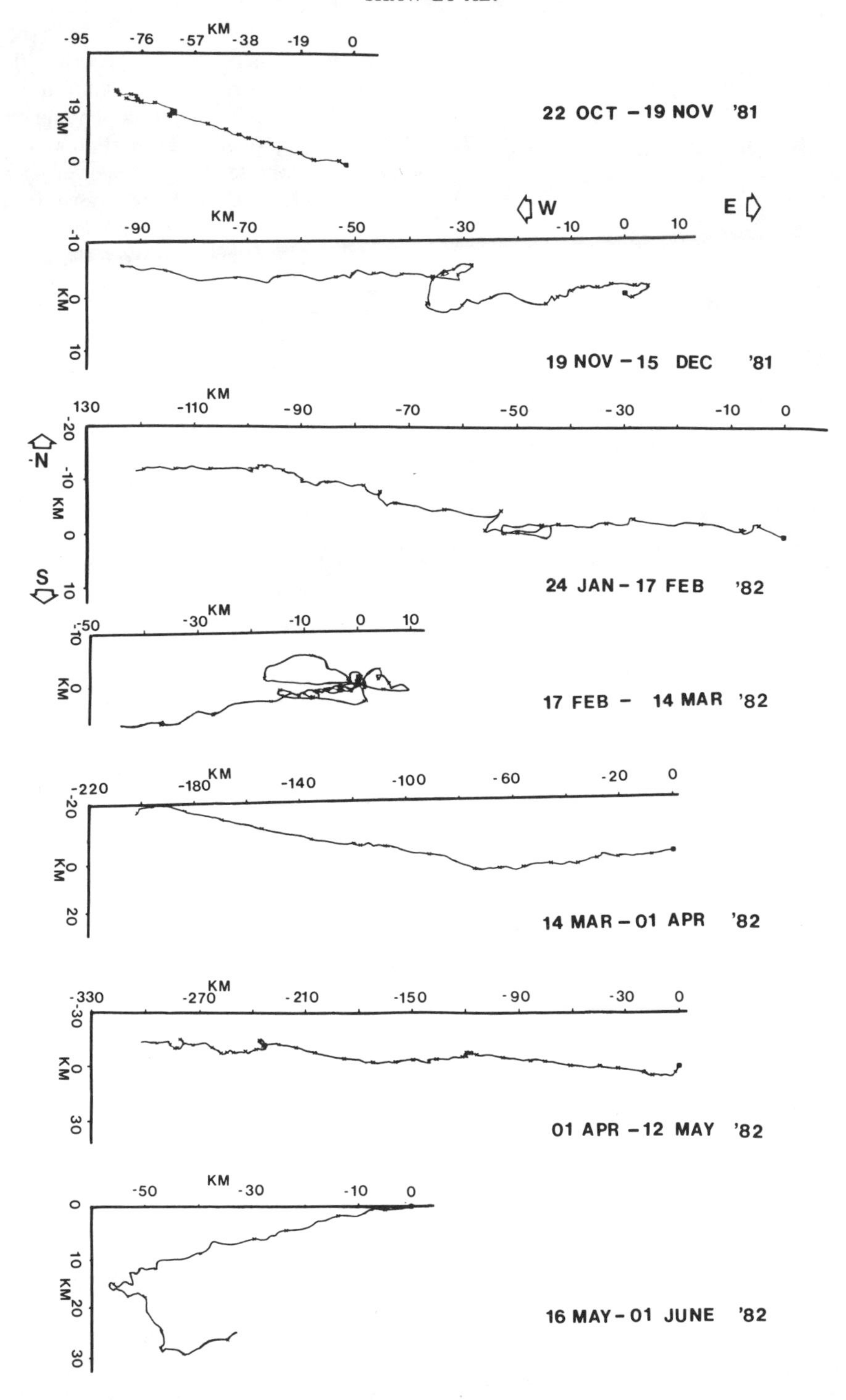

FIGURE 2.—Progressive vector diagrams constructed from raw current data taken every half hour and plotted by day from a moored current meter at site H (Figure 1) for the period 22 October 1981–1 June 1982. The current meter was located 3.7 m below the surface in 9.8 m of water, 11.4 km offshore. The abscissa is oriented west to east and negative values represent distance (km) in the westward direction. The ordinate is oriented north to south and negative values represent northerly flows. For each diagram, the solid black square (usually found on the right side of the plot) represents the start or first day of the current meter record (eulerian data). There were no data from 16 December 1981 to 23 January 1982.

represent, as it may first appear, an attempt to continuously trace the drift (lagrangian data) of a particular particle of water through time and space (Sverdrup et al. 1942).

Estuarine study.—Nekton samples were collected during the estuarine study from stations at each major water channel draining brackish marshland into Calcasieu Lake. West Cove Canal and Hog Island Gully stations were on the lake's west shore and Grand Bayou station was on the east shore (Figure 1). All stations were within the marshes of Sabine National Wildlife Refuge. Movement of late-stage larval and juvenile organisms was monitored by passive traps (Herke and Rogers 1984) similar to those used by King (1971). The box-like traps had plywood tops and bottoms and sides constructed of 1.2-mm-diameter monel wire screening (6 × 6 meshes/25.4 mm), which provided 3.0-mm-square openings. A V-shaped mouth of the same mesh, with a vertical slit 5 cm wide from top to bottom, allowed organisms easy ingress, but made egress difficult. Trap dimensions were 61 cm high, 61 cm wide, and 82 cm deep. These traps were mounted on pilings in midstream of the channels. Each station had traps oriented in both incoming and outgoing tidal directions for top, middle, and bottom depth levels. Between 19 March 1980 and 19 December 1982, the six traps at Hog Island Gully, and the six at West Cove Canal were fished approximately 22–26 h one or more times a week; Grand Bayou traps were fished similarly on the same day, but only from October 1981 through December 1982. All samples were held in ice water until processed. Organisms in every sample were identified to species and enumerated by midpoint 5-mm standard-length classes and then converted to total length for direct comparison with ichthyoplankton data. Catch was recorded by station, trap depth level, and direction of finfish movement. To make all trap catches more comparable, trap data were standardized as average catch per 24-h set of an individual trap. The catch data presented herein are reported as the mean of all traps fished during a given 2-week period. For the purposes of this paper, we present length-frequency data from the time interval between early November 1981 through late June 1982.

Normally, estimates of water volumes filtered were not made at trap stations. However, at the Hog Island Gully station, water volumes were measured with flowmeters (General Oceanics model 2030) during an ancillary marsh study between 8 February and 8 September 1982. Measurements were taken at each trap level every month during this interval. Flowmeter data were collected for 10 min each hour over a 24-h incoming tide only (during the late fall-to-spring period, water level and transport in shallow marshes often reflect the forces of atmospheric frontal passages rather than the mixed diurnal tides). These measurements are used as a rough estimate of the general ratio of water volumes filtered during incoming tides at the three trap levels. For more detailed information on sampling and subsampling methods, and the study area, see Herke and Rogers (1984) and Herke et al. (1984).

Results and Discussion

Vertical Distributions

In the offshore ichthyoplankton study, we found no significant differences between larval gulf menhaden densities from simultaneous surface and near-bottom horizontal tows taken during day or night (two-way analysis of variance, ANOVA; df = 46; $P = 0.28$), which may be a result of the characteristically intense vertical mixing of the water column during the winter period (Shaw et al. 1985b). However, catches of gulf menhaden in the marsh study showed significant vertical differences (two-way ANOVA and Duncan's multiple-range test; $P \leq 0.05$: Marotz 1984). Most gulf menhaden were caught in the midwater trap (51% of the cumulative catch), 39% were taken by the surface trap, and 10% by the bottom trap. Because the ratio of mean water volumes filtered for the surface, middle and bottom trap levels (at least during the ministudy at the Hog Island Gully station; $N = 388$ flowmeter observations) was approximately 2.0:1.9:0.9, respectively, it is possible that catch rates reflect differences in water volume filtered through a trap in a 24-h fishing period. However, the marsh vertical catch differences are in agreement with coastal and other estuarine studies, which have shown late-stage larval gulf menhaden to be distributed mostly in the mid to upper water column (Hoese 1965; Fore 1970a; Walker 1978; Vecchione et al. 1982).

In the offshore study, sand seatrout larvae appeared to be somewhat surface-oriented (the mean denstities from all horizontal tows were 9.2 larvae/100 m^3 at the surface and 1.9 near the bottom). However, this trend proved to be nonsignificant (at $P \leq 0.05$) in a four-way ANOVA involving time of day, horizontal tow type, month, and distance from shore (Cowan 1985). In

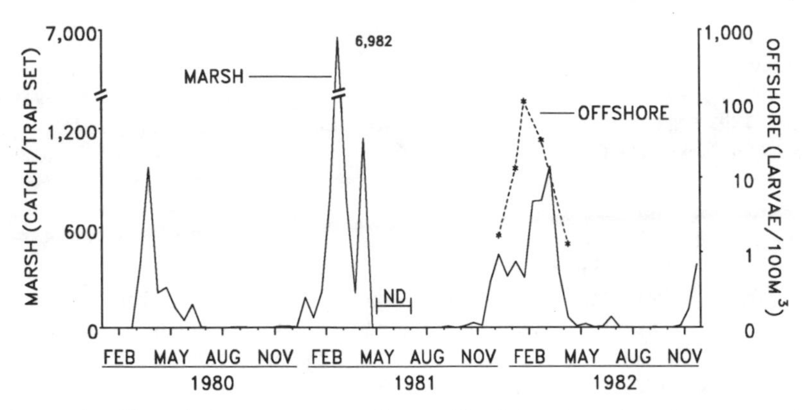

FIGURE 3.—Seasonal gulf menhaden occurrences for combined marsh stations (average catch per 24-h trap collection) for the period 19 March 1980–19 December 1982 (ND = no data) and for combined offshore ichthyoplankton collections (asterisks = average number of larvae/100 m^3) for five cruises between 15 December 1981 and 20 April 1982.

the marsh study, sand seatrout larvae and juveniles were taken in greater numbers in the midlevel and bottom traps (Rogers and Herke 1985a). For both gulf menhaden and sand seatrout, these differences in vertical catch distributions between shelf and marshland did not appear to be related to a change in water column stratification, because the three trap stations were also well mixed with respect to salinity, temperature, and dissolved oxygen.

The progressive vector diagrams, PVDs (Figure 2), from a near-surface current meter at site H (Figure 1) are representative of the trends seen at the near-surface or near-bottom records from the other two moored sites, N and S. During the winter–spring of 1981–1982, encompassing the spawning time of gulf menhaden and the spring spawn of sand seatrout, larvae were generally exposed to west–northwest advection. There were some instances of current reversals that were probably associated with atmospheric frontal passages and reversing wind fields (Van Heerden et al. 1983). The situation was very complicated during the periods of 17 February to 14 March and 16 May to 1 June 1982, yet, in both instances, there was still a net westward drift. Larvae in our offshore study area would not be transported directly across the shelf to the marsh study site but would most likely make landfall in Texas. We have assumed, therefore, that the observed ichthyoplankton shelf distribution (directly offshore of Calcasieu estuary) is representative of shelf spawning towards the east or "upstream." Previous Gulf-wide surveys (Fore 1970b; Christmas and Waller 1975) have encountered gulf menhaden spawning along the entire continental shelf from Texas to Florida but especially concentrated off Louisiana; eggs and larvae were distributed over similar water depths and distances from shore as in our shelf study. Similarly, to the east of our study area, Darnell et al. (1983) reported large concentrations of adult sand seatrout during the spawning season. Their synthesis of shelf fishery data suggests a source and potential shelf distribution of sand seatrout eggs and larvae similar to those we observed.

Temporal and Length-Frequency Distribution

Logistical problems prevented us from sampling shelf waters in October and November 1981, so we are unable to compare the time of first appearance of larval gulf menhaden offshore with the time of first arrival of postlarvae in the marsh study area. In the winter of 1981–1982, larval gulf menhaden densities peaked offshore in early February with a mean of 120 larvae/100 m^3 (Figure 3). By late April, densities were low and distributed only along the coast, and there was no indication of continued offshore spawning. March was the last month in which recently hatched larvae (< 5mm) were collected (Figure 4).

In 1982, the marsh abundance of larval and juvenile gulf menhaden peaked in late March, approximately 6 weeks after the offshore density peak (Figure 3). In Fourleague Bay, Louisiana, approximately 225 km east of our study area, Deegan (1985) found that gulf menhaden entered the estuary when 42–70 d old (mean, 8 weeks), based on otolith analyses. Growth conditions and

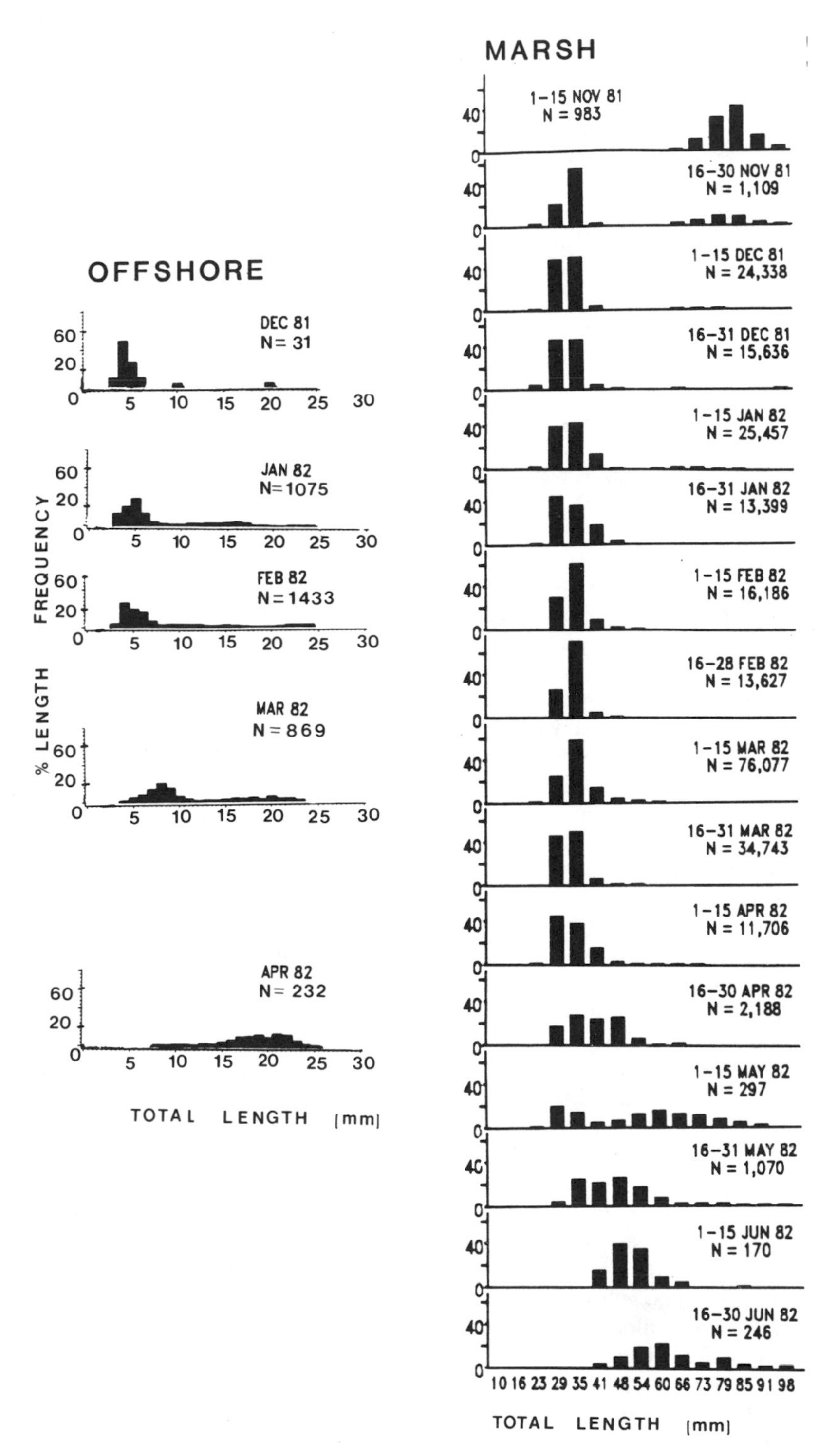

FIGURE 4.—Length-frequency data for gulf menhaden taken offshore (combined ichthyoplankton data) and in Calcasieu marsh (late larvae and early juveniles), 1981–1982. For the marsh data, the abscissa is in 5-mm midpoint size-classes, and N is the number of fish collected and either measured directly or estimated from subsamples.

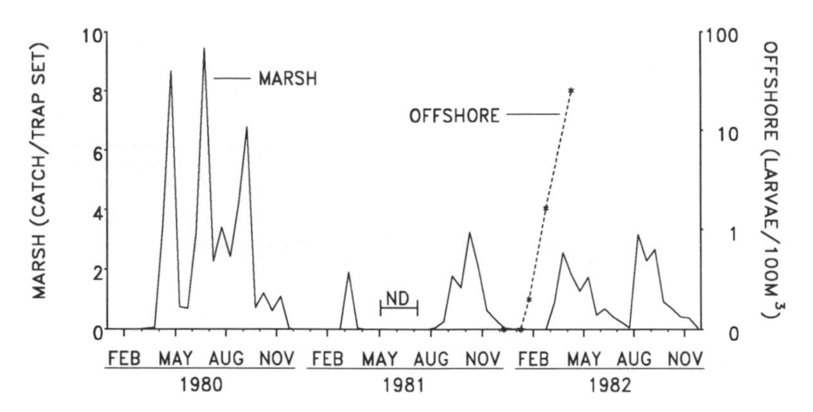

FIGURE 5.—Seasonal sand seatrout occurrences for combined marsh stations (average catch per 24-h trap collection) for the period 19 March 1980–19 December 1982 (ND = no data) and for combined offshore ichthyoplankton collections (asterisks = average number of larvae/100 m^3) for five cruises between 15 December 1981 and 20 April 1982.

transport dynamics may be different off Fourleague Bay, an estuarine system of the Atchafalaya River delta. However, these data and our estimated 6-week time lag between peak offshore and marsh abundances indicate that the transit time across the shelf to the estuary may be greater than the previous estimates of 3–5 weeks (Fore 1970a; Reintjes 1970; Christmas and Etzold 1977). If we apply Hettler's (1984) laboratory-determined growth rate of 0.3 mm/d to gulf menhaden that hatch offshore at approximately 3.0 mm and that later enter estuarine nurseries at 15–25 mm, we get a similar transit time of about 40–73 d.

In this analysis, we have focused on the timing of estuarine immigration based on the presence or absence of, or peak catch rates for, larvae in the marsh. More detailed analysis involving length frequencies is hazardous because both the offshore plankton nets and the marsh traps undersample larvae of 21–25 mm, the upper end of the length at which they recruit to the estuaries. However, larvae this large were present in offshore collections from December through April and in the marsh during at least November–January and April–May (Figure 4). There was a continuous supply of prejuvenile gulf menhaden one size-class larger (27–31 mm, plotted as the 29-mm midpoint in Figure 4) from late November 1981 through the end of May 1982. The length-frequency data also show the presence of overwintering juveniles in the marsh.

Sand seatrout evidently began offshore spawning during late January 1982, but larvae were not taken in any substantial number until February (Figure 5). Densities continued to increase through April, the last month of offshore sampling. Marsh length-frequency data (Figure 6) for early November 1981 show the presence of juveniles between 21 and 25 mm, the product of the previous late-summer spawn. In 1982, 14–18-mm sand seatrout, spawned during winter and spring, began to appear in the marsh in late March and continued through late May. There appeared to be a 1–2-month delay between offshore spawning and the appearance of the first marsh migrants. A shorter time lag might be expected for sand seatrout, because most spawning occurs near shore between the 5- and 25-m isobaths, but transit time across the shelf appears to be highly variable for sand seatrout, as seen below.

We estimated the ages of larval sand seatrout from Louisiana shelf waters by enumerating the growth increments in saccular otoliths from 149 fish, which ranged from 2.0 to 12 mm TL. The increment counts for these larvae ranged from 6 to 64, and most larvae had between 14 and 36 rings. If these growth increments were formed daily, the mean growth rate for all larvae was 0.17 mm/d. A daily rate of increment formation has not been validated for larval sand seatrout, but it has been for the related sciaenids red drum *Sciaenops ocellatus* and spot *Leiostomus xanthurus* (Peters et al. 1978; University of Texas Marine Science Institute 1982–1983; Warlen and Chester 1985). Also, otolith increments have been assumed to be daily in Atlantic croaker *Micropogonias undulatus* (Warlen 1982; Cowan, in press). Our estimate of growth rate for sand seatrout larvae is similar to the rates determined for other sciaenid larvae (Peters et al. 1978; Warlen 1982; University

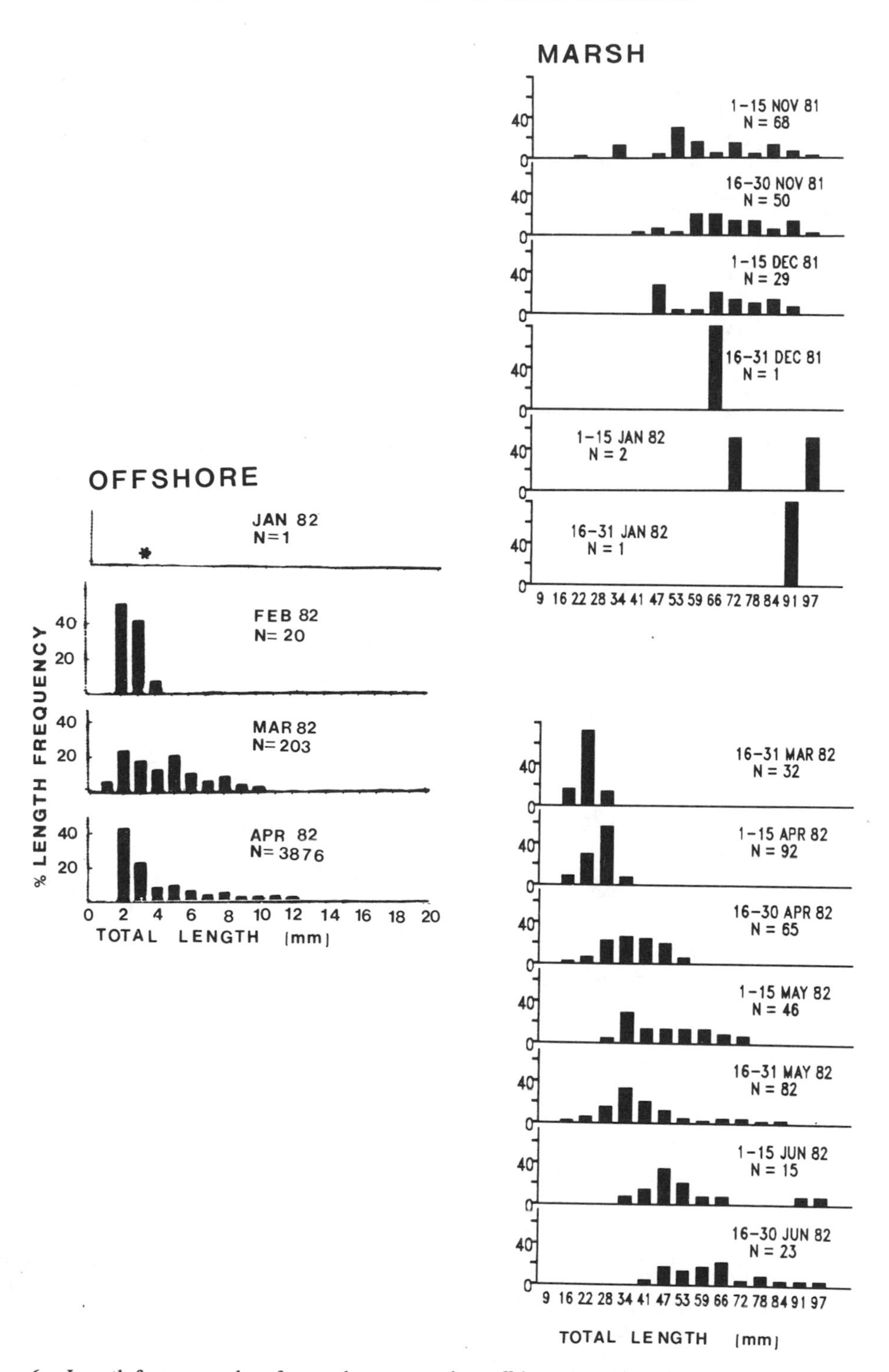

FIGURE 6.—Length-frequency data for sand seatrout taken offshore (combined ichthyoplankton data; asterisk indicates one 3.0-mm larva taken) and in Calcasieu marsh (late larvae and early juveniles), 1981–1982. For the marsh data, the abscissa is in 5-mm midpoint size-classes, and N is the number of fish collected and either measured directly or estimated from subsamples. No larvae or juveniles were taken from 1 February to 15 March 1982.

of Texas Marine Science Institute 1982–1983; Beckman and Dean 1984; Govoni et al. 1985). Therefore, we assumed the otolith increments in our larval sand seatrout to be daily. We added 5 d to the numbers of increments counted before the growth rates were calculated, because other sciaenids have been shown to begin increment deposition 4 to 5 d after hatching (Peters et al. 1978; University of Texas Marine Science Institute 1982–1983; Warlen and Chester 1985). Then, for sand seatrout larvae that hatch offshore at about 1.9 mm and enter the estuary at 10–17 mm, a growth rate of 0.17 mm/d implies transit times of 53–94 d. This estimate should be interpreted as the upper end of the range in transit time, because sand seatrout spawn relatively close to shore and may use the shallow Gulf as a nursery ground for a period of time (Shlossman and Chittenden 1981).

Finally, an independent estimate of transit time can be calculated directly from the moored current meter data. The best available estimate of shoreward transport on the mid to inner shelf is the N–S component of the mean vector from the near-surface meter at site H for the period 24 January–12 May 1982, which was 1.8 cm/s shoreward (Shaw et al. 1985b). Based on this rough estimate, it would take approximately 60 d for a passive buoyant particle to traverse 93 km of shelf, the offshore distance to our E-6 station or the 26-m isobath.

Critique and Future Direction

The implication of the current data analyses (Shaw et al. 1985b; Cochrane and Kelly 1986), that fish larvae recruiting to western Louisiana estuaries originate to the east, points out a difficulty facing fisheries oceanographers today. What appeared at first to be a fortunate juxtaposition of concurrent continental shelf and adjacent marshland studies turned out to be an informative, yet less than perfect, linkage of data. This illustrates the need for quantitative circulation data for the coastal ocean, and for shelf–estuary water mass exchange studies. On the basis of current meter data collected during three consecutive winters between October 1980 to April 1983, Shaw et al. (1985b) postulated that gulf menhaden larvae would be carried west-northwest along western Louisiana. Once near shore, they could be transported into the estuaries and marshes by astronomically and meteorologically driven flows or other shelf–estuary exchange mechanisms, which the larvae may select by behavioral means. We still do not know how close to shore larvae have to be to come under the influence of coastal ocean–estuary circulation and exchange mechanisms. This demonstrates the need for future tidal-pass or inlet studies, as well as for better information about vertical distribution, depth preferences, and changes in buoyancy or behavior of larvae. Such information is necessary to test hypotheses of passive or active transport and to further elucidate recruitment processes.

Acknowledgments

We thank Joanne Lyczkowski-Shultz and two anonymous reviewers for their constructive criticism of this manuscript. We acknowledge Frank J. Kelly, Jr., for use of his moored current meter data for the progressive vector diagrams and Linda A. Deegan and Bruce L. Marotz for use of their unpublished data. Data analysis was funded by the Louisiana Sea Grant College and the Louisiana State University Agricultural Center. The offshore data collection was supported by the Louisiana Department of Wildlife and Fisheries and the U.S. Department of Energy (cooperative agreement DE-FC96-81P010313). The marsh data collection was funded by the U.S. Fish and Wildlife Service (cooperative agreement 14-16-0009-79-1003) and the U.S. Soil Conservation Service (contract 53-7217-1-23). The area for the marsh study was provided by the Sabine National Wildlife Refuge. This is Louisiana State University's Marine and Coastal Fisheries Program contribution LSU-CFI-85-38 and Louisiana Cooperative Fish and Wildlife Research Unit contribution 47.

References

Ad Hoc Group of the Ocean Sciences Board. 1980. Fisheries ecology. Some constraints that impede advances in our understanding. National Academy of Sciences, National Research Council, Assembly of Mathematical and Physical Sciences, Washington, D.C.

Bakun, A., J. Beyer, D. Pauly, J. G. Pope, and G. D. Sharp. 1982. Ocean sciences in relation to living resources. Canadian Journal of Fisheries and Aquatic Sciences 39:1059–1070.

Beckman, D. W., and J. M. Dean. 1984. The age and growth of young-of-the-year spot, *Leiostomus xanthurus* Lacepede, in South Carolina. Estuaries 7:487–496.

Christmas, J. Y. 1960. Menhaden populations. Gulf Coast Research Laboratory Circular 92:26–28. (Ocean Springs, Mississippi.)

Christmas, J. Y., and D. J. Etzold. 1977. The menhaden fishery of the Gulf of Mexico United States: a regional management plan. Gulf Coast Research Laboratory, Technical Report 7, Ocean Springs,

Mississippi.
Christmas, J. Y., J. T. McBee, R. S. Waller, and F. C. Sutter III. 1981. Habitat suitability index model for gulf menhaden. Gulf Coast Research Laboratory, Ocean Springs, Mississippi.
Christmas, J. Y., and R. S. Waller. 1975. Location and time of menhaden spawning in the Gulf of Mexico. Gulf Coast Research Laboratory, Ocean Springs, Mississippi.
Cochrane, J. D., and F. J. Kelly. 1986. Low-frequency circulation on the Texas–Louisiana continental shelf. Journal of Geophysical Research 91:10645–10659.
Copeland, B. J. 1965. Fauna of the Aransas Pass Inlet, Texas: I. Emigration as shown by tide trap collections. Publications of the Institute of Marine Science, University of Texas 10:9–29.
Cowan, J. H., Jr. 1985. The distribution, transport and age structure of drums (family Sciaenidae) spawned in the winter and early spring in the continental shelf waters of western Louisiana. Doctoral dissertation. Louisiana State University, Baton Rouge.
Cowan, J. H., Jr. In press. Age and growth of Atlantic croaker, *Micropogonias undulatus,* larvae collected in the coastal waters of the northern Gulf of Mexico as determined by increments in saccular otoliths. Bulletin of Marine Science 42.
Daniels, K. L. 1977. Description, comparison, and distribution of larvae of *Cynoscion nebulosus* and *Cynoscion arenarius* from the northern Gulf of Mexico. Master's thesis. Louisiana State University, Baton Rouge.
Darnell, R. M., R. E. Defenbaugh, and D. Moore. 1983. Northwestern Gulf shelf bio-atlas: a study of the distribution of demersal fishes and penaeid shrimp on soft bottoms on the continental shelf from the Rio Grande to the Mississippi River delta. Gulf of Mexico OCS (Outer Continental Shelf) Regional Office, Minerals Management Service, Open File Report 84-04, Metairie, Louisiana.
Deegan, L. A. 1985. The population ecology and nutrient transport of gulf menhaden in Fourleague Bay, Louisiana. Doctoral dissertation. Louisiana State University, Baton Rouge.
Dugas, R. J. 1970. An ecological study of Vermilion Bay, 1968–1969. Master's thesis. University of Southwestern Louisiana, Lafayette.
Dunham, F. 1972. A study of commercially important estuarine-dependent industrial fishes. Louisiana Wildlife and Fisheries Commission, Technical Bulletin 4, New Orleans.
Finucane, J. H., L. A. Collins, and J. D. McEachran. 1977. Environmental studies of the south Texas continental shelf. 1976. Ichthyoplankton/mackerel eggs and larvae. Report to Bureau of Land Management, Washington, D.C. (National Technical Information Service, NTIS PB-283-873.)
Fore, P. L. 1970a. Life history of gulf menhaden *Brevoortia patronus*. U.S. Fish and Wildlife Service Circular 350:12–16.
Fore, P. L. 1970b. Oceanic distribution: eggs and larvae of the gulf menhaden. U.S. Fish and Wildlife Service Circular 341:11–13.
Gallaway, B. J., and K. Strawn. 1974. Seasonal abundance and distribution of marine fishes at a hot-water discharge in Galveston Bay, Texas. Contributions in Marine Science 18:71–137.
Govoni, J. J., A. J. Chester, D. E. Hoss, and P. B. Ortner. 1985. An observation of episodic feeding and growth of larval *Leiostomus xanthurus* in the northern Gulf of Mexico. Journal of Plankton Research 7:137–146.
Gunter, G. 1945. Studies on marine fisheries of Texas. Publications of the Institute of Marine Science, University of Texas 1:1–190.
Herke, W. H., and B. D. Rogers. 1984. Comprehensive estuarine nursery study completed. Fisheries (Bethesda) 9(6):12–16.
Herke, W. H., B. D. Rogers, and J. A. Grimes. 1984. A study of the seasonal presence, relative abundance, movements, and use of habitat types by estuarine-dependent fishes and economically important decapod crustaceans on the Sabine National Wildlife Refuge. Louisiana State University, Louisiana Cooperative Fishery Research Unit, Final Report, 3 volumes, Baton Rouge.
Hettler, W. F., Jr. 1968. Artificial fertilization among yellowfin and gulf menhaden (*Brevoortia*) and their hybrid. Transactions of the American Fisheries Society 97:119–123.
Hettler, W. F., Jr. 1984. Description of eggs, larvae, and early juveniles of gulf menhaden, *Brevoortia patronus,* and comparisons with Atlantic menhaden, *B. tyrannus,* and yellowfin menhaden, *B. smithi.* U.S. National Marine Fisheries Service Fishery Bulletin 82:85–95.
Hoese, H. D. 1965. Spawning of marine fishes in the Port Aransas, Texas, area as determined by the distribution of young and larvae. Doctoral dissertation. University of Texas, Austin.
Hoese, H. D., B. J. Copeland, F. N. Moseley, and E. D. Lane. 1968. Fauna of the Aransas Pass Inlet, Texas. III. Diel and seasonal variations in trawlable organisms of the adjacent area. Texas Journal of Science 20:33–60.
Kilby, J. D. 1955. The fishes of two Gulf coastal marsh areas of Florida. Tulane Studies in Zoology 2:175–247.
King, B. D., III. 1971. Study of migratory patterns of fish and shellfish through a natural pass. Texas Parks and Wildlife Department Technical Series 9:1–54. (Austin.)
Loesch, H. C. 1976. Observations of menhaden recruitment and growth in Mobile Bay, Alabama. Proceedings of the Louisiana Academy of Sciences 39:35–42.
Marotz, B. L. 1984. Seasonal movements of penaeid shrimp, Atlantic croaker, and gulf menhaden through three marshland migration routes surrounding Calcasieu Lake in southwestern Louisiana. Master's thesis. Louisiana State University, Baton Rouge.
Miller, J. M., J. P. Reed, and L. J. Pietrafesa. 1984.

Patterns, mechanisms and approaches to the study of migrations of estuarine-dependent fish larvae and juveniles. Pages 209–225 *in* J. D. McCleave, G. P. Arnold, J. J. Dodson, and W. H. Neill, editors. Mechanisms of migration in fishes. Plenum, New York.

Moffett, A. W., L. W. McEachron, and J. G. Key. 1979. Observations on the biology of sand seatrout (*Cynoscion arenarius*) in Galveston and Trinity bays, Texas. Contributions in Marine Science 22:163–172.

Norcross, B. L., and R. F. Shaw. (1984). Oceanic and estuarine transport of fish eggs and larvae: a review. Transactions of the American Fisheries Society 113:153–165.

Norden, C. R. 1966. The seasonal distribution of fishes in Vermilion Bay, Louisiana. Wisconsin Academy of Sciences, Arts and Letters 55:119–137.

NSF (National Science Foundation) Advisory Committee on Ocean Sciences. 1985. Emergence of a unified ocean science: a long-range plan for the ocean sciences program of the National Science Foundation. Washington, D.C.

Perret, W. S., and seven coauthors. 1971. Cooperative Gulf of Mexico estuarine inventory and study, Louisiana. Phase IV: biology, section II. Louisiana Wildlife and Fisheries Commission Technical Bulletin, New Orleans.

Peters, D. S., J. C. DeVane, Jr., M. T. Boyd, L. C. Clements, and A. B. Powell. 1978. Preliminary observations of feeding, growth, and energy budget of larval spot (*Leiostomus xanthurus*). Pages 377–397 *in* Annual report. U.S. National Marine Fisheries Service, Beaufort Laboratory, Beaufort, North Carolina.

Reid, G. K., Jr. 1955a. A summer study of the biology and ecology of East Bay, Texas. Part I. Texas Journal of Science 7:316–343.

Reid, G. K., Jr. 1955b. A summer study of the biology and ecology of East Bay, Texas. Part II. The fish fauna of East Bay, the Gulf beach and summary. Texas Journal of Science 7:430–453.

Reid, G. K., Jr. 1956. Ecological investigations in a disturbed Texas coastal estuary. Texas Journal of Science 8:296–327.

Reintjes, J. W. 1970. The gulf menhaden and our changing estuaries. Proceedings of the Gulf and Caribbean Fisheries Institute 22:87–90.

Roessler, M. A. 1970. Checklist of fishes in Buttonwood Canal, Everglades National Park, Florida, and observations on the seasonal occurrence and life histories of selected species. Bulletin of Marine Science 20:861–893.

Roithmayr, C. M., and R. A. Waller. 1963. Seasonal occurrence of *Brevoortia patronus* in the northern Gulf of Mexico. Transactions of the American Fisheries Society 92:301–302.

Rogers, B. D., and W. H. Herke. 1985a. Estuarine-dependent fish and crustacean movements and weir management. Pages 201–219 *in* C. F. Bryan, P. J. Zwank, and R. H. Chabreck, editors. Proceedings of the fourth coastal marsh and estuary management symposium. Louisiana State University, Baton Rouge.

Rogers, B. D., and W. H. Herke. 1985b. Temporal patterns and size characteristics of migrating juvenile fishes and crustaceans in a Louisiana marsh. Louisiana State University, School of Forestry, Wildlife, and Fisheries, Research Report 5, Baton Rouge.

Rothschild, B. J., and C. Rooth. 1982. Fish ecology III. A foundation for REX-recruiting experiment. University of Miami, Technical Report 82008, Miami, Florida.

Shaw, R. F., J. H. Cowan, Jr., and T. L. Tillman. 1985a. Distribution and density of *Brevoortia patronus* (gulf menhaden) eggs and larvae in the continental shelf waters of western Louisiana. Bulletin of Marine Science 36:96–103.

Shaw, R. F., W. J. Wiseman, Jr., R. E. Turner, L. J. Rouse, Jr., R. E. Condrey, and F. J. Kelly, Jr. 1985b. Transport of larval gulf menhaden, *Brevoortia patronus*, in continental shelf waters of western Louisiana: a hypothesis. Transactions of the American Fisheries Society 114:452–460.

Sherman, K., W. Smith, W. Morse, M. Berman, J. Green, and L. Eisymmont. 1984. Spawning strategies of fishes in relation to circulation, plankton production, and pulses in zooplankton off the northeastern United States. Marine Ecology Progress Series 18:1–19.

Shlossman, P. A., and M. E. Chittenden, Jr. 1981. Reproduction, movements, and population dynamics of the sand seatrout, *Cynoscion arenarius*. U.S. National Marine Fisheries Service Fishery Bulletin 79:649–669.

Simmons, E. G. 1957. An ecological survey of the upper Laguna Madre of Texas. Publications of the Institute of Marine Science, University of Texas 4:156–200.

Simoneaux, L. F. 1979. The distribution of menhaden, genus *Brevoortia* with respect to salinity, in the upper drainage basin of Barataria Bay, Louisiana. Master's thesis. Louisiana State University, Baton Rouge.

Springer, V. G., and K. D. Woodburn. 1960. An ecological study of the fishes of the Tampa Bay area. Florida Board of Conservation, Marine Laboratory Professional Papers Series 1.

Suttkus, R. D. 1956. Early life history of the gulf menhaden, *Brevoortia patronus*, in Louisiana. Transactions of the North American Wildlife and Natural Resources Conference 21:390–407.

Sverdrup, H. U., M. W. Johnson, and R. H. Fleming. 1942. The oceans: their physics, chemistry and general biology. Prentice-Hall, Englewood Cliffs, New Jersey.

Tagatz, M. E., and E. P. H. Wilkens. 1973. Seasonal occurrence of young gulf menhaden and other fishes in a northwestern Florida estuary. NOAA (National Oceanic and Atmospheric Administration) Technical Report NMFS (National Marine Fisheries Service) SSRF (Special Scientific Report Fisheries) 672:1–14.

Taubert, B. D., and D. W. Coble. 1977. Daily rings in

the otoliths of three species of *Lepomis* and *Tilapia mossambica*. Journal of the Fisheries Research Board of Canada 34:332–340.

Thompson, B. A., and S. J. Verret. 1980. Nekton of Lake Pontchartrain, Louisiana, and its surrounding wetlands. Pages 711–864 *in* J. H. Stone, editor. Environmental analysis of Lake Pontchartrain, Louisiana, its surrounding wetlands and selected land uses. Report (Contract DACW 29-77-C-0253) to U.S. Army Corps of Engineers, New Orleans District, New Orleans, Louisiana.

Turner, R. E., S. W. Woo, and H. R. Jitts. 1979. Estuarine influences on a continental shelf plankton community. Science (Washington, D.C.) 206:218–220.

University of Texas Marine Science Institute. 1982–1983. University of Texas mariculture program report. Port Aransas.

USNMFS (U.S. National Marine Fisheries Service). 1983–1986. Fisheries of the United States, 1982–1985. U.S. National Marine Fisheries Service Current Fisheries Statistics 8300, 8320, 8360, and 8385.

Van Heerden, I. Ll., J. T. Wells, and H. H. Roberts. 1983. River-dominated suspended-sediment deposition in a new Mississippi Delta. Canadian Journal of Fisheries and Aquatic Science 40 (Supplement 1):60–71.

Vecchione, M., C. E. Meyer, and C. L. Stubblefield. 1982. Zooplankton. Pages 8-1 to 8-69 *in* West Hackberry brine disposal project pre-discharge characterization. U.S. Department of Energy, Strategic Petroleum Reserve Project, Contract DE-AC96-80P010228, New Orleans, Louisiana.

Walker, H. J., Jr. 1978. Ichthyoplankton survey of nearshore Gulf waters between Barataria Bay and Timbalier Bay, Louisiana, during July, August, and December, 1973. Master's thesis. Louisiana State University, Baton Rouge.

Warlen, S. M. 1982. Age and growth of larvae and spawning time of Atlantic croaker larvae in North Carolina. Proceedings of the Annual Conference Southeastern Association of Fish and Wildlife Agencies 34:204–214.

Warlen, S. M., and A. J. Chester. 1985. Age, growth, and distribution of larval spot, *Leiostomus xanthurus,* off North Carolina. U.S. National Marine Fisheries Service Fishery Bulletin 83:587–599.

American Fisheries Society Symposium 3:90–103, 1988

Export and Reinvasion of Larvae as Regulators of Estuarine Decapod Populations

JOHN R. MCCONAUGHA

Department of Oceanography, Old Dominion University, Norfolk, Virginia 23529, USA

Abstract.—Larvae of decapod crustacean species residing or spawning in lower estuaries can be retained within the estuary, advected to the adjacent continental shelf (20–30 km offshore), or expelled and widely distributed across the shelf. For species whose larvae are advected from the estuary, year-to-year variations in transport processes can alter the year-class strength. Decapod larvae that are widely distributed across the continental shelf tend to be concentrated in the upper 3 m of the water column. For these species, larval distributions and transport are highly correlated with wind events. Larvae of other species that move only short distances from the estuary are concentrated at depth. Diurnal vertical migrations may be the basic mechanism by which these larvae are retained near the estuary. Selection pressures for retaining a long planktotrophic larval stage that is advected from the estuary include gene flow between estuaries, physiological requirements of the larvae (i.e., temperature and salinity tolerances), and reduced predation.

The success of any species is ultimately based on its ability to reproduce; hence, natural selection should favor reproductive patterns that minimize the energy expended by the adult while maximizing the number of offspring reaching sexual maturity. Greater than 70% of the marine benthic invertebrate species examined by Thorson (1946) had either a planktotrophic or a nonfeeding lecithotrophic larva during their early life history. Evolution towards lecithotrophy and direct development has been associated with low food availability, harsh environments, and conditions that could advect larvae away from the adult habitat (Thorson 1946, 1950; Mileikovsky 1971; Chia 1974).

For planktotrophy to be an advantageous reproductive strategy, larvae must be released during periods when abiotic and biotic factors are optimal for their survival (Giese and Pearse 1974). Retention of a planktonic larval stage may result in increased dispersal, colonization of new habitats (Thorson 1946, 1950; Mileikovsky 1971; Chia 1974), gene flow between populations (Scheltema 1971, 1975), and reduced maternal energy expenditure for production of individual eggs. Paleobiological evidence suggests that high dispersal potential associated with a planktotrophic larval stage has a direct positive correlation with species longevity (Jackson 1974; Scheltema 1977; Hansen 1980).

A negative aspect of planktonic development is the potential for advection of the larvae away from a suitable adult habitat. Although many species can extend the period of metamorphic competence until a suitable habitat is encountered, larvae advected long distances away from the adult habitat have generally been considered to be lost to the system. Because the net flow of most estuaries is seaward, retention of a planktonic larval phase in the reproductive pattern of estuarine species could result in advective loss of offspring, reduced recruitment success, and, ultimately, population failure. Because most estuarine species are continuously represented in their adult habitat, it is accepted that natural selection has led to adaptations that reduce larval export or enhance the return of late larvae or juveniles to the adult habitat. It is also logical that species that are most dependent on the estuary as adults would display an increased retention efficiency (see Sandifer 1975 and others), reduced larval duration, or both.

Larvae of many estuarine species exhibit behavioral traits that may increase the probability of their retention within the parental estuary (Bousfield 1954; Carriker 1967; Wood and Hargis 1971; Sandifer 1975; Cronin and Forward 1982; Johnson 1982). These behaviors generally fall into one of two categories: gradual change in vertical position throughout larval development (Bousfield 1954; Wood and Hargis 1971; Boicourt 1982; Johnson 1982), or vertical migration by the larvae in response to tidal or diurnal influences (Carriker 1967; Cronin and Forward 1982; Sulkin 1984). The former mechanism may be associated with stochastic processes within the estuary (de Wolf 1974; Boicourt 1982) as well as with deterministic behavioral traits of the larvae. The latter is clearly associated with deterministic behaviors.

Some species living or at least spawning in the lower reaches of the estuary show a net export of larvae from the estuary (Sandifer 1975; Scheltema 1975; Christy and Stancyk 1982; Epifanio and Dittel 1982; Johnson 1982; Provenzano et al. 1983; Epifanio et al. 1984; Sadler 1984). Scheltema (1975) suggested that larval export from the estuary might be an advantageous mechanism for colonization of adjacent estuaries and genetic exchange between estuaries. Burton and Feldman (1982) proposed that for a group such as decapod crustaceans whose larvae can substantially alter their vertical position in the water column, long-term planktotrophy is not highly correlated with genetic exchange. Because of environmental variability between estuaries, numerical models suggest that on a short time scale (≤100 years) there is little advantage to large-scale (>750 km) dispersal (Palmer and Strathman 1981; Strathman 1982). Thus the retention of a planktotrophic larval stage in the face of advection from the estuary may be associated with other adaptive aspects of the reproductive strategy.

The purpose of this paper is to review the mechanisms associated with advection of decapod crustacean larvae from the estuary and their subsequent reinvasions and to examine both the advantages and consequences of this reproductive pattern for estuarine population dynamics. Because brachyuran larvae have a definitive number of stages whose duration is known for ecologically relevant conditions, they represent an excellent model for the study of larval transport mechanisms.

Larval Export from the Estuary

Some species of brachyurans residing in the lower estuary display larval retention, but other species export larvae from the estuary onto the continental shelf. Mature female blue crabs *Callinectes sapidus* migrate to the lower reaches of the estuary to spawn (Van Engel 1958). Sandifer (1975) reported peak concentrations of stage-I *C. sapidus* larvae in the lower Chesapeake Bay and at an offshore station, and he suggested that larvae develop on the shelf. Megalopae and juveniles subsequently reinvade the estuary (Sandifer 1975). Recent sampling of decapod larvae along the main stem of Chesapeake Bay from Annapolis, Maryland, to 50 km offshore confirmed a strong advection of larvae offshore (McConaugha et al., unpublished). Low numbers of stage-I larvae were collected 60–80 km up the estuary (Figure 1), but larval densities in the lower bay never exceeded 2.2/m^3. The most advanced larvae collected within the estuary were at stage II. Maximum density of *C. sapidus* larvae was en-

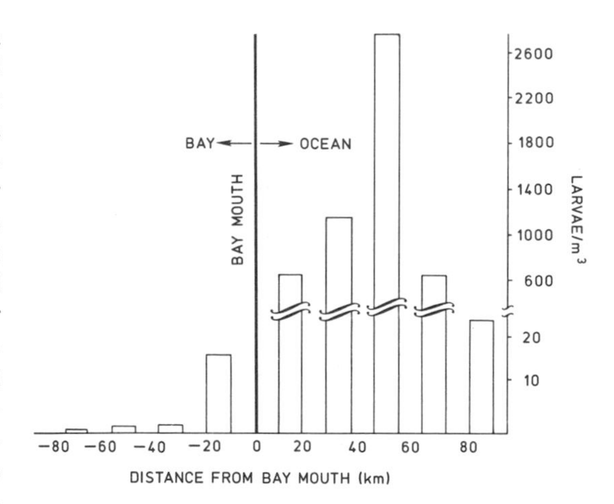

FIGURE 1.—Changes in density of *Callinectes sapidus* larvae (all developmental stages) with distance from the mouth of Chesapeake Bay. Data were collected in August 1982. (McConaugha, unpublished.)

countered 50 km offshore where all nine larval stages were collected. Greater than 80% of *C. sapidus* megalopae collected by D. F. Johnson (1982, 1985) along a transect from the York River to 18 km offshore were outside of the estuary.

A similar distribution of *C. sapidus* larvae has been reported for Delaware Bay (Epifanio and Dittel 1982; Epifanio et al. 1984). Nichols and Keney (1963) reported that, in a survey of 80 stations from Florida to Cape Hatteras, North Carolina, peak numbers of *Callinectes* larvae were found at stations 12 km offshore and larvae were present in samples up to 37 km seaward. Large numbers of stage-I *C. sapidus* larvae were collected in nearshore stations at Beaufort Inlet, North Carolina, but mid to late larval stages were not present (Dudley and Judy 1971). Substantial numbers of late-stage *Callinectes* larvae and megalopae have been reported 50–100 km off New Jersey and off the eastern shore of Virginia (Smyth 1980). The lack of mid- to late-stage larvae within the estuary and the high concentration of all larval stages on the shelf substantiate the hypothesis of larval advection from the estuary and subsequent reinvasion at the megalopal or juvenile stage for this species.

There is evidence to support the hypothesis of larval export for other estuarine decapod species (Dudley and Judy 1971; Sandifer 1975; Smyth 1980; Christy and Stancyk 1982; Dittel and Epifanio 1982; Johnson 1982; Johnson and Gonor 1982; Lambert and Epifanio 1982; Sadler 1984). Stage-I larvae of *Uca* sp. showed a net export from North

Inlet, a shallow, highly saline, well-mixed estuary in South Carolina (Christy and Stancyk 1982). Based on flux calculations, Christy and Stancyk (1982) suggested an export of 13–15% of stage-I larvae per tidal cycle. Given a stage-I duration of 3–4 d for *Uca* spp. (Christiansen and Yang 1976; O'Connor and Epifanio 1985), a flux rate of 15%/tidal cycle would be sufficient to transport all stage-I larvae out of the estuary. Given the flux rate and the relative lack of late-stage larvae in the estuary, the data are consistent with the hypothesis of offshore larval development. Significant ($P \leq 0.05$) export of crab larvae (species of *Uca, Sesarma,* and *Panopeus*) in this system occurred during spring tides, whereas invertebrate larvae characteristic of coastal waters showed significant influx during the same periods (Christy and Stancyk 1982). High concentrations of stage-I *Uca* larvae during nighttime ebb tides have been reported in the lower reaches of mixed (Dittel and Epifanio 1982; Lambert and Epifanio 1982; Brookins and Epifanio 1985) and stratified estuaries in the Middle Atlantic Bight (Goy 1976; Maris 1986). Based on the vertical distribution of stage-I larvae and megalopae in relation to tidal cycles, Lambert and Epifanio (1982) hypothesized a rapid flushing of *Uca* larvae from the estuary and a subsequent reinvasion by megalopae.

Maximum concentrations of *Pinnixa* larvae have been reported in surface waters during nighttime ebb tides in mixed estuaries (Dittel and Epifanio 1982; Brookins and Epifanio 1985). In North Inlet, South Carolina, *Pinnixa chaetopterana* displayed a semilunar nocturnal periodicity in larval density similar to that of *Uca* (Christy and Stancyk 1982), which suggested export of these larvae from the estuary.

Based on the distribution of megalopae, Johnson (1982) characterized larvae of *Uca* and *Pinnixa* as "expelled estuarine." Approximately 60% of the megalopae from these genera were collected outside the estuary, but their horizontal distribution was considerably more restricted than that of *C. sapidus*. Peak numbers of *Uca* and *Pinnixa* megalopae were collected within 10 km of the mouth of Chesapeake Bay (Johnson 1982). In comparison, peak numbers of *Callinectes sapidus* megalopae were collected 25 km, or farther, offshore (McConaugha, unpublished). Stage-I larvae of the anomuran crab *Callianassa californiensis* showed a net flux out of the Salmon River estuary, Oregon, whereas megalopae displayed a net flux into the estuary (Johnson and Gonor 1982). Hydrographic data indicate high flushing rates (<9 d) for the upper reaches of Pacific estuaries inhabited by *C. californiensis* (Johnson and Gonor 1982). Given the developmental duration for stage-I larvae (5 d) and the high flushing rates, most *C. californiensis* larvae appear to be exported to coastal waters.

Sadler (1984) reported high concentrations of *Pagurus longicarpus, P. pollicarius,* and *P. annulipes* larvae in shelf waters just outside Chesapeake Bay. Only larvae of *P. longicarpus* were found within the lower bay and these specimens accounted for less than 0.5% of the total larvae collected for this species. Larval distribution for all three species was centered around the Chesapeake Bay mouth and adjacent shelf. Although substantial numbers of larvae were collected up to 40 km seaward of the bay, peak concentrations of all three species were found approximately 25 km east of the mouth. Concentrations of *P. annulipes* were found farthest offshore, *P. longicarpus* larvae were closest to the bay, and *P. pollicarius* larvae were intermediate in distribution.

Two species of xanthid crabs, *Neopanope sayi* and *Panopeus herbstii,* occupy the lower estuary and presumably encounter the same physical environment as other expelled species. However, larvae of *N. sayi* and *P. herbstii* were concentrated in the lower Chesapeake Bay (Sandifer 1975; Goy 1976) and D. F. Johnson (1982, 1985) found over 90% of the megalopae of both species within the estuary. Similarly, larvae of *Hexapanopeus angustifrons* have been collected as far offshore as the Chesapeake Light Tower (40 km seaward), but greater numbers were collected near the bay's mouth (Maris and McConaugha 1983; Maris 1986), and greater than 75% of the megalopae were taken within the bay and estuary (D. F. Johnson 1985). These data suggest estuarine retention. However, unlike the other two xanthids, *H. angustifrons* has an adult habitat that includes the shallow shelf region (Williams 1984), and it is not clear whether the larvae collected on the shelf were advected from the lower estuary or resulted from spawning of the offshore populations. Larvae of *Rhithropanopeus harrisii,* a xanthid that inhabits the upper estuary, are retained in the adult habitat (Sandifer 1975; Cronin and Forward 1982; Epifanio 1988, this volume) apparently due to their vertical migration around the point of no net motion in response to an endogenous tidal rhythm (Cronin and Forward 1982).

Some brachyuran crustaceans, such as *Uca* spp., exhibit a hatching behavior in the laboratory and field that results in maximum larval release during nighttime ebb tides (Christy 1978, 1982; DeCoursey 1979; Bergen 1981; Christy and Stancyk 1982). Field evidence for similar synchronized hatching has been reported for *C. sapidus* (Provenzano et al. 1983; Epifanio et al. 1984), *Pinnixa* sp. (Brookins and Epifanio 1985; Salmon et al. 1985), *Panopeus herbstii,* and *Neopanope* sp. (Salmon et al. 1985). Such behavior may reduce predation on the larvae by visual predators and reduce, through advective dispersion, the probability that most offspring of one spawner will be killed by local predators. Nocturnal hatching does not appear to be directly related to estuarine retention or advection, because species with retained larvae and those with expelled larvae both display this behavior. However, maximum larval release during ebb tide could facilitate larval advection from the estuary.

Behavioral studies indicate that stage-I larvae of *C. sapidus,* as well as larvae of other species, are positively phototactic and negatively geotactic and have high barokinesis (Sulkin et al. 1980; Sulkin 1984). These behavioral characteristics insure that larvae hatched at depth move rapidly toward the neuston layer. Given the current velocities reported for the near-surface layer in lower estuaries and plumes during ebb tide (Boicourt 1981, 1982; D. R. Johnson 1985), larvae concentrated in this layer would be rapidly transported several kilometers seaward during the initial tidal cycle.

Horizontal Distribution

Available data suggest that distributions of decapod larvae in lower estuarine environments can be divided into three general patterns: retention within the lower estuary (*Panopeus, Neopanope,* and *Hexapanopeus?*), advection to the adjacent continental shelf within 20–30 km (*Uca, Pagurus, Pinnixa,* and *Hexapanopeus?*), and expulsion and wide distribution across the continental shelf (50–100 km) (*C. sapidus* and *Cancer magister*).

Dudley and Judy (1971) measured the abundances of brachyuran larvae seaward of the Beaufort Inlet, North Carolina (Table 1). Between 1.6 and 12 km from the estuary, the abundance of *Pinnixa* larvae increased and abundances of *Uca* and *Neopanope sayi* larvae decreased with distance offshore. Abundance of *Panopeus herbstii* larvae remained stable over the sampling area. Despite large reproductive populations of *N. sayi* and *P. herbstii* in the lower portions of this estuary (McConaugha, personal observation), the relative number of larvae collected outside the estuary was low, supporting the hypothesis of estuarine larval retention for these species. Lack of information regarding larval stage composition makes further analysis of these data difficult. Information on the distribution of *N. sayi* larvae in Chesapeake Bay supports the retention hypothesis for this species; peak abundance of zoeae (Sandifer 1973) and megalopae (D. F. Johnson 1985) occurred inside the bay mouth.

TABLE 1.—Abundances of decapod larvae (mean number of larvae/tow, standardized for unequal sample sizes) outside Beaufort Inlet, North Carolina. (From Dudley and Judy 1971.)

Species	Distance outside estuary (km)		
	1.6	6.5	12.0
Uca sp.	665	569	198
Neopanope sayi	93	72	17
Panopeus herbstii	81	43	60
Pinnixa sp.	228	334	420

Only stage-I and megalopae of *Uca* spp. were collected in the mouth of Delaware Bay (Lambert and Epifanio 1982) and North Inlet, South Carolina (Christy and Stancyk 1982). Goy (1976), who sampled along a transect northeast of Chesapeake Bay, found stage-I and megalopae of *Uca* adjacent to the mouth and an abundance of late-stage larvae at his most seaward station. If a similar distribution occurs at Beaufort Inlet, all available data support the hypothesis that larvae of some species of *Uca* are transported onto the continental shelf. Those species of *Uca* whose adults are restricted to the upper estuary appear to retain their larvae within the estuary (Sandifer 1973, 1975).

In a study in which sampling occurred in both the estuary and on the shelf, larvae of the genus *Pagurus* were concentrated in a narrow band approximately 25 km seaward (Sadler 1984). There were seasonal and species differences in distribution, but late-stage larvae were generally seaward of early-stage larvae and glaucothoe larvae.

Analysis of the distribution of *C. sapidus* larvae across the shelf indicated strong advection of larvae away from estuaries (Smyth 1980; Epifanio et al. 1984; McConaugha et al., unpublished). Samples taken across the continental shelf off Chesapeake Bay in 1982 and 1983 showed a progressive movement of larvae offshore as they matured and a shoreward movement of megalopae (Table 2). Stage-I larvae were distributed southeast of the bay whereas megalopae were to the north of the bay mouth (Figure 2), suggesting

TABLE 2.—Percentage distribution of selected stages of *Callinectes sapidus* offshore from Chesapeake Bay, 1982–1983 (McConaugha, unpublished data).

Distance from bay mouth (km)	Year	Larval stage			
		I	III	VI	Megalopa
0	1982	0	2	2	2
	1983	2	0.5	0	0.5
16	1982	66	4	0	3
	1983	68	0.5	0	0.5
34	1982	26	8	2	23
	1983	28	2	0	7
50	1982	2	15	30	60
	1983	1	72	2	33
65	1982	3	33	27	8
	1983	0.5	8	5	30
80	1982	3	42	39	4
	1983	0.5	17	93	29

a counterclockwise movement of blue crab larvae during development to megalopae.

Vertical Position

Various water masses are driven by different forcing factors so the vertical position (meso- and fine scales) of larvae in the water column can affect their horizontal transport and, ultimately, their recruitment success. Swimming speeds of decapod larvae, although insufficient to overcome horizontal currents, are generally sufficient for vertical positioning within the water column in response to environmental cues or innate behaviors that may vary ontogenetically (Mileikovsky 1971; Cronin 1979; Sulkin 1984). This observation has been the basis for the development of numerous crustacean larval retention and distribution models (Bousfield 1954; Sulkin et al. 1980; Cronin 1982; Kelly et al. 1982; Sulkin and Van Heukelem 1982; O'Connor and Epifanio 1985). Retention within an estuarine system has been associated with changes in vertical position in response to changes in tidal direction or ontogenetic changes in preferred depth; late-stage larvae typically are confined to deeper strata.

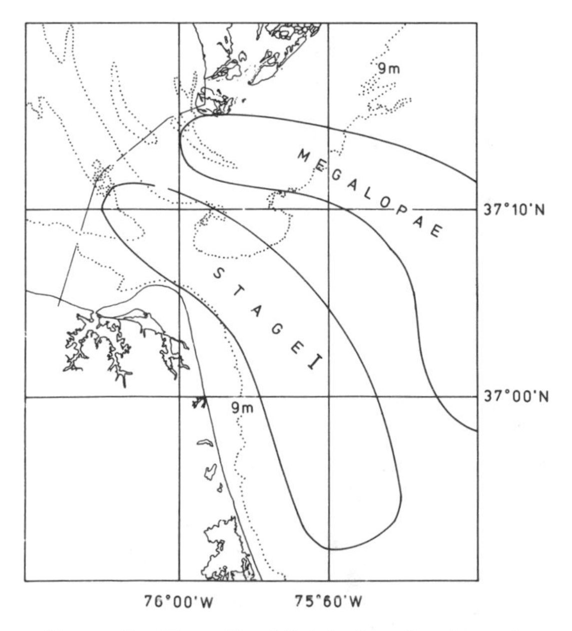

FIGURE 2.—Generalized distribution of peak numbers of *Callinectes sapidus* stage-I larvae and megalopae on the inner continental shelf adjacent to Chesapeake Bay.

In the case of brachyuran species with expelled larvae, there appears to be a dichotomy in the vertical distribution patterns observed in the field. Larvae of those species with a limited distribution on the inner shelf near the estuary are predominately found in bottom waters. Conversely, larvae of those species displaying a wide distribution across the shelf are concentrated in the upper 1 m of the water column.

All developmental stages of three *Pagurus* species were concentrated below 6 m (Sadler 1984). High concentrations of mid- to late-stage larvae of *Uca* and *Pinnixa* have been reported in bottom samples (Goy 1976; Dittel and Epifanio 1982). Therefore, larvae of these expelled species are concentrated in a water mass that has a net flux towards the estuary. This should result in retention or transport back into the estuary, yet the intermediate larval stages are the most seaward of all developmental stages. The majority of samples have been collected during daylight hours, so the possibility that diel vertical migration affects the observed distribution cannot be discounted. Maris (1986) reported that larvae of these species do undergo diel vertical migration. The extent of the diel migration varies by species and ontogenetically for a given species; earlier larval stages occupy a higher position in the water column than later stages. There is also some evidence that the degree of diel vertical migration varies at different geographical locations within the range of larval distributions (i.e., estuarine mouth or offshore; Maris 1986). These data suggest that diel vertical migration results in larval retention on the inner shelf. Reduction in the vertical distance of the diel migration at later developmental stages results in shoreward transport.

In contrast, larvae of *Callinectes sapidus,* which are distributed broadly across the continental shelf, are in general concentrated in the upper

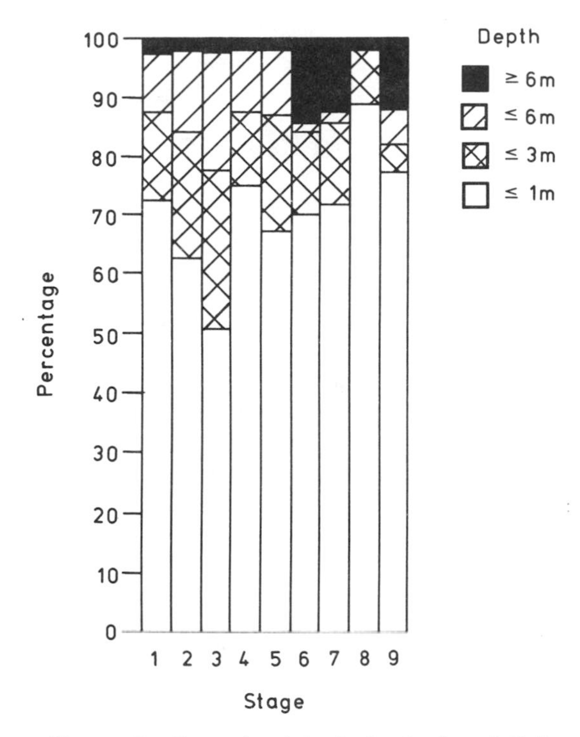

FIGURE 3.—Proportional depth distribution of *Callinectes sapidus* larvae on the inner shelf near Chesapeake Bay, summer 1980. Each value is a seasonal mean of eight stations sampled at 2-week intervals. (From McConaugha et al., unpublished.)

3 m of the water column at all developmental stages (Figure 3; McConaugha, unpublished). With few notable exceptions, the majority of the larvae captured during 1980 were concentrated between the surface and 1 m below the surface. There was a significant negative correlation ($P \leq 0.05$) between larval abundance and depth for all larval stages. Megalopae on the continental shelf were concentrated in the upper 1 m; megalopae in the bay mouth showed no consistent vertical stratification. Although seasonal means (Figure 3) indicate that all larval stages are generally concentrated in the upper layers, there are small-scale variations in larval vertical position on both spatial and temporal scales. Analysis of a 3-year data base indicated that larvae aggregated in dense patches are concentrated in the neuston layer; larvae in less-dense aggregations are more diffuse and are distributed, on average, deeper in the water column (Table 3). These differences in fine-scale vertical position are most likely due to biological interactions, especially trophic interactions. Larval densities similar to those reported in patches would rapidly consume all available food

TABLE 3.—Larval densities (numbers/m^3) and weighted mean depth (m, in parentheses) of *Callinectes sapidus* in shelf waters adjacent to Chesapeake Bay, 10 August 1982.

Stage	Distance offshore (km)				
	16	34	50	65	80
I	660 (1.1)	1,027 (2.0)			
II		24 (2.6)	437 (0.9)	58 (1.0)	0.3 (3.0)
III		1.8 (0.6)	560 (0.8)	285 (2.1)	0.4 (3.0)
IV			739 (0.6)	201 (2.2)	6 (1.4)
V			97 (0.6)	87 (2.4)	11 (1.5)
VI			131 (0.6)	41 (2.4)	2 (2.4)
VII			71 (0.6)	11 (2.5)	6 (2.0)
VIII			103 (0.6)	2 (3.0)	0.7 (2.6)
Megalopae			545 (0.9)	1 (3.0)	5 (3.0)

organisms (Anger and Nair 1979), and starved decapod larvae are consistently found higher in the water column than fed larvae (Cronin and Forward 1979). Our current understanding of mesoscale physical processes suggests that these differences in vertical position and the associated changes in biological interactions may play an important role in larval transport and recruitment. Further investigations of mesoscale physical processes are needed to confirm these hypotheses.

Two 72-h cruises, one at an offshore station and one in the bay mouth, were conducted to examine the role of tidal and diurnal vertical migration on megalopal transport. The offshore study confirmed that blue crab megalopae remain in the upper layer of the water column throughout the diurnal cycle (Figure 4). A small percentage of megalopae were found at depth during the nighttime sampling period and may reflect a change in behavioral characteristics associated with age or stage of the molt cycle. In contrast, the larval stages were found to be slightly deeper during the day than at night. This difference in vertical distribution was reduced as the larvae proceeded through development (Figure 4). In the bay-mouth study, there were significant changes in vertical position of megalopae over time (Maris 1986); statistical analysis confirmed a diel vertical migratory pattern, but there was no consistent vertical migration associated with changes in tidal phase (Figure 5).

Mechanisms of Dispersal

Development on the shelf raises the question of larval dispersal and reinvasion of the estuary. Although there are wide fluctuations in the abundance of *C. sapidus* within estuaries along the east coast of the USA, these interannual variations in year-class strength tend to fluctuate around a stable population base. Thus, it is reasonable to

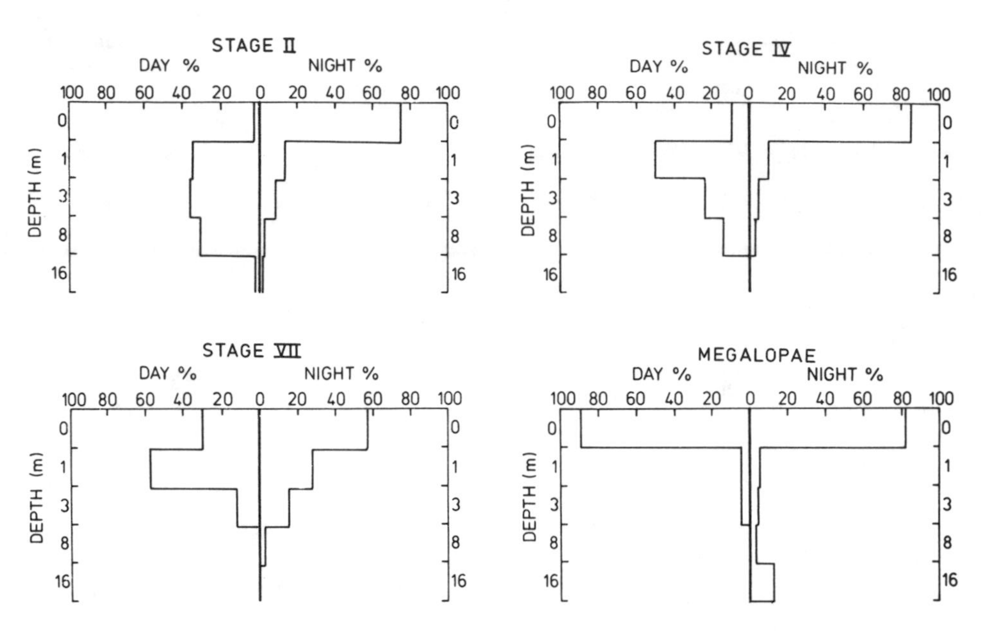

FIGURE 4.—Diurnal depth distribution of selected larval stages of *Callinectes sapidus* measured 50 km east of Chesapeake Bay. Samples were collected from five depths at 3-h intervals over 30 h in September 1980. (Modified from Maris 1986.)

assume that, despite interannual variability, transport mechanisms result in some recruitment annually. Evidence from long-term current measurements (Beardsley et al. 1976) and theoretical considerations (Bishop 1980) suggest a southerly flow at depth on the shelf of the Middle Atlantic Bight during all seasons. Larvae concentrated at depth would be entrained in this longshore current and would be transported southward toward Cape Hatteras and lost from the system. The prevailing winds along the Atlantic coast during the summer spawning period are characteristically from the southwest toward the northeast. Although generally weak in intensity, the wind stress in the Middle Atlantic Bight is of sufficient strength to develop a northward-flowing surface current that could retain larvae on the inner shelf (Johnson et al. 1984). Thus, the vertical position of the larvae on the shelf appears to be the critical factor in the retention mechanism. Discrete depth samples have shown that, offshore, all larval stages of *C. sapidus,* including the megalopal stage are concentrated in the upper 3 m of the water column. Their horizontal distribution on the shelf suggests that *C. sapidus* larvae exit the estuary in the plume; they are transported seaward during most of their development and then shoreward late in development. These vertical and horizontal distributions are consistent with the hypothesis that wind-generated surface currents are the driving force for larval transport.

Similar correlations between wind stress and recruitment have been reported for the dungeness crab *Cancer magister* along the west coast of the USA (Johnson et al. 1986). In this case, larval retention in the nearshore system and subsequent recruitment were associated with wind patterns that weakened or reversed the northward-flowing Davidson Current. This flow reversal reduced the mean north–south displacement that occurred during the 3- to 4-month developmental period. Megalopae of *C. magister* have been found in abundance in the neuston layer (Lough 1976) and attached to flotsam and to the hydroid *Vellella vellella* (Wickham 1979).

A similar relationship has been reported for larvae of the shrimp *Pandalus jordani* off the Pacific coast of North America (Rothlisberg and Miller 1983). Wind stress was not only associated with larval transport but also was a key factor in maintaining a temperature cold enough (12°C) for larval development by driving coastal up-welling processes.

Transport Model

A numerical model of the prevailing current patterns on the shelf in the Middle Atlantic Bight during the summer months, when most temperate

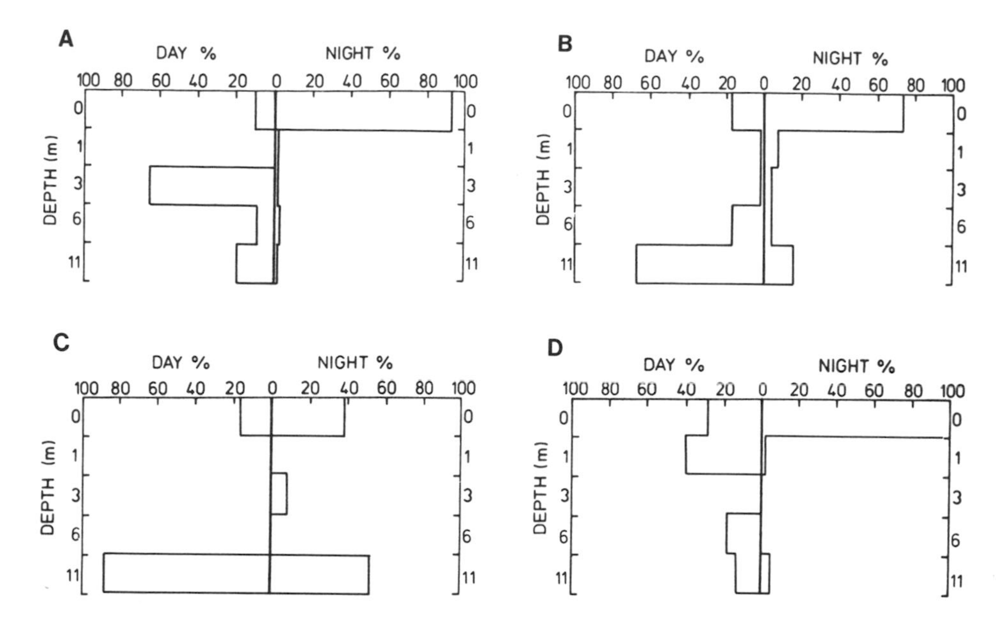

FIGURE 5.—Vertical position of *Callinectes sapidus* megalopae over diurnal and tidal cycles measured at the mouth of Chesapeake Bay, August 1983. Samples were collected at 3-h intervals over 72 h. A—slack before flood; B—maximum flood; C—slack before ebb; D—maximum ebb. (Modified from Maris 1986.)

invertebrate species spawn, has been developed (Johnson et al. 1984; D. R. Johnson 1985). The model assumes a strong pcynocline that allows the surface wind-driven layer to be decoupled from the lower friction-retarded layer. Data generated from model runs and the sparse current meter data available from the area around the mouth of Chesapeake Bay suggest the presence of a corridor of northward-flowing water in the near-shelf region. Model runs using wind data from the period of peak spawning for the blue crab and starting from the bay-mouth region indicate that larvae concentrated in the upper 1–2 m of the water column would be retained on the inner shelf (D. R. Johnson 1985). Predicted larval transport closely matches observed *Callinectes sapidus* larval distributions on the shelf during some periods. A good correlation between the wind index, which was generated from the model with local wind data, and commercial blue crab landings 1.5–2 years later was obtained (Hester 1983). This further supports the concept of wind transport and retention of larvae within the shelf system, because approximately 18 months are required for blue crabs recruited to Chesapeake Bay to reach commercial size. The basic model suggests that the degree of larval retention within the shelf system is fortuitous for any given year, but analysis of the wind index over 10 years indicates that some recruitment to the estuaries would occur annually (Hester 1983).

Surface current measurements taken in the plume and adjacent shelf waters of Chesapeake Bay during 10–18 June 1985 confirm the basic premise of a northward-flowing corridor in the near-shelf waters (D. R. Johnson, personal communication; Figure 6). The high velocities measured in the plume during this study (greater than 90 cm/s) and the average vectors at the shelf stations were consistent with observed larval distributions.

Recruitment Mechanisms

Although wind-generated currents appear sufficient to retain larvae on the shelf, they do not appear to be the major mechanism for recruitment to the estuary. For example, model runs with 1983 data on *C. sapidus* larval positions place the megalopae near the mouth of Chesapeake Bay but never within the bay (D. R. Johnson 1985). There is no doubt that recruitment from the shelf to the bay occurs; however, the exact transport mechanisms are not known. It is also unclear which life history stage, megalopa or juvenile crab or both, is recruited to the estuary.

The broad horizontal distribution of *C. sapidus* megalopae on the shelf raises questions regarding the fate of these larvae (i.e., whether they become recruited or wasted). Three hypotheses can be

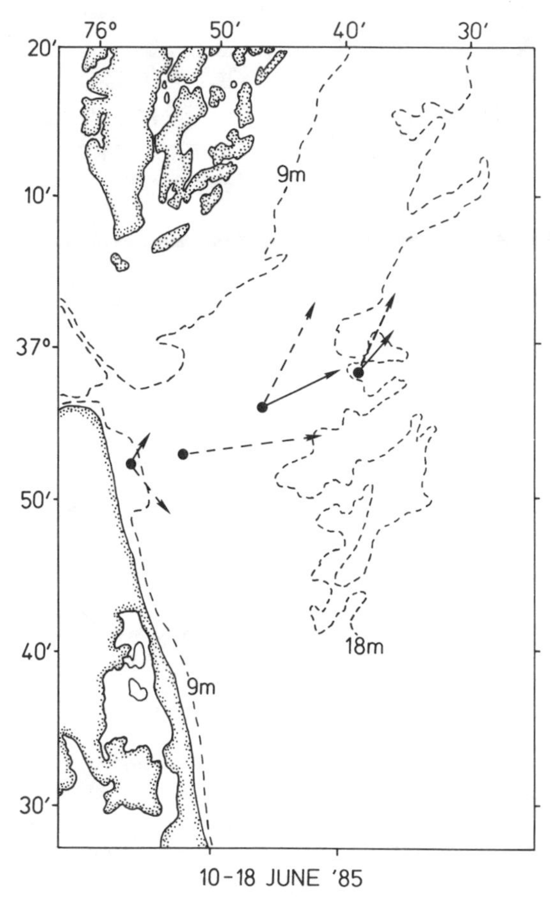

FIGURE 6.—Vectors of surface currents on the shelf adjacent to Chesapeake Bay, 10–18 June 1985. Solid lines represent the mean direction and relative speed for the 9-d period. Dashed lines indicate mean direction and relative speed for 10 June. (Reproduced with the permission of D. R. Johnson.)

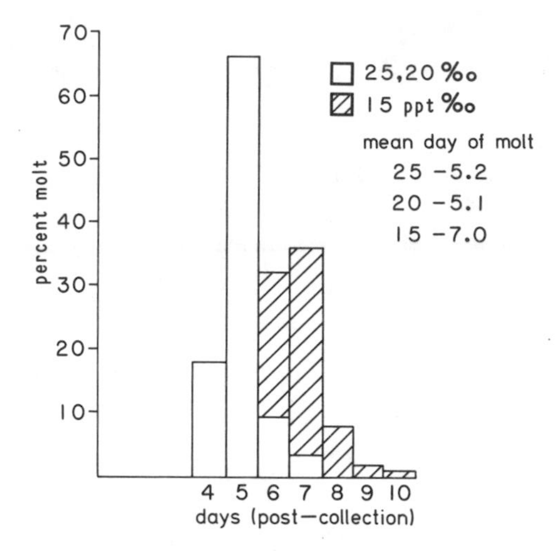

FIGURE 7.—Molt frequency of field-collected megalopae of *Callinectes sapidus* maintained in the laboratory at three salinities.

proposed to explain the observed distributions: (1) only a small proportion of the megalopae observed on the shelf are successfully recruited to estuarine habitats (i.e., there are high instantaneous mortality rates); (2) megalopae are capable of delaying ecdysis and metamorphosis until transported to a suitable habitat; or (3) the megalopal stage is not the major life history stage for reinvasion of the estuary. Available data are not sufficient to permit conclusions to be drawn, but they are adequate for informed speculation. Johnson (1982) reported that only 17% of the total number of blue crab megalopae collected in the Chesapeake Bay and adjacent shelf study area were taken within the bay mouth and only 2.2% were taken in the lower bay and York River area. R. J. Orth and J. Van Montfrans (personal communication) also reported small numbers of *Callinectes* megalopae (up to 70/trap) in Mobjack Bay, Virginia, just north of the York River and 50 km up-estuary from the Chesapeake Bay mouth. Olmi and Sandifer (in press) suggested that blue crab recruitment to open marshes in South Carolina was by megalopae. Thus, the megalopal stage must contribute to recruitment of the blue crab. In contrast, recruitment of blue crabs to South Carolina impoundments appeared to be by juvenile crabs. In the Chesapeake Bay, the sharp drop in megalopal densities at the lower bay and river stations compared to stations in the bay-mouth region suggests that early juvenile stages may play a major role in recruitment to the estuary.

The limited data available indicate that crustacean larvae can delay ecdysis and metamorphosis until a suitable habitat is encountered. For *C. sapidus,* this concept is based on the high degree of synchrony observed in the molt frequency of 600 field-collected megalopae (Figure 7). The 4–5-d delay between collection and ecdysis is consistent with time required to reinitiate a molt cycle that had been suspended at early premolt (McConaugha 1982). The type of stimulus that may have triggered this response is not clear. These megalopae were collected in a single 5-min plankton tow (28‰, 25°C). In the laboratory, megalopae were transferred to 25, 20, or 15‰ seawater at 25°C and fed freshly hatched *Artemia* nauplii. Therefore, a drop in ambient salinity or an increase in available food may have been the triggering stimulus. Feeding decapod larvae after they have been starved can reinitiate the molt cycle (Anger et al. 1981; McConaugha 1982, 1985).

A tidally induced vertical migration has been suggested for *C. sapidus* megalopae in mixed estuaries (Epifanio et al. 1984; Epifanio 1988). Recent evidence from the Chesapeake Bay, a bilayered system, does not support a tidally induced vertical migration as the reinvasion mechanism for this system (Maris 1986). Although there was a significant diel vertical migration by the larvae of several species, time-series analysis did not support a tidal migratory pattern. Larvae of most species, as well as *C. sapidus* megalopae, were found at middepths or lower during daylight hours in the bay mouth (Maris 1986). An alternative recruitment mechanism is that larvae use the general onshore transport found at these lower depths. Boicourt (1981) reported that the 28-d mean flow from the middle of the water column down was into the Chesapeake Bay. Because of the 16-h:8-h day:night cycle during the summer spawning months, the daily mean larval depth could be below the point of no net motion, resulting in transport into the estuary. A third alternative is wind-forced pumping of the lower estuary. Periodic wind events can lead to hydrographic conditions that result in large water masses being pumped into or out of estuaries (Wang and Elliott 1978; Wang 1979a, 1979b). Megalopae entrained in these water masses could be transported into (out of) the estuary.

Evolution of Larval Export

There is strong evidence to suggest that retention of a larval phase is advantageous over geological time (Strathman 1974; Hansen 1980). This, in conjunction with the difficulty of reacquiring a planktotrophic larval phase once it has been lost (Strathman 1978), may provide the selective advantage for the retention of a planktotrophic larval phase in estuarine species.

There are evolutionary arguments for retention of a planktotrophic larval phase, but arguments for the selective advantages for larval export from the estuary are less persuasive. As previously indicated, there appears to be little selective advantage over the short term (100 generations) for larval export when the classic criteria of gene flow between estuaries and the colonization of new habitats are considered (Palmer and Strathman 1981; Strathman 1982). The genus *Callinectes* is of tropical oceanic origin; thus, if developmental physiology were conservative, the larvae might require oceanic conditions for development. Laboratory studies on temperature and salinity effects on *C. sapidus* larval development support that hypothesis (Costlow and Bookhout 1959). Similarly, late-stage larvae of some species of *Uca* demonstrate a preference for the higher salinities found on the inner shelf adjacent to estuarine outfalls. Strathman (1982) suggested that export of larvae might be associated with predator–prey interactions and predicted that mortality rates for exported larvae would be greatly reduced when compared to those of retained estuarine species. Estimated instantaneous mortality rates *(Z)* for *C. sapidus* development were low (Table 4) for the developmental period between stage I and the megalopal stage. The rate of mortality is not constant throughout development; *Z* is high during the early stages and is reduced in the middle and late larval stages. There are insufficient data available for retained estuarine species to permit comparison of mortality rates.

TABLE 4.—Instantaneous mortality rates for *Callinectes sapidus* in shelf waters adjacent to Chesapeake Bay, 1980.

Developmental stage	Percent of total catch[a]	Instantaneous mortality rate (Z)
I–Megalopa	100.0	0.049
I–II	9.36	0.169
II–III	1.74	0.128
III–IV	0.73	0.066

[a]Total catch was 3.6 million larvae.

Although offshore development may reduce predation on the larvae, a possible disadvantage could be the density of prey items available to larvae. The estuary has been well documented as a nursery ground for larval and juvenile stages of many species. Numerous species of fish and penaeid shrimp spawn offshore and subsequently invade the estuarine nursery grounds as larvae or juveniles. Based on preliminary laboratory studies, it appears that up to 50% of late-stage *C. sapidus* larvae found on the shelf have undergone some period of starvation (McConaugha, unpublished data). Late-stage (VI, VII) laboratory-reared *C. sapidus* larvae were starved for periods of 4–5 d before feeding was resumed. These larvae eventually molted, but they were morphologically intermediate between the zoea and megalopa stages and were similar to stage-VIII larvae collected on the shelf. It it not currently known what percentage of stage-VIII larvae will complete development to the megalopal stage and become potential recruits to the estuary. Similar zoeal–megalopal intermediates have been reported among starved *Rhithropanopeus harrisii* larvae reared in the laboratory (McConaugha 1982). Yet, intermediates have not been identified in field studies in which larvae of *R. harrisii,* a retained estuarine species, were extensively collected (Pinschmidt 1963; Sandifer 1973; Goy 1976; Cronin 1979). These data support the concept that one cost of larval export to the shelf is a reduction in available prey and associated reduction in larval growth rate.

Summary

There is ample evidence to support the hypothesis of larval export by some decapod species inhabiting lower estuaries. There are two distinct dispersal patterns associated with larval advection: retention of advected larvae on the inner shelf adjacent to the mouth of the estuary, and cross-shelf transport with larval dispersal over a range of 100 km or greater. In all cases, maintenance of stable populations within the adult estuarine habitat depends on reinvasion by late-stage larvae or juveniles. Physical alterations of estuarine entrances and the associated changes in current pattern and velocity have the potential to affect recruitment of these ecologically and commercially important species. Although it is not yet possible to assess a priori the effect of any given alteration, those species with inner-shelf retention patterns (i.e., species of *Uca, Pinnixa,* and *Pagurus*) appear to be most susceptible to alterations of the physical characteristics of estuarine entrances. For the widely distributed larvae of *C. sapidus,* small changes in current patterns associated with structural modifications could have little effect on year-class strength compared with the annual fluctuations associated with cross-shelf transport. However, a more thorough knowledge of the mechanism(s) used by this species for reinvasion of the estuary is needed before an accurate assessment can be made.

Acknowledgments

I thank D. F. Johnson, R. C. Maris, D. R. Johnson, and P. W. Sadler for the use of unpublished data and many informative discussions regarding decapod larval transport and recruitment. Special thanks go to C. S. McConaugha for technical and clerical support. Original research presented here has been sponsored by grants NA79AA-D-00055, NA81AA-D-0025, NA85AA-D-SG016, and NA86AA-D-SG042 from the Office of Sea Grant, National Oceanic and Atmospheric Administration, to Virginia Graduate Marine Consortium and the Virginia Sea Grant College Program.

References

Anger, K., R. R. Dawirs, V. Anger, and J. D. Costlow. 1981. Effects of early starvation periods on zoeal development of brachyuran crabs. Biological Bulletin (Woods Hole) 161:199–212.

Anger, K., and K. K. Nair. 1979. Laboratory experiments on the larval development of *Hyas aranus* (Decapoda, Maijidae). Helgoländer Meeresuntersuchungen 32:36–54.

Beardsley, R. C., W. C. Boicourt, and D. V. Hansen. 1976. Physical oceanography of the Middle Atlantic Bight. American Society of Limnology and Oceanography Special Symposium 2:20–34.

Berge, M. E. 1981. Hatching rhythms in *Uca pugilator* (Decapoda: Brachyura). Marine Biology (Berlin) 63:151–158.

Bishop, J. M. 1980. A note on the seasonal transport on the middle Atlantic shelf. Journal of Geophysical Research 85:4933–4936.

Boicourt, W. C. 1981. Circulation in the Chesapeake Bay entrance region: estuary–shelf interactions. Pages 61–78 *in* J. W. Campbell and J. P. Thomas, editors. Chesapeake Bay plume study, 1980. National Aeronautics and Space Administration Science and Technology Information Branch, NASA Conference Publication 2188; NOAA/NEMP II; 81; ABCDEFG 0042, Washington, D.C.

Boicourt, W. C. 1982. Estuarine larval retention mechanisms on two scales. Pages 445–457 *in* Kennedy (1982).

Bousfield, E. L. 1954. Ecological control of the occurrence of barnacles in the Miramichi estuary. National Museum of Canada Bulletin 137.

Brookins, K. G., and C. E. Epifanio. 1985. Abundance of brachyuran larvae in a small coastal inlet over six consecutive tidal cycles. Estuaries 8:60–67.

Burton, R. S., and M. W. Feldman. 1982. Population genetics of coastal and estuarine invertebrates: does larval behavior influence population structure? Pages 537–551 *in* Kennedy (1982).

Carriker, M. R. 1967. Ecology of estuarine benthic invertebrates: a perspective. American Association for the Advancement of Science Publication 83: 442–487.

Chia, F. S. 1974. Classification and adaptive significance of developmental patterns in marine invertebrates. Thalassia Jugoslavica 10:121–130.

Christiansen, M. E., and W. T. Yang. 1976. Feeding experiments on the larvae of the fiddler crab *Uca pugilator* (Brachyura, Ocypodidae), reared in the laboratory. Aquaculture 8:91–98.

Christy, J. H. 1978. Adaptive significance of reproductive cycles in the fiddler crab *Uca pugilator*. Science (Washington, D.C.) 199:453–455.

Christy, J. H. 1982. Adaptive significance of semilunar cycles of larval release in fiddler crabs (genus *Uca*):

test of an hypothesis. Biological Bulletin (Woods Hole) 163:351–363.
Christy J. H., and S. E. Stancyk. 1982. Timing of larval production and flux of invertebrate larvae in a well-mixed estuary. Pages 489–503 *in* Kennedy (1982).
Costlow, J. D., Jr., and C. G. Bookhout. 1959. The larval development of *Callinectes sapidus* Rathbun reared in the laboratory. Biological Bulletin (Woods Hole) 116:373–396.
Cronin, T. W. 1979. Factors contributing to the retention of larvae of the crab, *Rhithropanopeus harrisii*, in the Newport River estuary, North Carolina. Doctoral dissertation. Duke University, Durham, North Carolina.
Cronin, T. W. 1982. Estuarine retention of larvae of the crab, *Rhithropanopeus harrisii*. Estuarine, Coastal and Shelf Science 15:207–220.
Cronin, T. W., and R. B. Forward, Jr. 1979. Tidal vertical migration: an endogenous rhythm in estuarine crab larvae. Science (Washington, D.C.) 204: 1020–1022.
Cronin, T. W., and R. B. Forward. 1982. Tidally timed behavior: effects on larval distribution in estuaries. Pages 505–520 *in* Kennedy (1982).
DeCoursey, P. J. 1979. Egg hatching rhythms in three species of fiddler crabs. Pages 399–406 *in* E. Naylor and R. G. Hartnol, editors. Cyclic phenomena in marine plants and animals. Pergamon, Oxford, England.
deWolf, P. 1974. On the retention of marine larvae in estuaries. Thalassia Jugoslavica 10:415–424.
Dittel, A. I., and C. E. Epifanio. 1982. Seasonal abundance and vertical distribution of crab larvae in Delaware Bay. Estuaries 5:197–202.
Dudley, D. L., and M. H. Judy. 1971. Occurrence of larval, juvenile and mature crabs in the vicinity of Beaufort Inlet, North Carolina. NOAA (National Oceanic and Atmospheric Administration) Technical Report NMFS (National Marine Fisheries Service) SSRF (Special Scientific Report Fisheries) 637.
Epifanio, C. E. 1988. Transport of invertebrate larvae between estuaries and the continental shelf. American Fisheries Society Symposium 3:104–114.
Epifanio, C. E., and A. I. Dittel. 1982. Comparison of dispersal of crab larvae in the Delaware Bay, U.S.A., and the Gulf of Nicoya, Central America. Pages 477–487 *in* Kennedy (1982).
Epifanio, C. E., C. C. Valenti, and A. E. Pembroke. 1984. Dispersal and recruitment of blue crab larvae in the Delaware Bay. Estuarine, Coastal and Shelf Science 18:1–12.
Giese, A. C., and J. S. Pearse. 1974. Reproduction of marine invertebrates. Academic Press, New York.
Goy, J. W. 1976. Seasonal distribution and the retention of some decapod crustacean larvae within the Chesapeake Bay, Virginia. Master's thesis. Old Dominion University, Norfolk, Virginia.
Hansen, T. A. 1980. Influence of larval dispersal and geographic distribution on species longevity in neogastropods. Paleobiology 6:193–207.
Hester, B. S. 1983. A model of the population dynamics of the blue crab in Chesapeake Bay. Doctoral dissertation. Old Dominion University, Norfolk, Virginia.
Jackson, J. B. C. 1974. Biogeographic consequences of eurytropy and stenotopy among marine bivalves and their evolutionary significance. American Naturalist 198:541–560.
Johnson, D. F. 1982. A comparison of recruitment strategies among brachyuran crustacean megalopae of the York River, lower Chesapeake Bay and adjacent waters. Doctoral dissertation. Old Dominion University, Norfolk, Virginia.
Johnson, D. F. 1985. The distribution of brachyuran crustacean megalopae in the waters of the York River, lower Chesapeake Bay and adjacent shelf: implications for recruitment. Estuarine, Coastal and Shelf Science 20:693–705.
Johnson, D. F., L. W. Botsford, R. D. Methot, Jr., and T. C. Wainwright. 1986. Wind stress and cycles in dungeness crab (*Cancer magister*) catch off California, Oregon, and Washington. Canadian Journal of Fisheries and Aquatic Sciences 43:838–845.
Johnson, D. R. 1985. Wind-force dispersion of blue crab larvae in the Middle Atlantic Bight. Continental Shelf Research 4:1–14.
Johnson, D. R., B. S. Hester, and J. R. McConaugha. 1984. Studies of a wind mechanism influencing the recruitment of blue crabs in the Middle Atlantic Bight. Continental Shelf Research 3:425–437.
Johnson, G. E., and J. J. Gonor. 1982. The tidal exchange of *Callianassa californiensis* (Crustacea, Decapoda) larvae between the ocean and the Salmon River estuary, Oregon. Estuarine, Coastal and Shelf Science. 14:501–515.
Kelly, P., S. D. Sulkin, and W. F. Van Heukelem. 1982. A dispersal model for larvae of the deep sea red crab *Geryon quinquedens* based upon behavioral regulation of vertical migration in the hatching stage. Marine Biology (Berlin) 72:35–43.
Kennedy, V., editor. 1982. Estuarine comparisons. Academic Press, New York.
Lambert, R., and C. E. Epifanio. 1982. A comparison of dispersal strategies in two genera of brachyuran crabs in a secondary estuary. Estuaries 5:182–188.
Lough, R. G. 1976. Larval dynamics of the dungeness crab, *Cancer magister*, off the central Oregon coast, 1970–71. U.S. National Marine Fisheries Service Fishery Bulletin 74:353–376.
Maris, R. C. 1986. Patterns of diurnal vertical distribution and dispersal—recruitment mechanisms of decapod crustacean larvae and postlarvae in the Chesapeake Bay, Virginia and adjacent offshore waters. Doctoral dissertation. Old Dominion University, Norfolk, Virginia.
Maris, R. C., and J. R. McConaugha. 1983. Effects of diurnal vertical distribution and patchiness of decapod larvae on dispersal along the inner continental shelf. American Zoologist 23:982.
McConaugha, J. R. 1982. Regulation of crustacean metamorphosis in larvae of the mud crab, *Rhithropanopeus harrisii*. Journal of Experimental Zoology 223:155–163.
McConaugha, J. R. 1985. Nutrition and larval growth. Pages 127–154 *in* A. M. Wenner, editor. Crustacean

issues, volume 2. Larval growth. Balkema Press, Boston.
Mileikovsky, S. A. 1971. Types of larval development in marine bottom invertebrates, their distribution and ecological significance: a reevaluation. Marine Biology (New York) 10:193–213.
Nichols, P. R., and P. M. Keney. 1963. Crab larvae (*Callinectes*) in plankton collections from M/V *Theodore N. Gill,* south Atlantic coast of the United States, 1953–54. U.S. Fish and Wildlife Service Special Scientific Report—Fisheries 448.
O'Connor, N. J., and C. E. Epifanio. 1985. The effect of salinity on the dispersal and recruitment of fiddler crab larvae. Journal of Crustacean Biology 5: 137–145.
Olmi, E. J., and P. A. Sandifer. In press. Recruitment of blue crab, *Callinectes sapidus,* in open and impounded marsh systems in South Carolina. Journal of Shellfish Research.
Palmer, A. R., and R. R. Strathman. 1981. Scale of dispersal in varying environments and its implications for life histories of marine invertebrates. Oecologia (Berlin) 43:308–318.
Pinschmidt, W. C. 1963. Distribution of crab larvae in relation to some environmental conditions in the Newport River estuary, North Carolina. Doctoral dissertation. Duke University, Durham, North Carolina.
Provenzano, A. J., J. R. McConaugha, K. B. Phillips, D. F. Johnson, and J. Clark. 1983. Vertical distribution of first stage larvae of the blue crab, *Callinectes sapidus,* at the mouth of the Chesapeake Bay. Estuarine, Coastal and Shelf Science 16: 489–499.
Rothlisberg, P. C., and C. B. Miller. 1983. Factors affecting the distribution, abundance and survival of *Pandalus jordani* (Decapoda, Pandalidae) larvae off the Oregon coast. U.S. National Marine Fisheries Service Fishery Bulletin 81:455–472.
Sadler, P. W. 1984. The spatial and temporal distribution of the larvae of sympatric pagurid hermit crabs (Decapoda Anomura) in Virginian estuarine and coastal waters. Master's thesis. Old Dominion University, Norfolk, Virginia.
Salmon, M., W. H. Seiple, and S. G. Morgan. 1985. Hatching rhythms of fiddler crabs and associated species at Beaufort, North Carolina. Journal of Crustacean Biology 6:24–36.
Sandifer, P. A. 1973. Distribution and abundance of decapod larvae in the York River estuary and adjacent lower Chesapeake Bay, Virginia, 1968–1969. Chesapeake Science 14:235–257.
Sandifer, P. A. 1975. The role of pelagic larvae in recruitment to populations of adult decapod crustaceans in the York River estuary and adjacent lower Chesapeake Bay, Virginia. Estuarine and Coastal Marine Science 3:269–279.
Scheltema, R. S. 1971. Larval dispersal as a means of genetic exchange between geographically separated populations of shallow-water benthic marine gastropods. Biological Bulletin (Woods Hole) 140: 284–322.
Scheltema, R. S. 1975. Relationship of larval dispersal, gene flow and natural selection to geographic variation of benthic invertebrates in estuarine and along coastal regions. Estuarine Research 1:372–391.
Scheltema, R. S. 1977. Dispersal of marine invertebrate organisms: palebiogeographic and biostratigraphic implications. Pages 73–108 *in* E. G. Kauffman and J. F. Hazel, editors. Concepts and methods of biostratigraphy. Dowden, Hutchinson and Ross, Stroudsburg, Pennsylvania.
Smyth, P. O. 1980. *Callinectes* (Decapoda: Portunidae) larvae in the Middle Atlantic Bight, 1975–77. U.S. National Marine Fisheries Service Fishery Bulletin 78:251–265.
Strathman, R. R. 1974. The spread of sibling larvae of sedentary marine invertebrates. American Naturalist 108:29–44.
Strathman, R. R. 1978. The evolution and loss of feeding larval stages of marine invertebrates. Evolution 32:894–906.
Strathman, R. R. 1982. Selection for retention or export of larvae in estuaries. Pages 521–536 *in* Kennedy (1982).
Sulkin, S. D. 1984. Behavioral basis of depth regulation in the larvae of brachyuran crabs. Marine Ecology Progress Series 15:181–205.
Sulkin, S. D., and W. Van Heukelem. 1982. Larval recruitment in the crab, *Callinectes sapidus* Rathbun: an amendment to the concept of larval retention in estuaries. Pages 459–475 *in* Kennedy (1982).
Sulkin, S. D., W. Van Heukelem, P. Kelly, and L. Van Heukelem. 1980. The behavioral basis of larval recruitment in the crab *Callinectes sapidus* Rathbun: a laboratory investigation of ontogenetic changes in geotaxis and barokinesis. Biological Bulletin (Woods Hole) 159:402–417.
Thorson, G. 1946. Reproductive and larval development of Danish marine bottom invertebrates. Meddelelser fra Kommissionen for Danmarks Fisheri-og Havundersogelser Serie Plankton 4:1–523 (Charlottenlund, Denmark.)
Thorson, G. 1950. Reproductive and larval ecology of marine bottom invertebrates. Biological Reviews of the Cambridge Philosophical Society 25:1–25.
Van Engel, W. A. 1958. The blue crab and its fishery in Chesapeake Bay. Part 1—Reproduction, early development, growth and migration. Commercial Fisheries Review 20(8):6–17.
Wang, D.-P. 1979a. Subtidal sea level variations in the Chesapeake Bay and relations to atmospheric forcing. Journal of Physical Oceanography 9:413–421.
Wang, D.-P. 1979b. Wind-driven circulation in the Chesapeake Bay, winter 1975. Journal of Physical Oceanography 9:564–572.
Wang, D.-P., and A. J. Elliott. 1978. The effect of meteorological forcing on the Chesapeake Bay: the coupling between and estuarine system and its adjacent coastal waters. Pages 127–145 *in* J. C. J. Nihoul, editor. Hydrodynamics of estuaries and fjords. Elsevier, Amsterdam.
Wickham, D. E. 1979. The relationship between megalopae of the Dungeness crab, *Cancer magister,* and the hydroid, *Velella velella,* and its influence on abundance estimates of *C. magister* megalopae.

California Fish and Game 65:184–186.
Williams, A. B. 1984. Shrimps, lobsters, and crabs of the Atlantic coast of the eastern United States, Maine to Florida. Smithsonian Institution Press, Washington, D.C.
Wood, L., and W. J. Hargis. 1971. Transport of bivalve larvae in a tidal estuary. Pages 29–44 *in* D. J. Crisp, editor. Fourth European marine biology symposium. Cambridge University Press, Cambridge, England.

American Fisheries Society Symposium 3:104–114, 1988

Transport of Invertebrate Larvae between Estuaries and the Continental Shelf

CHARLES E. EPIFANIO

College of Marine Studies, University of Delaware, Lewes, Delaware 19958, USA

Abstract.——Marine and estuarine invertebrates evidence a spectrum of reproductive strategies. These include (1) direct development, whereby the larval stage is bypassed in the egg; (2) brooding, whereby larvae are held within the body of the adult female; (3) nonfeeding lecithotrophic development, whereby larvae are free in the plankton but rely on self-contained yolk for nourishment; and (4) planktotrophic development, whereby larvae are free in the plankton and prey on other planktonic species for nourishment. Planktotrophic development is by far the most common reproductive strategy among tropical and temperate species. Advantages of this mode of development are several and include high potential to take advantage of newly available habitat, extensive gene flow among dispersed populations, and exploitation of different habitats during larval and adult life. Concomitant disadvantages include increased vulnerability to predation or starvation and the possibility of advection away from habitats suitable for adult or juvenile existence. This latter problem is most acute for species living in estuaries where subtidal flow (vertically averaged) is seaward. The mechanisms for retention of these larvae in estuaries have been of great interest to marine scientists for nearly a century. This paper summarizes present theory concerning mechanisms of dispersal and recruitment of estuarine larvae and discusses these in light of anthropogenic effects on estuaries.

There exist two schools of thought concerning factors that control the dispersive transport of invertebrate larvae. Proponents of one school maintain that horizontal movements of estuarine larvae are largely controlled by physical factors and that larval swimming plays little role in the process; this is generally termed the "passive hypothesis." Support for this hypothesis is drawn from the results of de Wolf (1973), who studied dispersal of barnacle larvae in the Wadden Sea, a well-mixed estuary on the coast of the Netherlands. De Wolf's field measurements suggested that barnacle larvae were generally distributed in the lower regions of the water column due to passive sinking. These larvae would be mixed up into the water column by turbulent processes during periods of flooding or ebbing tidal currents and would sink to near-bottom at slack water. Because flood current velocities exceed ebb velocities in bottom estuarine waters, this would provide a mechanism for transport upstream.

More recent supporters of the passive hypothesis include Boicourt (1982) and Seliger et al. (1982), who have brought more sophisticated physical oceanographic approaches to the problem. Boicourt utilized both eulerian (current meter) and lagrangian (dye deployment) techniques to explain differences in settlement of oyster larvae in two apparently similar creeks connected to the Choptank River, a secondary estuary in the Chesapeake Bay system in Maryland. Seliger et al. employed a different methodology to address the same problem in the Choptank by making a series of synoptic measurements of the vertical distribution of salinity isopleths and larvae in the water column. In both studies, the authors explained upstream transport of larvae as a generally passive physical process, although Seliger et al. pointed out that tidally rhythmic vertical migration of the larvae might augment the process.

This caveat by Seliger et al. (1982) reflects the other transport school, which invokes volitional movements by larvae. Wood and Hargis (1971) tested the passive hypothesis for oyster larvae in the James River estuary, Virginia. They collected larvae every 2 h over several tidal cycles from two depths at three stations located along a transect near a large oyster bar. They found larvae higher in the water column on flooding than on ebbing tidal currents, which would result in upstream transport. However, they found that the vertical distribution of larvae was different from that of coincidentally occurring coal particles of the same size and density as the larvae (Figure 1). The authors concluded that larval swimming was an important effector of the vertical distribution and of the resultant horizontal advection of larvae. These results were anticipated by earlier investigators (Nelson and Perkins 1930; Carriker 1951,

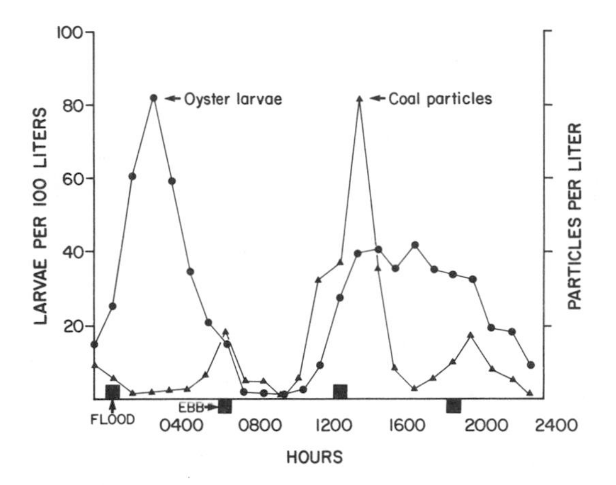

FIGURE 1.—Concentration of oyster larvae and coal particles sampled hourly over two tidal cycles at a depth of 7–10 m in the James River, Virginia. Tide marks along the abscissa mark time of maximum flow velocity during flood and ebb tides. (Modified from Wood and Harris 1971.)

1961), who found molluscan larvae higher in the water column during flood tidal phases.

A variation of active transport for retention of larvae in estuaries involves a change in vertical distribution as ontogenetic development proceeds. Bousfield (1955) studied the vertical and horizontal distribution of barnacle larvae in the Miramichi estuary on the east coast of Canada. Extensive sampling at several depths along an upstream–downstream transect indicated that early-stage larvae maintained a position high in the water column and were thus transported downstream, whereas late-stage larvae took up position deeper in the water column and were transported back upstream, thus effecting retention in the estuary. Presumably, changes in larval swimming behavior were important in maintenance of different vertical positions at different points in larval development.

The consensus conclusion from field evidence, then, is that larvae of many species of estuarine invertebrates are actively involved in the retentive process. Changes in vertical position result in differential horizontal advection in estuaries where velocity and even direction of current may vary greatly with depth. The "clever" larva can influence its horizontal dispersal to a large degree and can manage an upstream transport in the face of an average downstream movement of estuarine water. But what are the factors that elicit these vertical movements? Do vertically migrating larvae respond to environmental cues or is the behavior endogenous?

The behavioral basis for retentive mechanisms (Sulkin 1984) of estuarine larvae has been best studied in the brachyuran crabs (Figure 2). Anesthetized crab larvae sink rapidly in the water column, and their rate of sinking hastens with increases in size that occur during larval development. Accordingly, these larvae must swim vertically to regulate their position in the water column and their consequent horizontal transport in stratified estuarine systems.

Crab larvae respond behaviorally to a varicty of stimuli. These responses can be broadly divided into two categories, "taxis responses" and "kinesis responses" (Fraenkel and Gunn 1961). A taxis response occurs when an animal orients toward a stimulus; a kinesis response occurs when the level of some activity, e.g., swimming, increases as a consequence of a stimulus. Clearly both taxis and kinesis responses are necessary for a larva to undertake a directed movement. Kinesis responses occur without regard to direction of the stimulus. The principal orienting stimuli in the vertical plane are light and gravity. Kinesis stimuli most important in the marine environment are light, temperature, salinity, and hydrostatic pressure (Sulkin 1984).

Larvae of most species of crabs show positive phototaxis in the laboratory, but light is not an important cue for vertical orientation related to retention of larvae in estuaries. This is due to the high turbidity of most estuaries and the extreme diurnal variation in visible radiation reaching the surface of estuaries. These conditions restrict the importance of light as an orienting stimulus to the near-surface layer during daylight hours and to the very-near surface during periods of full moon. In contrast, gravity is a conservative stimulus for orientation throughout the water column regardless of time of day. Positive response is directed toward the center of the earth, negative response away from it.

Several experiments have been concerned with the taxis and kinesis responses of crab larvae to gravity (e.g., Sulkin 1973; Latz and Forward 1977; Sulkin et al. 1980, 1983; Schembri 1982). Although experiments have not always allowed separation of passive sinking from active swimming, early-stage larvae are generally negatively geotactic whereas late stages and postlarval megalopae are more often positively geotactic (Sulkin 1984). This pattern of response would place early-stage larvae high in the water column, thus effecting a down-

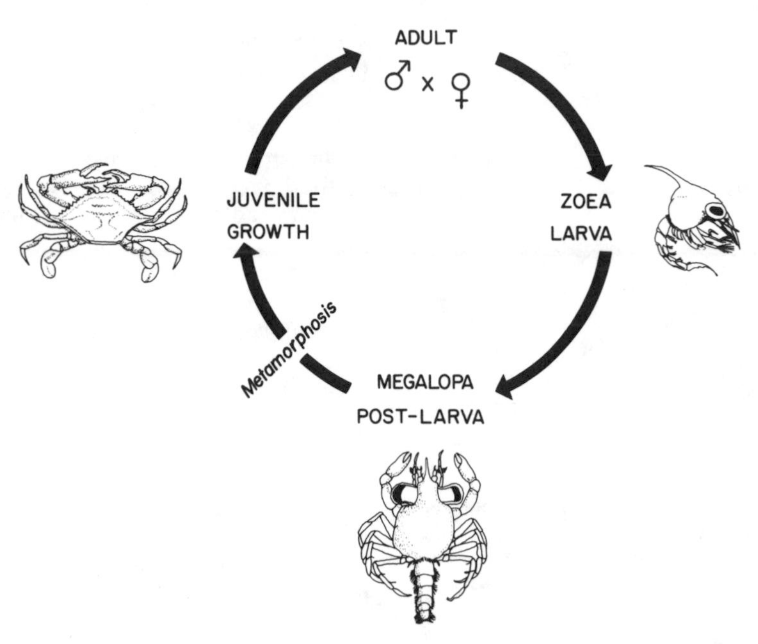

FIGURE 2.—Life cycle of a typical brachyuran crab. Planktonic zoeal stages vary from two to eight, depending on species. The period of planktonic existence is usually less than 30 d, but may extend as long as 90 d for some species. Megalopae show planktonic and benthic behavior. This postlarval phase may last 30 d.

stream transport, and the late-stage larvae and megalopae deep in the water, resulting in upstream transport.

Other studies have investigated effects of a variety of stimuli on locomotor activity (kinesis responses) in crab larvae. Generally, directed swimming speeds under "optimal environmental conditions" range from approximately 0.1 to 0.3 cm/s for early-stage larvae. Several investigations have dealt with effects of changes in salinity and temperature on swimming activity (Ott and Forward 1976; Sulkin 1980, 1982, 1983; Kelly et al. 1982; McConnaughey 1983; O'Connor and Epifanio 1985). Both of these variables vary with depth in estuaries; salinity generally increases whereas temperature decreases (during warm seasons) as vertical position deepens. Salinity has minor effects on locomotor activity. Sharp haloclines such as those occurring in highly stratified estuaries may slow vertical movement, but they do not prevent it. Not surprisingly, swimming speed increases with temperature, but sharp thermoclines do not prevent upward or downward migration of larvae. Crab larvae, then, appear to be able to migrate vertically in the water column regardless of the degree of stratification.

Another effector of increased locomotor activity by crab larvae is hydrostatic pressure (Hardy and Bainbridge 1951; Naylor and Isaac 1973; Sulkin 1973; Bentley and Sulkin 1977; Wheeler and Epifanio 1978; Sulkin and Van Heukelem 1982). This factor is more conservative than either salinity or temperature because it increases predictably with depth regardless of the degree of stratification of an estuary. Once a threshold level of pressure is exceeded, early-stage larvae increase swimming speed as pressure increases further, even in very small increments. This response is called barokinesis. Thresholds vary among species, but levels as low as 2×10^4 Pa have been reported (Sulkin and Van Heukelem 1982). Later-stage larvae are either insensitive to pressure change or exhibit low barokinesis. However, megalopae may show a greater sensitivity to pressure changes than late larvae, while maintaining a negative geotaxis. Additionally, megalopae show a much increased rate of sinking, so during periods of low swimming activity, megalopae might sink to the bottom of relatively shallow estuarine water columns. Increases in hydrostatic pressure associated with the passage of the tidal wave would induce these larvae to swim upward from the bottom during flooding tidal phases. During the ensuing ebb phase, they would decrease their swimming speed and sink to the bottom. Indeed, this has been proposed as a

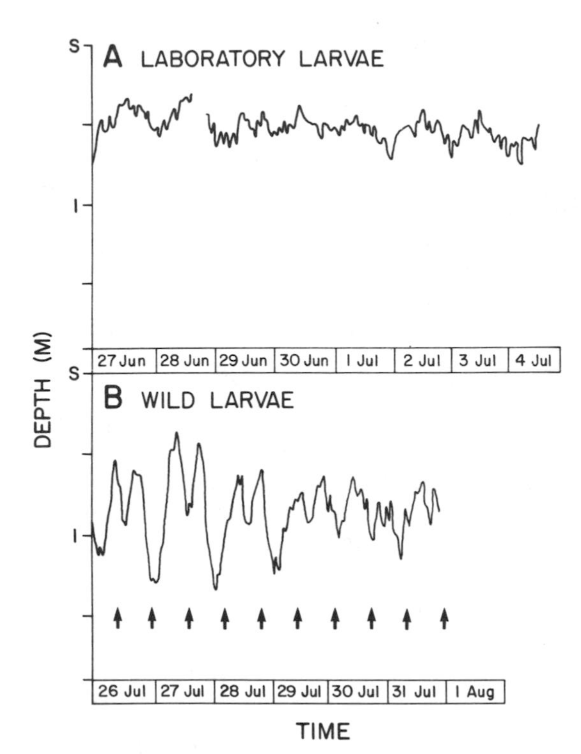

FIGURE 3.—Vertical distribution of mud crab larvae *Rhithropanopeus harrisii* in a 2-m-high plexiglass column: (A) laboratory-reared larvae; (B) field-collected larvae. S = water surface; arrows represent time of low tide at the collection site. (Modified from Cronin and Forward 1979.)

mechanism for upstream transport by megalopae (Knight-Jones and Qasim 1966; Epifanio et al. 1984).

There has been much less study of endogenous behavior by crab larvae. Sulkin et al. (1979) showed a circadian rhythm in swimming speeds of laboratory-cultured blue crab larvae *Callinectes sapidus*, but found it difficult to relate this to natural environmental conditions. In a later investigation, Cronin and Forward (1979) compared the patterns of vertical migration by laboratory-reared and field-caught larvae of the mud crab *Rhithropanopeus harrisii* (Figure 3). Field-caught larvae showed a clear cycle of tidal migration whereas laboratory-reared groups exhibited only a weak circadian rhythm of swimming activity. Field-caught larvae moved deeper in the water column at times coinciding with low tide at the site of collection and up in the water column at times coincident with high tide. This pattern of migration was maintained in the laboratory for several tidal cycles. The experiment was conducted in darkness at constant salinity and temperature, so it seems clear that the cycle of migration was endogenous, i.e., entrained in the natural environment. While it is not clear what cues are necessary to entrain this rhythm in the field, the pattern of migration would result in inhibition of net downstream transport of the larvae.

Export of Larvae from Estuaries

Retention of larval forms of estuarine species is not perfect, and there is considerable leakage of even those species that show retentive behavioral traits. Larvae of several estuarine crabs are relatively common in the waters of the inner continental shelf in the Middle Atlantic and South Atlantic bights on the east coast of the USA (Nichols and Keney 1963; Dudley and Judy 1973). (Similar data for larvae of estuarine forms other than crustaceans are rare.) However, some estuarine species appear to export their larvae preferentially. Extensive field investigations in the mouth of Chesapeake Bay and on the adjacent shelf have allowed grouping of species by the degree of retention of their larvae (Sandifer 1975; Johnson 1982). Retained species include all those in the family Xanthidae (mud crabs) whose adults are incapable of swimming and often live in the mid to upper reaches of the estuary. Expelled species included not only swimming crabs such as the blue crab *Callinectes sapidus*, but also the fiddler crabs *Uca* spp. This latter case is noteworthy, because adult fiddler crabs are semiterrestrial, and the habitat of one of the indigenous species extends far up the estuary into fresh water. Juveniles are poor swimmers and, due to their small size, are probably incapable of long, overland migrations. It is most probable that adult habitat is determined when the postlarval megalopae give up planktonic existence at metamorphosis.

Early-stage fiddler crab larvae are clearly adapted for export from the marsh creek habitat of the adults. Larvae are hatched near the time of high tide around new and full moons, assuring their downstream transport during maximum ebb flow (Wheeler 1978; Christy 1982; Lambert and Epifanio 1982). Early-stage larvae maintain a position high in the water column, resulting in their transport into the mouths of primary estuaries such as Delaware Bay (Dittel and Epifanio 1982) and even onto the adjacent continental shelf (Christy and Stancyk 1982). However, late-stage larvae take a position deep in the water column, thus taking advantage of subtidal landward flow back into the primary estuary (Dittel and Epifanio 1982). Movement into the marsh environments required by adults appears to be facilitated by tidally rhythmic vertical migration (Meredith 1982).

Adult and juvenile forms of the blue crab are also restricted to a largely estuarine habitat, and classical treatments of the life history assumed that larvae developed in estuaries as well (Churchill 1919; Porter 1956; Van Engel 1958). However, recent work in Chesapeake and Delaware bays indicates that larvae are not retained in the estuary (Epifanio et al. 1984), but rather are dispersed onto the continental shelf. Results of these investigations show that gravid females migrate to the estuarine mouth where the eggs hatch near the time of nocturnal high tide (Provenzano et al. 1983; Epifanio et al. 1984). Newly hatched larvae migrate to the surface and are carried onto the shelf by the subsequent ebbing tide. As only newly hatched zoeae and megalopae are collected in the estuary, there is little retention of zoeae in the estuary. To the contrary, the larvae undergo their entire period of zoeal development on the continental shelf and reinvade the estuary as megalopae or small juvenile crabs (Dittel and Epifanio 1982; Epifanio et al. 1984).

Larval development of blue crabs includes seven (occasionally eight) zoeal stages and a megalopal stage. Zoeal development takes 5–6 weeks in the laboratory at 25° C (Sulkin 1975), and field evidence suggests that similar rates of development may pertain under natural conditions in the Middle Atlantic Bight, because peak abundances of newly hatched zoeae and megalopae were separated by 5–6 weeks in the mouth of Delaware Bay (Epifanio et al. 1984).

Based on the results of laboratory behavior studies, it was expected that early larvae would be transported seaward in surface waters. More advanced zoeae would then take up a deeper position in the water column, resulting in transport back to the estuary via landward, subtidal drift (Sulkin et al. 1980; Sulkin and Van Heukelem 1982). However, field investigations on the inner continental shelf of the Middle Atlantic Bight showed that zoeae maintain a surface distribution throughout development (Smyth 1980; McConaugha et al. 1983). A substantial body of oceanographic data indicated that subtidal movement of inshore surface water in the Middle Atlantic Bight was generally to the south with only brief reversals (Bumpus 1965, 1969, 1973; Pape and Garvine 1982); there was no apparent mechanism for retention of surface-dwelling zoeae in the vicinity of the parent estuary. Rather, it appeared that zoeae might be transported long distances from adult habitat. Indeed, calculations for the Delaware Bay system indicated that surface larvae could be transported at rates approaching 10 km/d, resulting in total southward advection of 350 km over 5 weeks of zoeal development (Pape and Garvine 1982; Epifanio, unpublished data).

However, such long-distance transport does not seem to occur, because large concentrations of early- and late-stage zoeae have been reported on the inner continental shelf within 70 km of the mouth of Chesapeake Bay, suggesting that the larvae were produced within that system (McConaugha et al. 1983). Additionally, there are stable populations of blue crab in the lagoonal estuaries along the New Jersey coast north of Delaware Bay indicating that these estuaries do not depend on occasional recruitment from Delaware Bay during periods of surface-flow reversal on the continental shelf.

It appears, then, that circulation along the inner continental shelf of the Middle Atlantic Bight is more complicated than previously supposed. For example, preliminary measurements of surface currents off the mouth of Chesapeake Bay suggested the presence of a wind-generated, northward-flowing current during the summer months (Boicourt 1981, 1982). A subsequent theoretical study (Johnson et al. 1984) supported this finding. An extension of that work showed a relationship between reported distributions of blue crab larvae off Chesapeake Bay and theoretical surface-current patterns predicted from wind records (Johnson 1985). Additional results from current meters moored off Delaware Bay (R. W. Garvine, unpublished data) have yielded unequivocal evidence for a northward flowing current during the summers of 1983 and 1984. This current exists as a relatively narrow band (Figure 4) between the southward surface current right along the coast (Pape and Garvine 1982) and the southward surface current dominating the outer shelf (Bumpus 1965).

This system of surface currents provides a potential mechanism for retention of blue crab larvae on the inner continental shelf near their parental estuaries. Early-stage larvae are flushed from middle Atlantic estuaries along with the general flow of surface water (Epifanio et al. 1984). In theory, the larvae are then carried southward in the near-shore estuarine plume. As

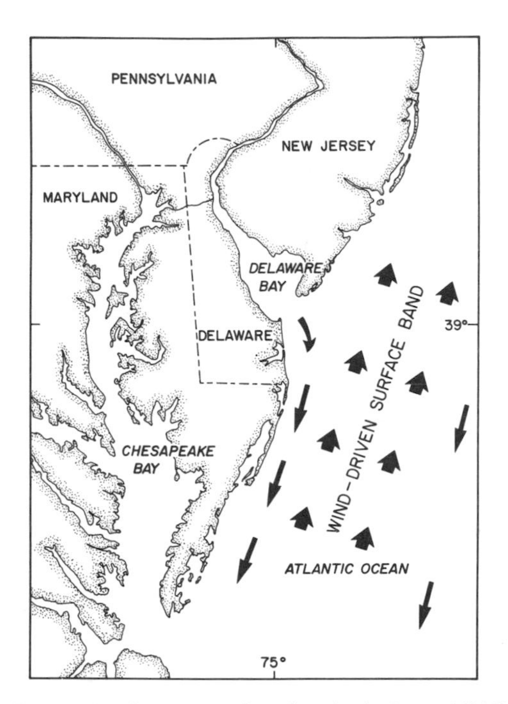

FIGURE 4.—Summer surface flow in the lower Middle Atlantic Bight adjacent to Delaware and Chesapeake bays. Southward flows along the beach and in the midshelf region are separated by a wind-driven northward flow important for the retention of estuarine larvae near parent estuaries. Interannual variation in strength of northward flow may be related to interannual variation in estuarine recruitment.

the momentum of the plume dissipates with increasing distance from the estuary, the larvae are mixed with the northward-flowing band of wind-driven water and carried to the vicinity of the parent estuary. Recent field data from the continental shelf adjacent to Chesapeake Bay provide strong support for this model (McConaugha et al. 1983).

Although the model predicts high retention of larvae during spawning seasons when winds over the inner continental shelf are strong and persistently from the south, it provides no mechanism for transport of advanced larvae and megalopae in an onshore direction and back into the estuary. Johnson (1985) found maximum concentrations of blue crab megalopae in the near-surface layer over the inner continental shelf adjacent to Chesapeake Bay; only 12–25% of the megalopae were found in near-bottom water. Citing the local predominance of onshore winds during early autumn (Boicourt 1973), Johnson proposed a model of wind-driven transport of the surface-dwelling megalopae back into the estuary. However, there is a strong and persistent outflow of surface water from large estuaries such as the Chesapeake and Delaware bays. Therefore, wind-driven advection of surface-dwelling megalopae into these estuaries would be expected only during periods of very strong easterly winds associated with major cyclonic disturbances.

An alternative model for transport of megalopae into estuaries (Sulkin and Epifanio 1986) considers that 12–25% of megalopae that Johnson found in near-bottom waters. This model assumes that megalopae move deeper in the water column as metamorphosis approaches, and thus interprets those larvae collected near the bottom in Johnson's study as those closest to metamorphosis. Less-mature megalopae, then, would be transported onshore by wind-driven currents as in Johnson's model, while more mature megalopae would be entrained in the subtidal, shoreward movement of the near-bottom waters of the inner shelf. This is consistent with behavioral characteristics of megalopae in laboratory experiments (Sulkin and Van Heukelem 1982).

One problem with this most recent model is the slow movement of near-bottom water back into the estuary (Pape and Garvine 1982). However, additional field evidence indicates that megalopae may exploit tidal currents to increase their speed of transport back into the estuary. Meredith (1982), working in a shallow, secondary estuary near the mouth of Delaware Bay, reported that megalopae were absent from the water column during ebb phases, but were common during periods of flooding tidal currents. Epifanio et al. (1984) and Brookins and Epifanio (1985) reported similar results over four to six consecutive tidal cycles in the mouths of Delaware Bay and Indian River Bay, Delaware. These results were interpreted as tidally rhythmic vertical migration and are consistent with reports of megalopal swarming at discrete depths in the water column (Chace and Barnish 1976; Rice and Kristensen 1982). Furthermore, Sulkin and van Heukelem (1982) described behavioral responses of blue crab megalopae that could account for precise depth regulation and resultant change in vertical position as a function of periodic stimuli. Estuarine invasion via tidally rhythmic vertical migration has been reported for postlarvae of one other crustacean (Johnson and Gonor 1982).

This model, combining wind-driven transport of zoeae and early megalopae with subtidal transport and vertical migration of mature megalopae, provides a plausible mechanism for dispersal and

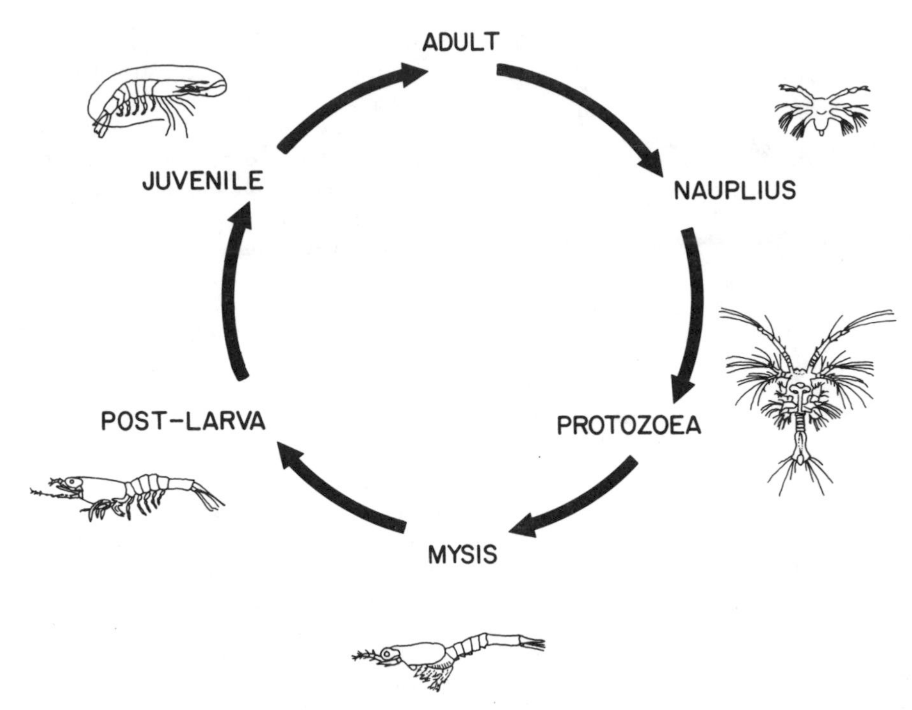

FIGURE 5.—Life cycle of a typical panaeid shrimp. Adults spawn on the inner continental shelf and larval development through the mysid stage procedes on the shelf. Postlarvae are transported into estuaries where juveniles develop. Advanced juveniles migrate back to the continental shelf.

recruitment of blue crab larvae. However, it does not exclude the possibility that some substantial fraction of the megalopae undergoes metamorphosis on the continental shelf and reinvades the estuary as juveniles (Johnson 1985). Juvenile blue crabs grow very rapidly and, soon after metamorphosis, reach a size at which their swimming abilities can be important in effecting their movement into the estuary.

Import of Larvae to Estuaries

Larvae of species living on the inner continental shelf are commonly transported into estuaries via tidal currents (Sandifer 1975; Goy 1976; Dittel and Epifanio 1982). However, investigation of the dynamics of this transport suggests that the net flux is zero (Brookins and Epifanio 1985). Nevertheless, there are some species whose larvae are spawned on the continental shelf and then transported into estuaries where subsequent juvenile development occurs. This is relatively common among fish (e.g., Nelson et al. 1977); among invertebrates, it has been best studied for penaeid shrimp.

Female penaeid shrimp do not brood their eggs externally (as do the brachyuran crabs), but rather release them near the bottom shortly before hatching. Usually there are five naupliar stages, three protozoeal stages, and three mysid stages (Figure 5). The mysid phase is followed by the postlarval stage and subsequent juvenile stages. Some species are either completely marine or completely estuarine, but most migrate between the two habitats at different times in their life history (Kutkuhn 1966). Many species spawn on the inner continental shelf, and the larvae develop in the inshore plankton. Postlarvae are then transported onshore and into estuaries, where subsequent juvenile stages remain for 3–4 months before they migrate back to the shelf (Edwards 1977). Migration of juveniles out of the estuary is probably an active process involving swimming in the horizontal plane.

The weak-swimming postlarvae maintain a position deep in the water column, thus facilitating their landward transport in the subtidal estuary–shelf circulation (Garcia and le Reste 1981). Moreover, extensive sampling in the mouths of several tropical and subtropical estuaries points to a distinct diel–tidal pattern of vertical migration (Young and Carpenter 1977) wherein postlarvae are more common in the water column during nighttime flood tides. Presumably, the postlarvae sink to the bottom during daylight hours and during nighttime ebb tides. Although the significance of the diel component in this periodicity is unclear (possibly it reduces visual predation on the postlarvae), the tidal component undoubtedly augments upstream transport. Additionally, postlarval invasion of estuaries appears to be maximal during spring tides (Roessler and Rehrer 1971; Edwards 1977). This would require a lunar rhythm in reproductive patterns, and would further accelerate the upstream transport of the tidally migrating postlarvae.

Summary and Conclusions

Planktonic larval development is a common component of the life cycles of invertebrates in temperate, subtropical, and tropical estuaries throughout the world. These larval forms are minute, and the effect of their sustained horizontal swimming is inconsequential compared to horizontal advection by tidal and subtidal motion. As the net, subtidal flow of most estuaries is seaward, these larvae face the risk of transport out of the estuary and away from required adult habitat.

A large body of laboratory and field evidence suggests that transport of larvae in estuaries is not a passive phenomenon, but that larval behavior indeed influences this process. It appears that three general responses to the problem of retention of larvae in estuaries have evolved. As a result of the first response, early larvae maintain a position high in the estuarine water column, and later stages sink to the near-bottom water (Figure 6A). This allows downstream dispersal of early stages with some leakage onto the continental shelf and the possibility of transport to other nearby estuaries. Later stages would be entrained in the subtidal, upstream flow of estuarine bottom water and would be effectively retained in the estuary.

A second mechanism for retention in estuaries is evidenced by species that migrate vertically in the water column in rhythm with tidal cycles (Figure 6B). Larvae move up in the water column during flooding tidal phases and sink to deeper levels during ebb phases. This allows the larvae to maximize their upstream transport during flood tide, while minimizing downstream movement during ebbing periods. Tidally rhythmic vertical migration maximizes larval retention in even the most upstream of estuarine locations.

A third group of species maintains stable estuarine populations without retention of larvae in

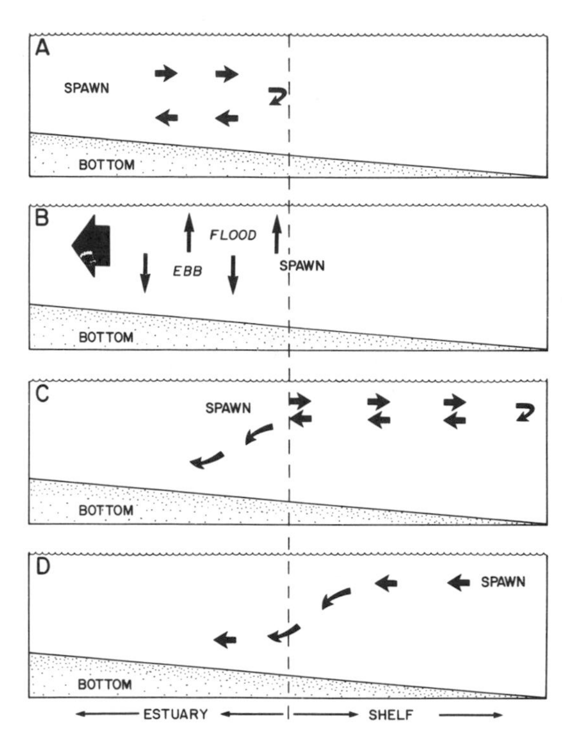

FIGURE 6.—Comparison of four mechanisms for dispersal and recruitment of estuarine larvae. (A) Spawning occurs in the estuary and young larvae are transported downstream by surface flows; older larvae sink and are transported upstream in the residual bottom flow. (B) Spawning occurs near the estuary mouth; larvae migrate up in the water column on flood tides and sink on ebb tides, resulting in their net upstream transport. (C) Larvae are spawned near the estuary mouth and are flushed onto shelf; larvae are retained near the estuarine mouth by current patterns. (D) Larvae are spawned on the shelf and are transported onshore by currents; postlarvae sink and are transported into the estuary by residual bottom flow.

the estuary (Figure 6C). Larvae of these species show behavioral characteristics that accelerate transport to the inner continental shelf. Some species in this group return to the estuary by sinking deep in the water column during the later stages of larval development, whereas others maintain a surface position and exploit gyres or wind-driven countercurrents for retention on the inner shelf near the parent estuary. In either case, there is great potential for transport away from the parent estuary, and there probably is a large amount of gene flow among dispersed populations of species in this group. Large interannual fluctuations in the magnitude of larval recruitment also would be expected from these species, because transport to unfavorable habitats must occur frequently as a function of interannual variations in patterns of surface currents on the inner shelf.

A final group of invertebrate species uses the inner continental shelf for adult and larval existence, but their postlarvae invade the estuary and remain there through juvenile development (Figure 6D). Upstream transport of these postlarvae is against the net seaward flow of estuarine water, and is accomplished through general maintenance of a deep position in the water column and enhanced by tidally rhythmic vertical migration.

An understanding of the ways in which the different groups have evolved is important in the assessment of effects of natural or human perturbations on reproductive success. For example, the extreme rains of Hurricane Agnes had extensive effects on recruitment of bivalve molluscs in Chesapeake Bay in 1972, but had little effect on recruitment of blue crabs (Andrews 1973). This is easily explained in light of the retention strategy for larvae employed by many estuarine bivalves compared to the export strategy evidenced by blue crabs. In contrast, blue crab recruitment might be expected to be poor during years when northerly winds generate southward-flowing currents in the Middle Atlantic Bight (Johnson et al. 1984), while estuarine bivalve reproduction would be unaffected.

Similarly, human-induced changes in estuarine flow patterns, (e.g., oil spills, sewage outfalls, dredging activity, agricultural runoff, entrainment of water for industrial cooling purposes) should affect species with retained larvae more than those with exported larvae. On the other hand, activities that (1) alter flow patterns in the mouths of estuaries (e.g., stabilization of inlets, construction of breakwaters, dredging of navigational channels) or (2) affect the inner continental shelf (ocean sewage outfalls, sludge dumping, construction of floating power plants or industrial islands) would have the greater effects on species with exported larvae or those whose larvae develop on the continental shelf before postlarval invasion of estuarine nursery areas.

Acknowledgments

This work is a result of research sponsored by National Oceanic and Atmospheric Administration Office of Sea Grant, under NA85AA-D-SG033 (project R/M-4).

References

Andrews, J. D. 1973. Effects of tropical storm Agnes on epifaunal invertebrates in Virginia estuaries. Chesapeake Science 14:223–234.

Bentley, E., and S. D. Sulkin. 1977. The ontogeny of barokinesis during the zoeal development of the xanthid crab *Rhithropanopeus harrisii* (Gould). Marine Behavior and Physiology 4:275–282.

Boicourt, W. C. 1973. The circulation of water on the continental shelf from Chesapeake Bay to Cape Hatteras. Doctoral dissertation. The Johns Hopkins University, Baltimore, Maryland.

Boicourt, W. C. 1981. Circulation in the Chesapeake Bay entrance region: estuary–shelf interaction. Pages 61–78 *in* G. Campbell and J. Thomas, editors. Chesapeake Bay plume study, 1980. National Aeronautics and Space Administration Science and Technology Information Branch, NASA Conference Publication 2188, NOAA/NEMP II; 81; ABCDEFG 0042, Washington, D. C.

Boicourt, W. C. 1982. Estuarine larval retention mechanisms on two scales. Pages 445–458 *in* V. Kennedy, editor. Estuarine comparisons. Academic Press, New York.

Bousfield, E. L. 1955. Ecological control of the occurrence of barnacles in the Miramichi estuary. National Museum of Canada Bulletin 137.

Brookins, K. G., and C. E. Epifanio. 1985. Abundance of brachyuran larvae in a small inlet over six consecutive tidal cycles. Estuaries 8:60–67.

Bumpus, D. F. 1965. Residual drift along the bottom on the continental shelf in the Middle Atlantic Bight area. Limnology and Oceanography (supplement) 10:R50–R53.

Bumpus, D. F. 1969. Reversals in the surface drift in the Middle Atlantic Bight area. Deep Sea Research and Oceanographic Abstracts (supplement) 16:17–23.

Bumpus, D. F. 1973. A description of the circulation on the continental shelf of the east coast of the United States. Progress in Oceanography 6:111–157.

Carriker, M. R. 1951. Ecological observations on the distribution of oyster larvae in New Jersey estuaries. Ecological Monographs 21:19–38.

Carriker, M. R. 1961. Interrelation of functional morphology, behavior and autecology in early stages of the bivalve, *Mercenaria mercenaria*. Journal of the Elisha Mitchell Scientific Society 77:168–241.

Chace, F. A., and G. Barnish. 1976. Swarming of a raninid megalopa at St. Lucia, West Indies (Decapoda, Brachyura). Crustaceana (Leiden) 31:105–107.

Christy, J. H. 1982. Adaptive significance of semilunar cycles of larval release in fiddler crabs (Genus *Uca*): test of an hypothesis. Biological Bulletin (Woods Hole) 163:251–263.

Christy, J. H., and S. E. Stancyk. 1962. Timing of larval production and flux of invertebrate larvae in a well-mixed estuary. Pages 489–503 *in* V. Kennedy, editor. Estuarine comparisons. Academic Press, New York.

Churchill, E. P., Jr. 1919. Life history of the blue crab. U.S. Bureau of Fisheries Bulletin 36:91–128.

Cronin, T. W., and R. B. Forward, Jr. 1979. Tidal vertical migration: an endogenous rhythm in estuarine crab larvae. Science (Washington, D.C.) 7:1020–1021.

de Wolf, P. 1973. Ecological observation on the mechanisms of dispersal of barnacle larvae during planktonic life and settling. Netherlands Journal of Sea Research 6:1–129.

Dittel, A. I., and C. E. Epifanio. 1982. Seasonal abundance and vertical distribution of crab larvae in Delaware Bay. Estuaries 5:197–202.

Dudley, D. L., and M. D. Judy. 1973. Seasonal abundance and distribution of juvenile blue crabs in Cove Sound, N.C., 1965–68. Chesapeake Science 14:51–54.

Edwards, R. R. C. 1977. Field experiments on growth and mortality of *Penaeus vannamei* in a Mexican coastal lagoon complex. Estuarine and Coastal Marine Science 5:107–121.

Epifanio, C. E., C. C. Valenti, and A. E. Pembroke. 1984. Dispersal and recruitment of blue crab larvae in the Delaware Bay, USA. Estuarine, Coastal and Shelf Science 18:1–12.

Franekel, G. S., and D. L. Gunn. 1961. The orientation of animals. Dover, New York.

Garcia S., and L. le Reste. 1981. Life cycles, dynamics, exploitation and management of coastal penaeid shrimp stocks. FAO (Food and Agriculture Organization of the United Nations) Fisheries Technical paper 203.

Goy, J. W. 1976. Seasonal distribution and the retention of some decapod crustacean larvae within the Chesapeake Bay, Virginia. Masters thesis. Old Dominion University, Norfolk, Virginia.

Hardy, A. C., and R. Bainbridge. 1951. Effect of pressure on the behavior of decapod larvae. Nature (London) 167:354–355.

Johnson, D. F. 1982. A comparison of recruitment strategies among brachyuran crustacean megalopae of the York River, lower Chesapeake Bay and adjacent waters. Doctoral dissertation. Old Dominion University, Norfolk, Virginia

Johnson, D. R. 1985. Wind-forced dispersion of blue crab larvae in the Middle Atlantic Bight. Continental Shelf Research 4:1–14.

Johnson, D. R., B. S. Hester, and J. R. McConaugha. 1984. Studies of a wind mechanism influencing the recruitment of blue crabs in the Middle Atlantic Bight. Continental Shelf Research 3:425–437.

Johnson, G. E., and J. J. Gonor. 1982. The tidal exchange of *Callianassa californiensis* (Crustacea, Decapoda) larvae between the ocean and the Salmon River estuary, Oregon. Estuarine, Coastal and Shelf Science 14:501–515.

Kelly, P., S. D. Sulkin, and W. F. Van Heukelem. 1982. A dispersal model for larvae of the deep sea red crab *Geryon quinquedens* Smith based upon behavioral regulation of vertical migration in the hatching stage. Marine Biology (Berlin) 72:35–43.

Knight-Jones, E. W., and S. Z. Qasim. 1966. Response of Crustacea to change in hydrostatic pressure. Pages 1132–1150 *in* Proceedings of the symposium on crustacea, part 3. Marine Biological Association of India, Madras.

Kutkuhn, J. H. 1966. Dynamics of a penaeid shrimp population and managment implications. U.S. Fish and Wildlife Service Fishery Bulletin 65:313–338.

Lambert, R., and C. E. Epifanio. 1982. A comparison of dispersal strategies in two genera of brachyuran crab in a secondary estuary. Estuaries 5:182–188.

Latz, M. I., and R. B. Forward, Jr. 1977. The effect of salinity upon phototaxis and geotaxis in a larval crustacean. Biological Bulletin (Woods Hole) 153:163–179.

McConaugha, J. R., D. F. Johnson, A. J. Provenzano, and R. C. Maris. 1983. Seasonal distribution of larvae of *Callinectes sapidus* in the waters adjacent to Chesapeake Bay. Journal of Crustacean Biology 3:582–591.

McConnaughey, R. 1983. The influence of thermoclines on vertical migration of stage I blue crab *Callinectes sapidus* larvae and the implications to recruitment. Master's thesis. University of Maryland, College Park.

Meredith, W. A. 1982. The dynamics of zooplankton and micronekton community structure across a salt-marsh estuarine interface of lower Delaware bay. Doctoral dissertation. University of Delaware, Newark.

Naylor, E., and M. J. Isaac. 1973. Behavioral significance of pressure responses in megalopa larvae of *Callinectes sapidus* and *Macropipus* sp. Marine Behavior and Physiology 1:341–350.

Nelson, T. C., and E. B. Perkins. 1930. The reactions of oyster larvae to currents and to salinity gradients. Anatomical Record 40:288. (Abstract.)

Nelson, W. R., M. C. Ingham, and W. E. Schaef. 1977. Larval transport and year-class strength of Atlantic menhaden, *Brevoortia tyrannus*. U.S. National Marine Fisheries Service Fishery Bulletin 75:23–41.

Nichols, P., and P. M. Keney. 1963. Crab larvae (*Callinectes*) in plankton collections from cruises of M/V *Theodore N. Gill*, south Atlantic coast of the United States 1953–54. U.S. Fish and Wildlife Service Special Scientific Report—Fisheries 448.

O'Connor, N. J. and C. E. Epifanio. 1985. The effect of salinity on the dispersal and recruitment of fiddler crab larvae. Journal of Crustacean Biology 5:137–145.

Ott, F. S., and R. B. Forward, Jr. 1976. The effect of temperature on phototaxis and geotaxis by larvae of the crab *Rhithropanopeus harrisii*. Journal of Experimental Marine Biology and Ecology 23:97–107.

Pape, E. H., III, and R. W. Garvine. 1982. The subtidal circulation in Delaware Bay and adjacent shelf waters. Journal of Geophysical Research 87:7955–7970.

Porter, H. H. 1956. Delaware blue crab. Estuarine Bulletin 2:3–5.

Provenzano, A. J., Jr., J. R. McConaugha, K. B. Phillips, D. F. Johnson, and J. Clark. 1983. Vertical distribution of first stage larvae of the blue crab, *Callinectes sapidus*, at the mouth of Chesapeake Bay. Estuarine, Coastal and Shelf Science 16:486–499.

Rice, A. L., and I. Kristensen. 1982. Surface swarm of swimming crab megalopae at Curacao (Decapoda, Brachyura). Crustaceana (Leiden) 42:233–240.

Roessler, M. A., and R. C Rehrer. 1971. Relation of

catches of postlarval pink shrimp in Everglades National Park, Florida, to the commercial catches on the Tortugas grounds. Bulletin of Marine Science 21:790–805.

Sandifer, P. A. 1975. The role of pelagic larvae in recruitment to populations of adult decapod crustaceans in the York River estuary and adjacent lower Chesapeake Bay, Virginia. Estuarine and Coastal Marine Science 3:269–279.

Schembri, P. J. 1982. Locomotion, feeding, grooming, and the behavioral responses to gravity, light, and hydrostatic pressure in the stage I zoea larvae of *Ebalia tuberosa* (Crustacea:Decapoda:Leucosiidae). Marine Biology (Berlin) 72:125–134.

Seliger, H. H., J. A. Boggs, R. B. Rivkin, W. H. Biggley, and K. R. H. Aspden. 1982. The transport of oyster larvae in an estuary. Marine Biology (Berlin) 71:57–72.

Smyth, P. O. 1980. *Callinectes* (Decapoda: Portunidae) larvae in the Middle Atlantic Bight, 1975–77. U.S. National Marine Fisheries Service Fishery Bulletin 78:251–265.

Sulkin, S. D. 1973. Depth regulation of crab larvae in the absence of light. Journal of Experimental Marine Biology and Ecology 13:73–82.

Sulkin, S. D. 1975. The influence of light in the depth regulation of crab larvae. Biological Bulletin (Woods Hole) 148:333–343.

Sulkin, S. D. 1984. Behavioral basis of depth regulation in the larvae of brachyuran crabs. Marine Ecology Progress Series 15:181–205.

Sulkin, S. D., and C. E. Epifanio. 1986. A conceptual model for recruitment of the blue crab, *Callinectes sapidus* Rathbun, to estuaries of the Middle Atlantic Bight. Canadian Special Publication of Fisheries and Aquatic Science 92:117–123.

Sulkin, S. D., C. E. Epifanio, and A. Provenzano. 1982. Proceedings of a workshop on recruitment of the blue crab in Middle-Atlantic Bight estuaries. University of Maryland Sea Grant Program Technical Report UM-SG-TS-82-04.

Sulkin, S. D., I. Phillips, and W. Van Heukelem. 1979. On the locomotory rhythm of brachyuran crab larvae and its significance in vertical migration. Marine Ecology Progress Series 1:331–335.

Sulkin, S. D., and W. Van Heukelem. 1982. Larval recruitment in the crab, *Callinectes sapidus* Rathbun: an amendment to the concept of larval retention in estuaries. Pages 459–475 *in* V. Kennedy, editor. Estuarine comparisons. Academic Press, New York.

Sulkin, S. D., W. Van Heukelem, and P. Kelly. 1983. Behavioral basis of depth regulation in the hatching and post-larval states and of the mud crab *Eurypanopeus depressus* Hay and Shore. Marine Ecology Progress Series 11:157–164.

Sulkin, S. D., W. Van Heukelem, P. Kelly, and L. Van Heukelem. 1980. The behavioral basis of larval recruitment in the crab, *Callinectes sapidus* Rathbun: a laboratory investigation of ontogenetic changes in geotaxis and barokinesis. Biological Bulletin (Woods Hole) 159:402–417.

Van Engel, W. 1958. The blue crab and its fishery in Chesapeake Bay. Part I. Reproduction, early development, growth, and migration. Commercial Fisheries Review 20(8):6–17.

Wheeler, D. 1978. Semilunar hatching periodicity in the mud crab *Uca pugnax* (Smith). Estuaries 1:268–269.

Wheeler, D., and C. E. Epifanio. 1978. Behavioral response to hydrostatic pressure in larvae of two species of xanthid crabs. Marine Biology 46(Berlin):167–174.

Wood, I., and W. J. Hargis, Jr. 1971. Transport of bivalve larvae in a tidal estuary. Pages 29–44 *in* D. Crisp, editor. Proceedings fourth European marine biology symposium. Cambridge University Press, Cambridge, England.

Young, P. C., and S. M. Carpenter. 1977. Recruitment of postlarval Penaeid prawns to nursery areas in Moreton Bay, Queensland. Australian Journal of Marine and Freshwater Research 28:745–773.

American Fisheries Society Symposium 3:115-131, 1988

Modeling of Physical and Behavioral Mechanisms Influencing Recruitment of Spot and Atlantic Croaker to the Cape Fear Estuary

JOHN P. LAWLER, MICHAEL P. WEINSTEIN, HWANG Y. CHEN, AND THOMAS L. ENGLERT

Lawler, Matusky & Skelly Engineers, One Blue Hill Plaza, Pearl River, New York 10965, USA

Abstract.—A series of mathematical models was developed to simulate spot *Leiostomus xanthurus* and Atlantic croaker *Micropogonias undulatus* transport to, and accumulation in, primary nurseries of the Cape Fear estuary, North Carolina. The ultimate product of this effort was the Cape Fear fish population model (FPM) used to describe the physical and behavioral mechanisms influencing the recruitment process. Because recruitment seemed to be a two-stage phenomenon, influenced initially by advective (hydrodynamic) processes and later by behavioral traits of the organisms themselves, a salt-budget model was developed first. This model evaluated the hydrodynamics by estimating the net nontidal flows in the estuary under various freshwater flow conditions. The second model, the FPM, incorporated these net nontidal flows with the life cycle parameters and behavioral mechanisms to simulate the distribution and growth of spot and Atlantic croakers inside the estuary. Larvae and early juveniles of both species were perceived to be transported into the estuary by the ocean exchange rate at the estuary mouth. Once inside the estuary, juvenile spot concentrated near the bottom during the day and were transported to the upriver nursery areas by the lower-layer flow. At night, spot were evenly distributed in the water column, and thus could reach the tributary creek and marsh nursery areas via the flood tide into the creeks. Most Atlantic croakers, in contrast, concentrated near the bottom at all times and therefore were found primarily in the upriver nursery areas. Thus, physical mechanisms—represented by the ocean exchange rate, tidal flows, and net nontidal flows—and behavioral mechanisms—exhibited by diurnal vertical migration patterns—played major roles in the recruitment and distribution of these two species in the Cape Fear estuary. Construction activities at and around inlets can interfere with recruitment rates, recirculation, behavioral cues, and other components of estuarine recruitment; the models described here have the potential to predict the outcome of inlet modification.

Many coastal fishes have estuarine-dependent early life stages. This life history trait is particularly prominent in the southeastern United States, where spawning by adult fishes occurs offshore and larvae (and juveniles) are recruited to estuaries throughout the year. Two of the most abundant coastal species, spot *Leiostomus xanthurus* and Atlantic croaker *Micropogonias undulatus*, exhibit these characteristics; adults migrate offshore in the fall, and spawning occurs in the population from November through March.

Estuaries provide an abundant supply of food for young fish and shellfish and serve as predation refugia for many species. The young of some species reach high abundance in deep, brackish water near the heads of estuaries. Spot tend to concentrate in the tributary creeks of the salt marsh drainage system, whereas young Atlantic croakers are most abundant in the deep shoals and channels of many estuaries (Weinstein 1979; Weinstein et al. 1980a, 1980b). How much of their habitat selection is active and how much is due to passive concentration mechanisms, or perhaps to differential mortality due to predation, is currently under debate (Boehlert and Mundy 1988, this volume; Miller 1988, this volume).

Despite what is known of the distributional ecology of these species, general mechanisms by which larval fishes recruit to, and concentrate in, estuaries are poorly understood. Larval juvenile fishes are poor swimmers, and much of the early migration toward shore from the spawning grounds is probably passive, i.e., controlled by wind-driven transport and advection (Norcross and Shaw 1984). Larvae requiring estuarine nursery areas may benefit from onshore Ekman transport. Spawning may also be associated with semipermanent coastal gyres that bring larvae to the coast where they are entrained in estuarine flows (Norcross and Shaw 1984). By the time most larvae reach estuaries they have grown considerably and are able to exert greater control over their vertical position in the water column. Such behavioral traits allow young spot and Atlantic croakers to take advantage of the two-layered circulation pattern characteristic of stratified estuaries. The exact cueing mechanisms that allow juveniles to position themselves properly in the water column are not known, but are believed to involve a suite of factors associated with tidal flux (at a particular location) and an endogenous rhythm with a tidal periodicity (Boehlert and Mundy 1988).

The success of estuarine-dependent species in reaching the nursery areas within the estuary is partly governed by the ability of young fish to negotiate coastal inlets. Construction around such inlets has the potential to reduce the rate of recruitment to interior marshes. Serious alterations in flows around inlets, combined with other anthropogenic influences (pollutants, dredge and fill, power plant operation), could substantially reduce the standing crops of young-of-year fishes and shellfish.

As part of an impact assessment associated with a power plant situated on the Cape Fear River estuary, North Carolina, a series of mathematical models was developed to simulate the transport of spot and Atlantic croakers to their primary nurseries and their accumulation there. The purpose of this paper is to describe the evolution of the Cape Fear fish population model and its use in simulating the physical and behavioral mechanisms influencing the recruitment process. Because recruitment is a two-stage process, influenced first by advective (hydrodynamic) processes and later by behavioral traits of the organisms themselves, the salt-budget model, which evaluated the hydrodynamics by estimating the net nontidal flows in the estuary under various freshwater flow conditions, was the first one developed. The second model, the FPM, then incorporated these net nontidal flows with the life cycle parameters and behavioral mechanisms to simulate the distribution and growth of these species inside the estuary. Models such as these can be readily adapted to considerations of flux through inlets with and without human-induced perturbations, e.g., jetty construction or channel dredging.

Cape Fear Estuary

The Cape Fear River basin encompasses one of the major North Carolina river systems entering the Atlantic Ocean. The 45-km reach affected by the tide, between Wilmington and the mouth of the river, is the area generally referred to as the Cape Fear estuary (Figure 1).

The estuary is relatively shallow, much of it less than 5 m deep. The exception is a narrow, 12-m-deep shipping channel maintained by the U.S. Army Corps of Engineers. Extensive tidal salt marshes exist along the shoreline of the river, and the shallow, open-water areas contain islands, tidal flats, and tidal streams. The shipping channel has a top width of 122–152 m, a volume of approximately 7.4×10^7 m^3, and a surface area of about 607 hectares at mean low water. The shallow open water, by comparison, has a combined volume of approximately 1.73×10^8 m^3 and a surface area of about 7,200 hectares at mean low water. In general, the river width decreases progressively upstream, ranging from about 4,572 m at its widest point to less than 610 m near Wilmington. The average daily freshwater inflow to the estuary for the period 1952–1978 was estimated at about 283 m^{3}s^{-1}. Monthly average flows, averaged over the period 1952–1975, show a peak value of about 450 m^{3}s^{-1} in March and a low of 116 m^{3}s^{-1} in November. The lower Cape Fear River, through oceanic influence, exhibits regular lunar tides, with two complete tidal cycles approximately every 25 h. The average tidal flow at the mouth of the Cape Fear River has been estimated at 5,664 m^{3}s^{-1} during the ebb tide (J. H. Carpenter, U.S. Nuclear Regulatory Commission, unpublished data).

Fresh water enters the Cape Fear estuary and flows downstream to the ocean. At the same time, heavier seawater intrudes upstream from the ocean entrance and flows landward beneath the lighter estuarine water. Because of tidal motion, turbulent eddies mix the fresh water downward and the heavier seawater upward. As the seawater moves toward the surface as a result of vertical mixing, more ocean-derived water intrudes upstream in the lower layer. Thus, a two-layer flow pattern is developed in which the net water movement is downstream in the upper layer and upstream in the lower layer on a tidally averaged basis. These upper and lower flows are termed the net nontidal flows.

Under most freshwater conditions, the two-layer flow pattern is predominant in the Cape Fear River above Sunny Point (Figure 1). The flow pattern between Sunny Point and the estuary mouth is more variable. Tidal currents in this reach are strong, and the estuary is highly complex and irregular in shape. When the strong tidal currents interact with the irregular estuary shape, rapid mixing of water occurs and the reach is generally well mixed. However, under high fresh-

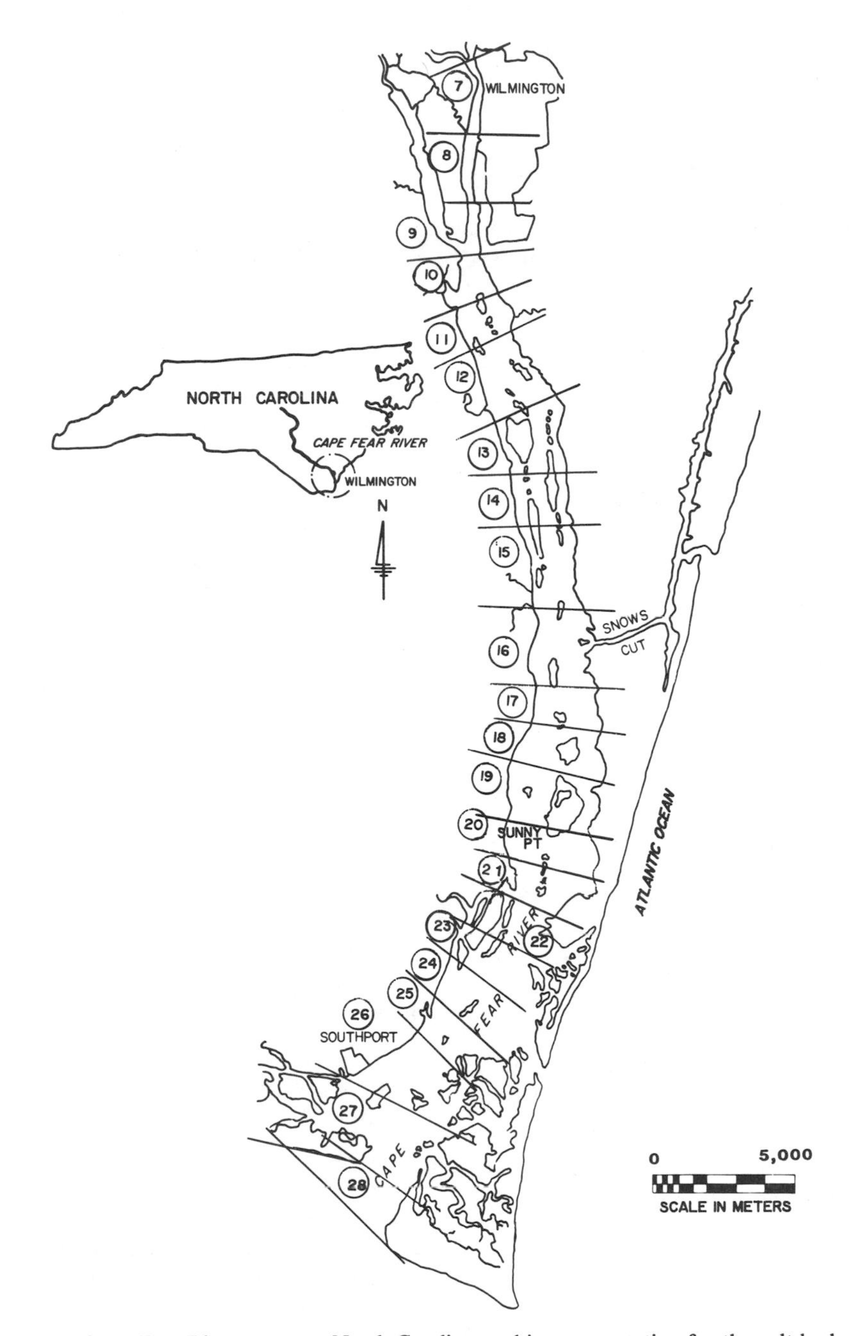

FIGURE 1.—The Cape Fear River estuary, North Carolina, and its segmentation for the salt-budget model.

water flow conditions, some stratification may still occur in this reach and the two-layer flow is better developed.

Salt-Budget Model

To quantify the circulation patterns and hydrodynamic characteristics of the Cape Fear estuary, a three-dimensional, steady-state salt-budget model was developed. The model was based on the principles of flow continuity and conservation of mass as applied to salinity to compute the net nontidal flows. In the model, as shown in Figure 1 and summarized in Table 1, the Cape Fear estuary was divided into 28 longitudinal segments from the estuary mouth to approximately 60 km upstream. By choice, the geometry within each segment was relatively similar. Because the river cross section at most segments consisted of shoals on both sides of a deep, dredged midchannel, a two-layer (one upper, one lower) system was adopted. Only one upper layer was designated for the east and west shoal regions. This "four-box" scheme (Figure 2) was modified for some segments with different types of geometry.

For a conservative substance such as salt, the steady-state material balance in each box can be simplified to

$$\text{input} - \text{output} = 0, \tag{1}$$

and the material balance of salt, incorporating all input and output terms, yields

$$\left(Q_xS - E_xA_x\frac{\partial S}{\partial x}\right)_{\text{in}} - \left(Q_xS - E_xA_x\frac{\partial S}{\partial x}\right)_{\text{out}} + \left(Q_yS - E_yA_y\frac{\partial S}{\partial y}\right)_{\text{in}} - \left(Q_yS - E_yA_y\frac{\partial S}{\partial y}\right)_{\text{out}} + \left(Q_zS - E_zA_z\frac{\partial S}{\partial z}\right)_{\text{in}} - \left(Q_zS - E_zA_z\frac{\partial S}{\partial z}\right)_{\text{out}} = 0; \tag{2}$$

Q_x, Q_y, Q_z = longitudinal, vertical, and lateral flows, respectively;
E_x, E_y, E_z = longitudinal, vertical, and lateral dispersion coefficients;
A_x, A_y, A_z = areas perpendicular to longitudinal, vertical, and lateral axes;
S = salinity;
$\frac{\partial S}{\partial x}, \frac{\partial S}{\partial y}, \frac{\partial S}{\partial z}$ = salinity gradients in the longitudinal, vertical, and lateral directions.

The longitudinal (E_x) and lateral (E_z) dispersion coefficients in equation (2) are determined from empirical expressions defined by Sayre (1968) and Abood (1977):

$$E_x = 0.265 \cdot D \cdot u; \tag{3}$$

$$E_z = 0.0085 \cdot D \cdot u; \tag{4}$$

D = average cross-sectional depth (m);
u = tidal velocity (m/s).

The flow continuity equation yields

$$Q_{x(\text{in})} + Q_{y(\text{in})} + Q_{z(\text{in})} - Q_{x(\text{out})} - Q_{y(\text{out})} - Q_{z(\text{out})} = 0. \tag{5}$$

Thus, two equations can be written for each box and eight simultaneous equations are available for a typical four-box scheme. If all input terms from the upstream end are known, the model can solve for the eight unknown variables:

four longitudinal outflows, Q_x;
two lateral flows, Q_z;
vertical flow, Q_y;
vertical dispersion coefficient, E_y.

Considering freshwater flow as the longitudinal inflow to the upper layer, the model starts its computation at the upstream segment where the salt front is located and proceeds downstream. To make the computations, tidally averaged salinity profiles are required within the salt intrusion region and at several cross sections.

Three intensive salinity surveys were conducted in 1977–1978 (Carolina Power & Light Company 1980) to provide salinity data for the Cape Fear under low, medium, and high freshwater flow conditions. Salinity data were collected at the east, middle, and west regions of five to six transects from surface to bottom and over a full tidal cycle. The river freshwater flows during

TABLE 1.—Segmentation of the Cape Fear River for the salt-budget model.

Segment	Kilometers from river mouth	Segment length
1	60.00–57.28	2.72
2	57.28–54.71	2.57
3	54.71–51.68	3.02
4	51.68–48.78	2.90
5	48.78–46.15	2.64
6	46.15–43.60	2.54
7	43.60–40.42	3.19
8	40.42–38.18	2.24
9	38.18–36.43	1.75
10	36.43–34.63	1.80
11	34.63–32.71	1.92
12	32.71–30.47	2.24
13	30.47–27.77	2.70
14	27.77–26.21	1.56
15	26.21–23.20	3.01
16	23.20–20.40	2.80
17	20.40–18.92	1.48
18	18.92–18.81	0.11
19	18.81–17.44	1.37
20	17.44–15.16	2.28
21	15.16–13.11	2.04
22	13.11–11.26	1.85
23	11.26–9.73	1.53
24	9.73–8.09	1.64
25	8.09–6.55	1.54
26	6.55–4.07	2.48
27	4.07–0.00	4.07
28	0.00–−2.00	2.00

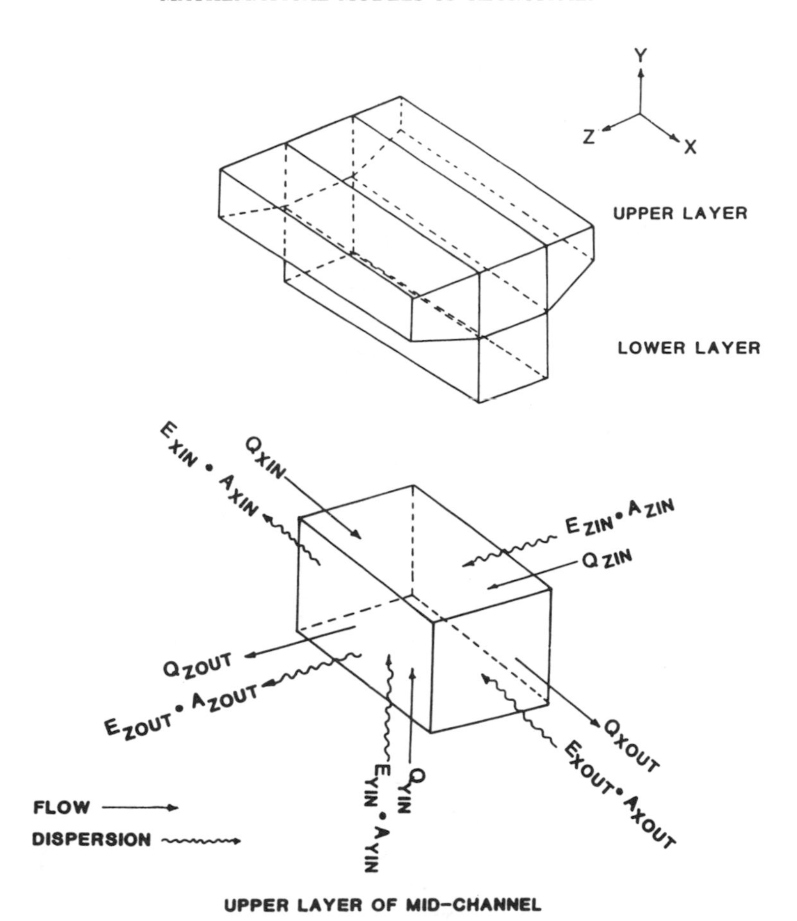

FIGURE 2.—"Four-box" schematization of Cape Fear River segments for use in the three-dimensional, steady-state, salt-budget model. Upper and lower layers are defined as the volumes above and below the level of no net motion. A = cross-sectional area; E = dispersion, Q = flow.

these surveys were 33.3 m^3s^{-1} on 14 July 1977, 459.8 m^3s^{-1} on 29 March 1978, and 103 m^3s^{-1} on 7 June 1978. Tidally averaged salinities in the upper and lower layers during these surveys are shown in Figure 3.

The net nontidal upper-layer flows computed by the salt-budget model from the measured salinity profiles (Figure 4) indicate that these flows generally increased, as did the salinity stratification, when the freshwater flows increased. Between the ocean entrance and Sunny Point, the net nontidal flow increased sharply during high freshwater flow conditions. As noted above, because of the complex geometry and the strong ocean influence, the hydrodynamics in this reach are very complex.

For the lateral flows computed by the salt-budget model from midchannel to the shallows in various segments, no consistent (east to west or west to east) trend in the flow direction was found. The lateral flows computed were sensitive to small changes in the lateral salinity gradient. However, because the lateral flows were lower than the longitudinal flows by more than an order of magnitude, the overall net nontidal flows were not affected significantly.

Ocean Exchange Rate

Dye tracer and current studies were conducted to measure the rate of exchange of the Cape Fear estuary water with adjacent ocean water (Carpenter, unpublished). Based on five surveys with freshwater flows ranging from 13.9 to 231 m^3s^{-1}, the net ocean exchange rate was estimated to range from 991 to 1,699 m^3s^{-1} (Table 2). The exchange rates appeared to depend more on local wind and tidal conditions than on the rate of freshwater inflow. During these surveys, the

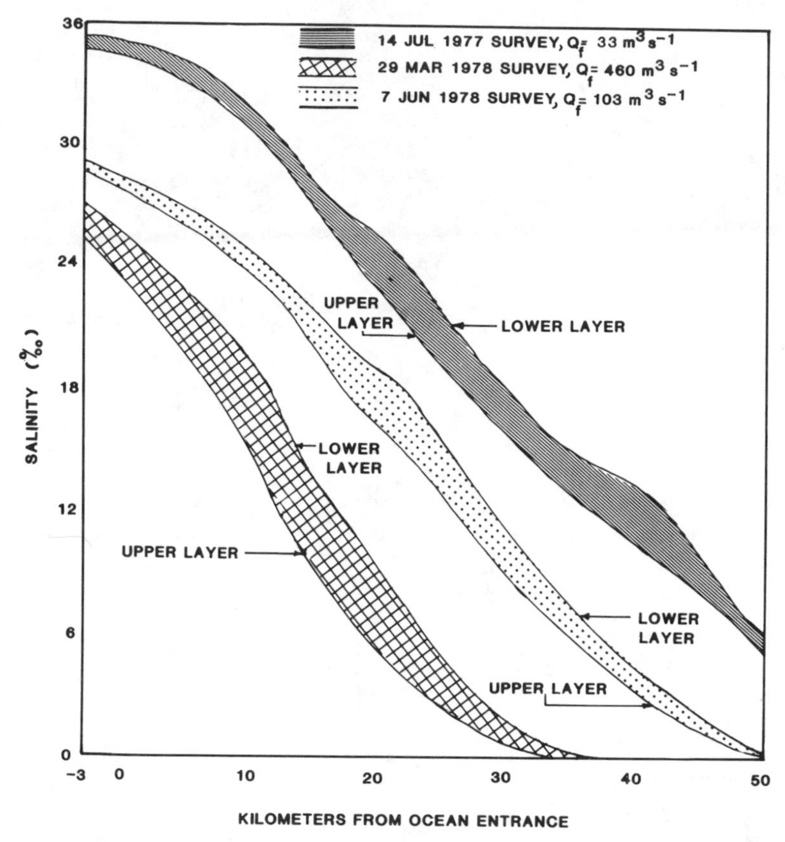

FIGURE 3.—Tidally averaged salinity profiles, Cape Fear River estuary, North Carolina. Upper and lower layers are volumes above and below the level of no net motion. Q_f = freshwater inflow.

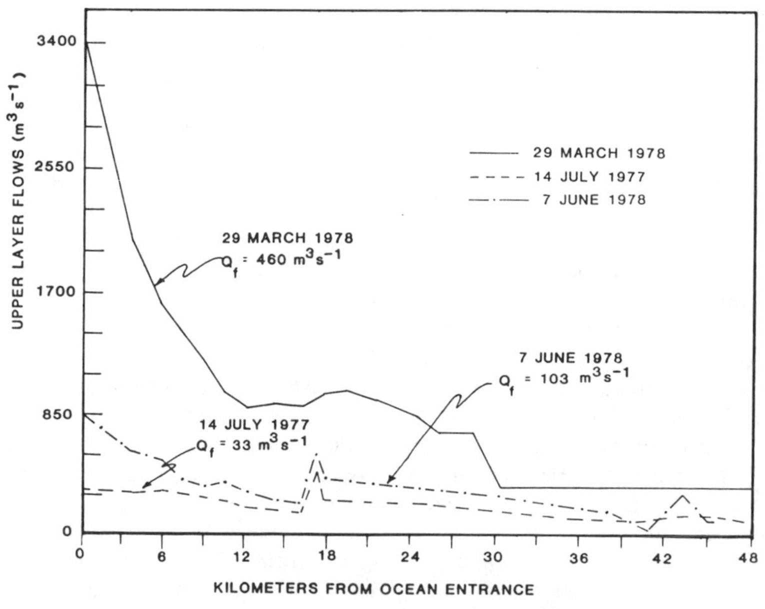

FIGURE 4.—Net nontidal upper-layer flows in the Cape Fear River estuary as computed by the salt-budget model. Upper layer is the volume above the point of no net motion. Q_f = freshwater inflow.

TABLE 2.—Ocean exchange rates in the Cape Fear estuary estimated from dye tracers and current measurements.

Month, year	Tide	Fresh inflow (m^3s^{-1})	Exchange rate (m^3s^{-1})
May 1976	Spring, 1.5 m	110	1,189
Oct 1976	Neap, 1.2 m	30	1,699
May 1977	Neap, 1.1 m	33	991
Jul 1977	Spring, 1.5 m	14	1,359
Apr 1978	Average, 1.4 m	231	1,473

reach of the Cape Fear estuary seaward of the Sunny Point area (Figure 1) did not show the classic two-layer circulation pattern of a partially stratified estuary. Instead, the exchange rate appeared to be the result of tidal flows, drift of these waters toward the west, and inflow of new ocean water from the east on the ensuing flood tide. Approximately 50% of the estuarine water that had moved out through the entrance during an ebb tide was recirculated into the estuary during the ensuing flood tide. Thus, the total outflow through the entrance may be estimated by

$$Q_{out} = \frac{Q_e + Q_f}{1 - r}; \qquad (6)$$

Q_{out} = total outflow;
Q_e = net ocean exchange rate;
Q_f = freshwater inflow;
r = recirculation ratio.

During the salinity survey on 29 March 1978, freshwater inflow (Q_f) was estimated at 460 m^3s^{-1}. For an assumed net exchange rate (Q_e) of 1,416 m^3s^{-1} and recirculation ratio of 0.5, the total computed outflow was

$$Q_{out} = \frac{1{,}416 + 460}{1 - 0.5} = 3{,}752 \ m^3s^{-1}.$$

The total outflow computed above is similar to that projected by the salt-budget model, which showed a total upper-layer outflow of about 3,512 m^{3-1}. The net exchange rate estimated by Carpenter (unpublished) was coupled with the concentrations of fish larvae in the eastern offshore areas to compute the ocean recruitment rate.

Fish Population Model

The defining equations for the fish population model (FPM) are similar to those used to simulate the transport and decay of a nonconservative substance in physics and engineering. To simulate the changing behavioral characteristics of fish

TABLE 3.—Life-stage designations used in the Cape Fear River fish population model.

Life stage	Length range (mm)	Characteristics
1	6–12	Mostly planktonic, but exhibit diurnal migration
2	13–20	
3	21–30	Swimming ability sufficient to migrate to preferred habitat areas
4	31–60	Juveniles actively seek preferred habitat
5	61–90	
6	91–140	

organisms as they mature, individual equations were written for six life stages and defined in terms of the length of the organisms (Table 3). These designations were applied to both spot and Atlantic croaker. In addition to accounting for the natural mortality of the organisms, the model equations must include growth of the organisms from one life stage to the next. Separate equations were also written for each habitat considered in the model: the main channel, tributary creeks, and the high-marsh areas.

For life stages 1 and 2, the defining equation for the concentration of organisms in the lower layer of the main stem of the estuary, C_E, was

$$\frac{\partial C_E}{\partial t} = \pm \frac{Q}{A}\frac{\partial C_E}{\partial x} \pm K_g C_E - K_D C_E - \frac{Q_v}{V_L} C_E; \qquad (7)$$

$\frac{\partial C_E}{\partial t}$ = advection;
$\frac{Q}{A}\frac{\partial C_E}{\partial x}$ = accumulation;
$K_g C_E$ = growth;
$K_D C_E$ = natural mortality;
$\frac{Q_v}{V_L} C_E$ = transport of organisms between upper and lower layers;
Q = tidally averaged flow;
K_g = maturation rate;
K_D = natural mortality rate;
Q_v = vertical flow;
x = distance in the longitudinal dimension;
V_L = volume of lower layer;
t = time;
A = cross-sectional area.

The equation for the upper layer of the main stem, which includes the upper water column of the ship channel as well as shoal areas, is analogous, but includes an additional term, $-f_T Q C_E$, that accounts for transport of the organism from the main stem into the tributary creeks. The equations for the tributary creeks are simpler than those for the main stem, because they do not include spatial variation and can be written as ordinary differential equations having the form

$$\frac{dC_T}{dt} = f_T \frac{Q_f}{V_T} C_E \pm K_g C_T \pm K_D C_T - f_m \frac{Q_f}{V_T} C_T; \quad (8)$$

$\frac{dC_T}{dt}$ = accumulation;

$f_T \frac{Q_f}{V}$ = transport of larvae from upper layer to tributary creeks;

$K_g C_T$ = growth;

$K_D C_T$ = natural mortality;

$f_m \frac{Q_f}{V} C_T$ = transport of larvae from the tributary creek to the high marsh;

V_T = volume of tributary creeks;

Q_f = flood flow into tributary creeks;

f_T = transfer coefficient of larvae from the upper layer into tributary creeks;

f_m = transfer coefficient of larvae from tributary creeks into the high marsh.

The equation for the high marsh is nearly identical except that there is no transport out of the marsh.

The simulation of a given year class of larvae begins immediately before the appearance of the first larvae in the estuary, so the initial conditions for equations (7) and (8) are all $C(0)$ equal to 0. The boundary condition for equation (7) is discussed below, along with the description of the solution scheme.

The equations for life stages 3 and above were identical to those described for life stages 1 and 2, except that the advective transport terms were not used. Instead, the longitudinal distribution of organisms in the model was adjusted to agree with the distributions computed from the field data. This approach was based on the assumption that, by the time the organisms reached 20 mm in length, they were no longer susceptible to transport by the currents in the estuary but were able to control their position in the estuary by behavioral means.

Solution Scheme

Equations (7) and (8) were solved by a solution scheme tailored to the particular system being modeled. Application of the numerical accuracy criterion $\Delta t < \Delta x/u$, u being the tidally averaged velocity and Δx the preliminary spatial segmentation, resulted in a time step of about 3 h. With Δt set at 3 h, the spatial segmentation was adjusted so that Δx was greater than $u\Delta t$. This resulted in seven segments ranging from 5.3 to 8.8 km long (Figure 5). The coastal sector was used to define the source concentrations of larvae entering the estuary. Each of the segments was divided into two layers in the vertical direction (Figure 6). Where appropriate, tidal creeks and marsh areas were added to the upper layer. Material balance was then performed on the number of organisms in each of the control volumes by solving a series of coupled differential equations by the finite-difference method.

In the salinity-intruded reach of the estuary, the tidally averaged flows moved upstream in the lower layer and downstream in the upper layer. Thus, specification of the concentration of ocean-spawned larvae at the mouth of the estuary permitted the solution of equation (7) sequentially, beginning at the mouth of the estuary and proceeding upstream. Net flows in the upper layer, both in the salinity-intruded reach and above it, were always in the downstream direction. The equation for the upper layer could thus be solved sequentially, beginning at the most upstream segment and proceeding toward the mouth following the solution of the lower-layer equation. Equations for the tributary creeks and marshes were solved simultaneously with the upper-layer equations for each given segment. After each pass through this solution scheme, the organisms were redistributed between the upper and lower layers according to their diurnal migration patterns as shown by the field data.

Evaluation of Model Input Parameters

The input parameters required by the FPM can be divided into mass transport and biological parameters. The former include net nontidal flows evaluated by the salt-budget model and the ocean exchange rates determined from Carpenter's (unpublished) dye and current studies. Although mass transport parameters also included power plant effects in our original assessment, they are not reproduced here. Major biological parameters include ocean recruitment rates, temporal and

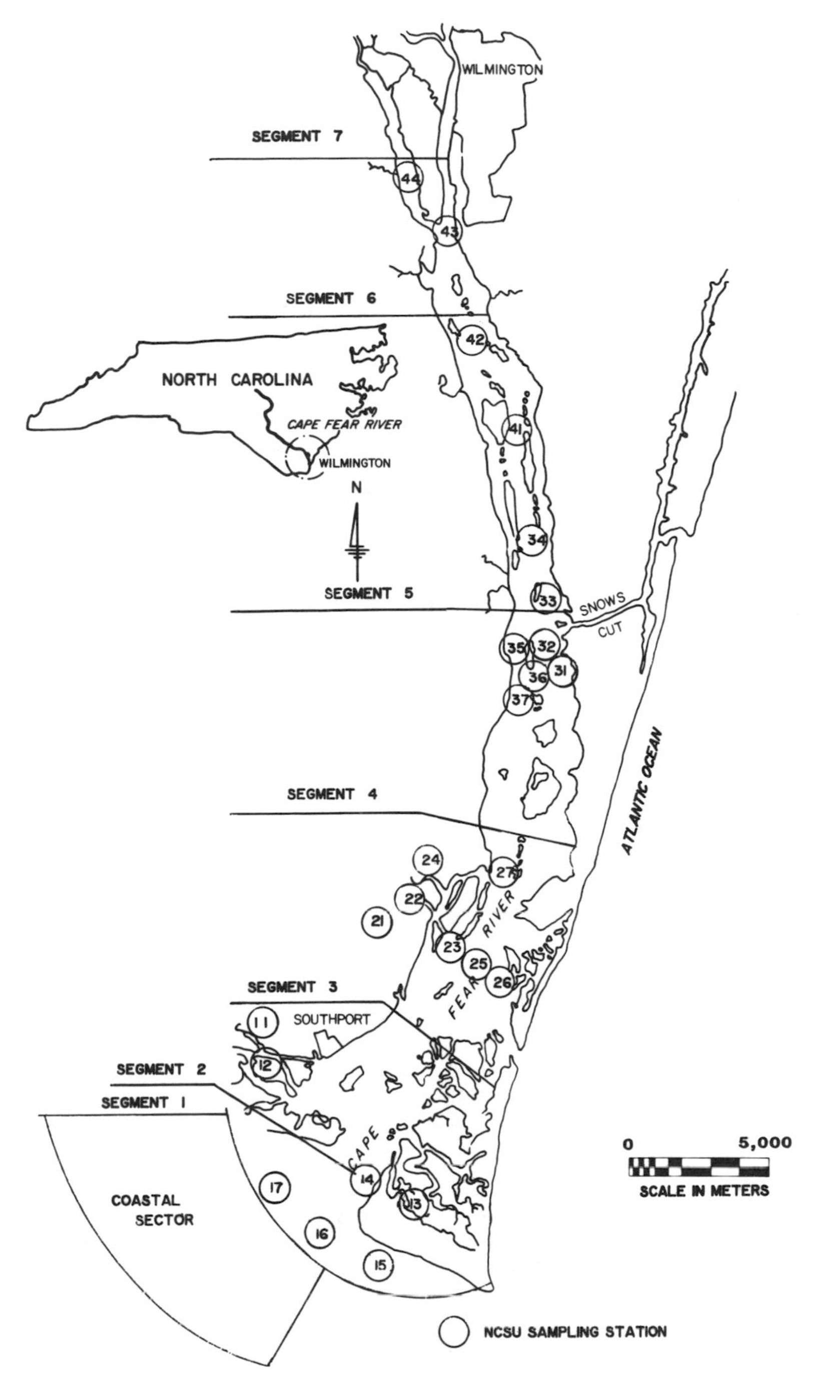

FIGURE 5.—River segments established for the fish population model were based on a 3-h time step and ranged in length from 5.3 to 8.8 km. The coastal sector extended approximately 4.8 km outside the estuary mouth. NCSU = North Carolina State University.

spatial distributions, vertical and longitudinal migration patterns, growth rates, and natural mortality rates. As discussed later, most of these input parameters were derived from the available field data. Juvenile spot and Atlantic croakers were collected in the Cape Fear estuary by Cope-

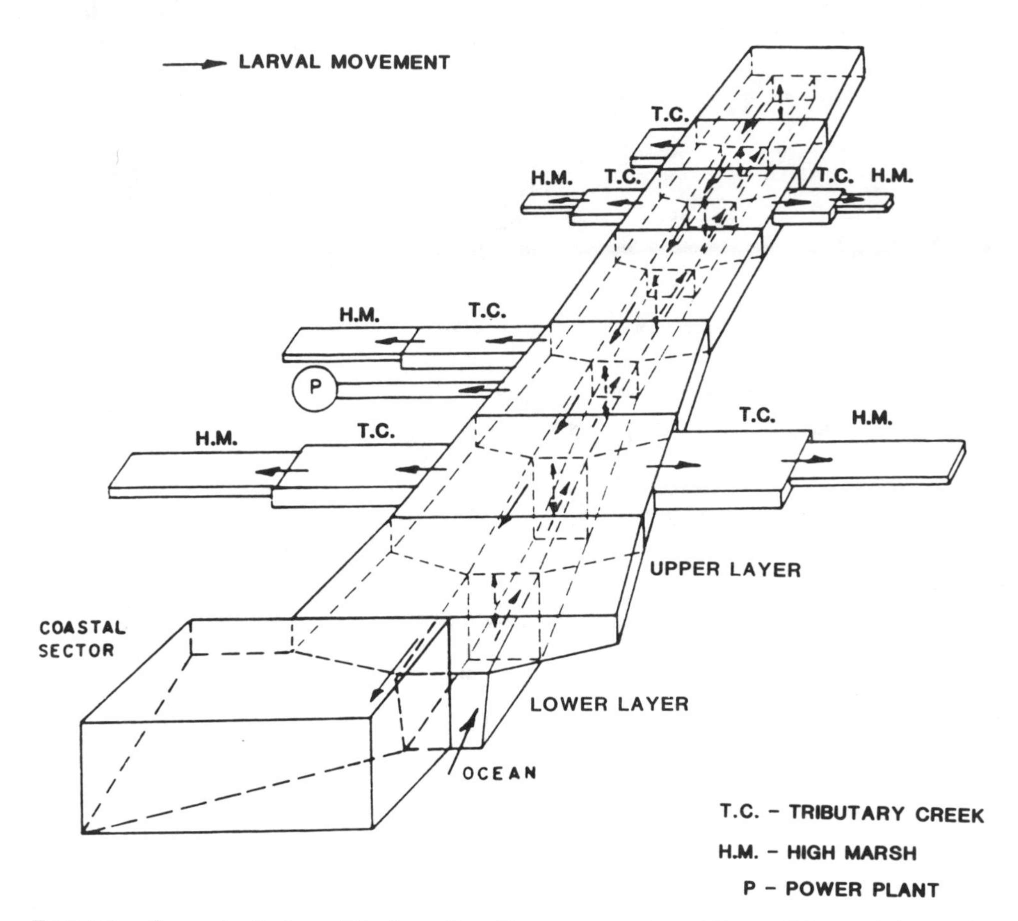

FIGURE 6.—Conceptualization of the Cape Fear River estuary adopted for the fish population model.

land and Hodson (1980) and by Weinstein et al. (1980a). The former collections probably represent one of the most comprehensive data sets for depicting spatial and temporal trends in the distributional ecology and behavior of these taxa in the water column. Paired surface and bottom tows were collected semimonthly at 44 stations distributed throughout the estuary and nearshore ocean during the entire period of recruitment in 1976–1977 and 1977–1978. The distributions of spot and Atlantic croakers were depicted along the long axis of the estuary, along major tributaries, and vertically in the water column (at surface and bottom during day and night) from 21,120 samples (Tables 4, 5). These data were supplemented by the studies of Weinstein and Davis (1980), Weinstein et al. (1980a), and Weinstein and Walters (1981), in which additional information was collected on response to tides, gear efficiency, and mortality rates. Figure 7, taken from Weinstein et al. (1980a), summarizes behavioral responses for spot and Atlantic croaker

TABLE 4.—Ratios of spot densities in surface and bottom samples from the Cape Fear River main stem, day and night.

Date	Day		Night	
	Surface (%)	Bottom (%)	Surface (%)	Bottom (%)
18 Jan 1977	9.6	90.4	53.9	46.1
1 Feb	18.6	81.4	49.1	50.9
15 Feb	19.1	80.9	47.1	52.9
28 Feb	15.9	84.1	43.8	56.2
15 Mar	4.8	95.2	55.6	44.4
29 Mar	1.0	99.0	48.3	51.7
13 Apr	3.2	96.8	46.5	53.5
26 Apr	10.0	90.0	50.0	50.0
7 Dec	7.4	92.6	73.7	26.3
4 Jan 1978	4.0	96.0	67.3	32.7
17 Jan	20.7	79.3	62.2	37.8
31 Jan	10.9	89.1	33.7	66.3
1 Feb	3.3	96.7	49.2	50.8
14 Feb	13.3	86.7	48.2	51.8
28 Feb	26.0	74.0	36.9	63.1
1 Mar	14.5	85.5	62.1	37.9
15 Mar	17.7	82.3	4.1	95.9
29 Mar	12.4	87.6	57.7	42.3
11 Apr	1.8	98.2	50.9	49.1

TABLE 5.—Ratios of croaker densities in surface and bottom samples from the Cape Fear River main stem, day and night.

Date	Day		Night	
	Surface (%)	Bottom (%)	Surface (%)	Bottom (%)
27 Oct 1976	0.0	100.0	67.4	32.6
2 Nov	0.1	99.9	30.2	69.8
9 Nov	0.0	100.0	44.5	55.5
23 Nov	1.1	98.9	28.4	71.6
7 Dec	2.1	97.9	12.4	87.6
21 Dec	15.0	85.0	28.7	71.3
4 Jan 1977	0.7	99.3	7.6	92.4
18 Jan	18.4	81.6	44.0	56.0
1 Feb	16.5	83.5	43.9	56.1
15 Feb	7.3	92.7	28.2	71.8
1 Mar	1.4	98.6	12.9	87.1
15 Mar	0.6	99.4	25.0	75.0
29 Mar	0.1	99.9	21.3	78.7
13 Apr	0.5	99.5	24.7	75.3
26 Apr	0.3	99.7	3.8	96.2
10 May	0.0	100.0	21.9	78.1
13 Sep	0.7	99.3	6.5	93.5
11 Oct	1.0	99.0	30.2	69.8
8 Nov	0.3	99.7	38.0	62.0
7 Dec	1.9	98.1	37.1	62.9
4 Jan 1978	5.1	94.9	44.0	56.0
17 Jan	6.2	93.8	19.4	80.6
31 Jan	1.4	98.6	16.5	83.5
1 Feb	7.5	92.5	17.6	82.4
14 Feb	4.5	95.5	20.4	79.6
28 Feb	5.9	94.1	17.0	83.0
1 Mar	14.9	85.1	51.8	48.2
15 Mar	5.8	94.2	1.4	98.6
29 Mar	0.3	99.7	27.2	72.8

derived from these studies. When information was not available from our studies to assess parameters, literature values were used. If necessary, these values were adjusted during the model calibration procedures, but only after they had been judged biologically realistic.

Evaluation of Input Parameters for Spot

When they arrived at the estuarine plume, spot were from 8 to 20 mm long. Their swimming ability was assumed to be limited, and they were largely perceived to be drifting with the ocean currents. Thus, the ocean recruitment rate of spot into the Cape Fear River could be computed from densities at the estuary mouth (the ocean source concentration) and the associated ocean exchange rate. Dye and current studies showed the ocean exchange at the mouth to consist of a net inflow from the east and a net outflow to the west, so spot densities in eastern offshore areas were assumed to be the source of immigrants.

In addition to the ocean recruitment, some spot were assumed to be flushed from the estuary into the coastal sector by the tidal flow at the mouth. Some of these spot were further assumed to be retained within the coastal sector and returned to the estuary by the recirculation flow or by behavioral mechanisms. No loss of juveniles was assumed to occur from the coastal sector to the ocean.

Once spot were transported into the estuary, they appeared to migrate diurnally in the water column, concentrating near the bottom during the day, and becoming almost evenly distributed in the water column at night (Copeland and Hodson 1980; Weinstein et al. 1980a). Density splits (Table 4) between surface and bottom compartments for both day and night periods were calculated from field data collected during 1976–1978 (Copeland and Hodson 1980).

The daytime density ratio for spot in 1977 was divided into two periods. Before 15 March, an average 15% of the juveniles were in the daytime surface compartment; after 15 March, 5%. In 1978, the daytime surface:bottom ratio averaged 12:88% for the whole season. For both years, a nighttime ratio of 50:50 was adopted. These ratios were based on the assumptions that densities calculated from surface samples represented densities throughout the upper layer and that those from bottom samples represented densities throughout the lower layer. Based upon the results of a larval retention study (Weinstein et al. 1980a), spot concentrated in the lower layer during the day; concentrations from middepth to the bottom were about the same. However, near the bottom in the shallows, which are also part of the upper layer, spot concentrations were as high as at the channel bottoms. Therefore, the mean upper-layer concentrations were adjusted to reflect the higher spot concentrations in the shallow bottoms. We assumed that 40% of the upper-layer volume was relatively shallow (and therefore showed the same densities as the lower layer), and recomputed the daytime surface:bottom ratios of 15:85 and 5:95 as 34:66 and 30:70, respectively, for the periods before and after 15 March 1977.

A special Tucker trawl study was conducted in March 1979 (Copeland and Hodson 1980) to determine the density profile for juvenile spot from surface to bottom. Based on the data collected, the density ratios between the upper and lower layers were 35:65 during the day and 56:44 at night, indicating that the 40% adjustment used above was a reasonable estimate. Thus, the daytime density ratios of 34:66 and 30:70 were used

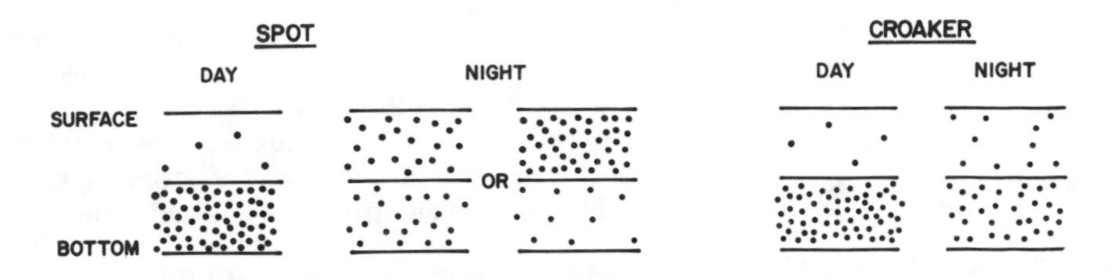

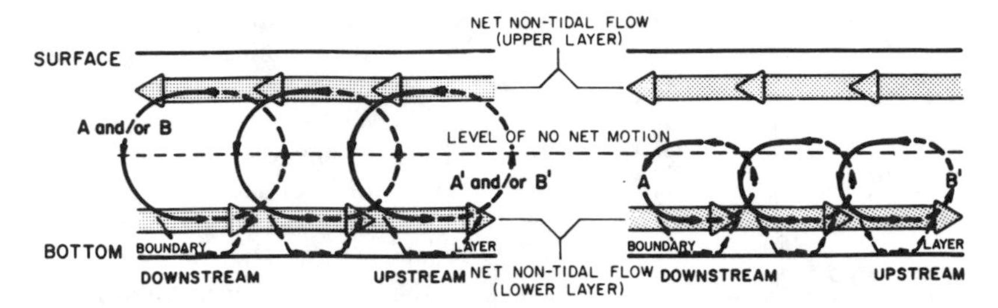

FIGURE 7.—Conceptualized behavioral responses of juvenile spot and Atlantic croakers to photoperiod and tide. (From Weinstein et al. 1980a)

for the model runs in 1977 and 1978; the nighttime ratios for both years were assumed to be 50:50. After each time step, the model adjusted the vertical distributions so that concentrations in the upper and lower layers conformed with the measured density splits.

In addition to the vertical migration parameter described above, other life cycle parameters required by the FPM included natural mortality rates and growth rates. These were evaluated from the biological data collected at four marsh systems (Weinstein and Walters 1981) during 1977 and 1978 (Table 6).

The sizes of spot used in the computation ranged from 20 to 51 mm (life stages 3 and 4) in 1977 and from 12 to 51 mm (life stages 2, 3, and 4) in 1978. Because the computed mortality rate represents the average decay rate within the size range, it was decided to apply mortality rates of 0.05, 0.03, and $0.02 \cdot d^{-1}$ to life stages 2, 3, and 4. The mortality rate of life stage 1 could not be computed directly, so the estimated value of $0.08 \cdot d^{-1}$ was obtained through model calibration. For purposes of this demonstration an added mortality component associated with power plant operation was not considered.

Length-frequency data collected in various marsh systems were also used to estimate the growth rates (Weinstein and Walters 1981). During the recruitment season, actual growth rates were masked by the large number of new recruits. After recruitment was over, however, growth rates could be estimated with greater accuracy. For the period April through June, the growth rates of spot were estimated to be 0.31 and 0.28 $mm \cdot d^{-1}$ for 1977 and 1978, respectively. These growth rates were combined with the length span of each life stage to compute the transfer rate from one life stage to the next.

Different sampling methods were used to measure larval densities in the main stem and tributary creek (Copeland and Hodson 1980) and in the high marsh (Weinstein 1979; Weinstein et al. 1980a). To compare the measured densities on an

TABLE 6.—Mortality estimates for spot populations in four marsh systems, Cape Fear estuary.

	Natural mortality rate (per day)	
Marsh system	1977[a]	1978[b]
Baldhead	0.061	0.052
Dutchman	0.038	0.037
Walden	0.030	0.030
Upriver	0.035	0.051

[a]Size range, 20–51 mm.
[b]Size range, 12–51 mm.

equal basis, the data were corrected to account for possible differences in gear efficiencies among the sampling methods.

Four special studies were conducted to evaluate the gear efficiency of the seine and rotenone sampling for spot in the high marsh (Weinstein and Davis 1980). These studies showed the average capture efficiency was 61% for the seine and 58% for rotenone. Accordingly, an average efficiency of 60% was adopted for the high marsh data. The gear efficiency for the river sampling was evaluated by comparing densities at a power plant discharge (considered 100% efficient) with the densities at the same plant's intake canal, located on the estuary near Southport. Both were measured with the same gear used in the river sampling. Based on these comparisons, a 45% gear efficiency was estimated for life stages 1 and 2. The data further indicated that the efficiency dropped off sharply for spot juveniles above 20 mm and reached zero at 25 mm (Copeland and Hodson 1980). Therefore, for life stage 3 (21–30 mm), gear efficiency values of 15 and 30% were adopted in two separate runs.

Model Calibration for Spot

The calibration procedure began with a set of input parameters derived from available field data. Because field data were not available for some parameters, reasonable values were assumed in making initial model runs to predict the temporal and spatial distributions of spot larvae. If the model output was not in satisfactory agreement with the field observations, the values of the assumed input parameters were adjusted and the model was rerun.

The input parameters evaluated through the calibration procedure included (1) lateral transfer coefficients associated with flood flows to the tributary creeks and marshes, and (2) larval transfer rates between life stages. The lateral transfer coefficients, f_T and f_m (multiplied by the flood flows, Q_f, and upper-layer concentration to compute the rate of transfer of larvae to the tidal creeks and marshes), were adjusted individually for various tidal creeks and marshes until the field-measured buildup of larval concentrations in these areas was reproduced by the model. The larval transfer coefficients, originally estimated from the growth rates and length span of each life stage, were adjusted so that the life stage distributions computed by the model agreed reasonably well with the field data.

An important consideration in the calibration process was the definition of "satisfactory agreement" between the field data and the model output. In the present study, two standard errors on either side of the mean were used to define the statistical error in field measurement. This resulted in a range of values approximating the 95% confidence interval about each data point. The goal of the calibration procedure was to adjust model input parameters until the model results fell within or close to the 95% interval. This was tantamount to having no statistically significant difference between the model result and the field measurements.

Calibration of the FPM with the 1976 and 1977 spot data collected by Copeland and Hodson (1980) and by Weinstein (1979) and Weinstein et al. (1980b) is shown in Figures 8–11 for the whole main channel, channel segments, tidal creeks, and marshes. Model outputs were almost always within the 95% confidence intervals of observed data. Thus, we concluded that the FPM reasonably simulated the spatial and temporal distribution of spot in the Cape Fear estuary from the time they were recruited until they became young of the year and emigrated.

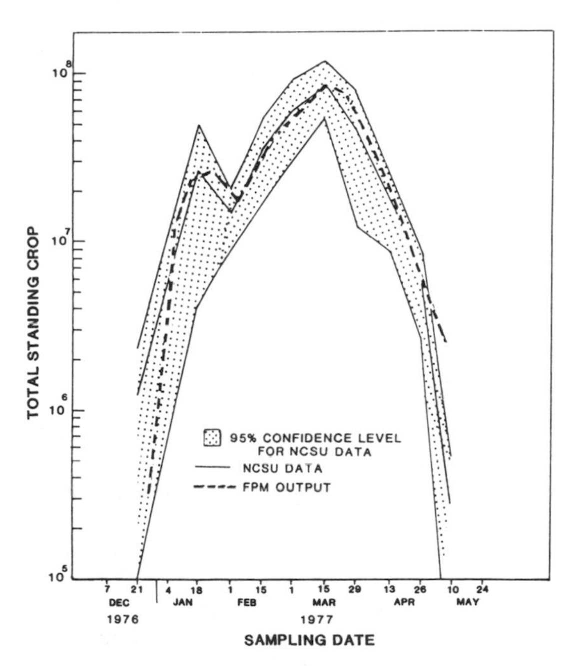

FIGURE 8.—Total standing crop and 95% confidence intervals for spot in the main channel of the Cape Fear River estuary. NCSU = North Carolina State University; FPM = fish population model.

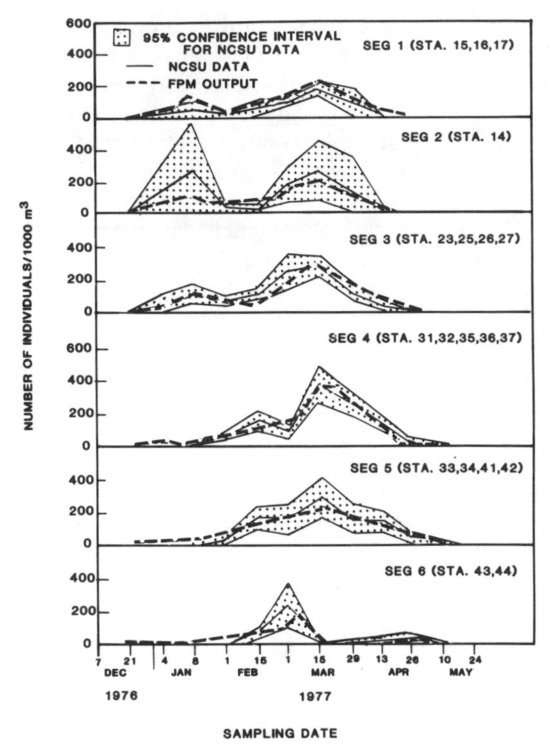

FIGURE 9.—Concentrations of spot and 95% confidence intervals in the main Cape Fear River channel of each model segment (SEG). NCSU = North Carolina State University; FPM = fish population model. Segments and stations (STA) are shown in Figure 5.

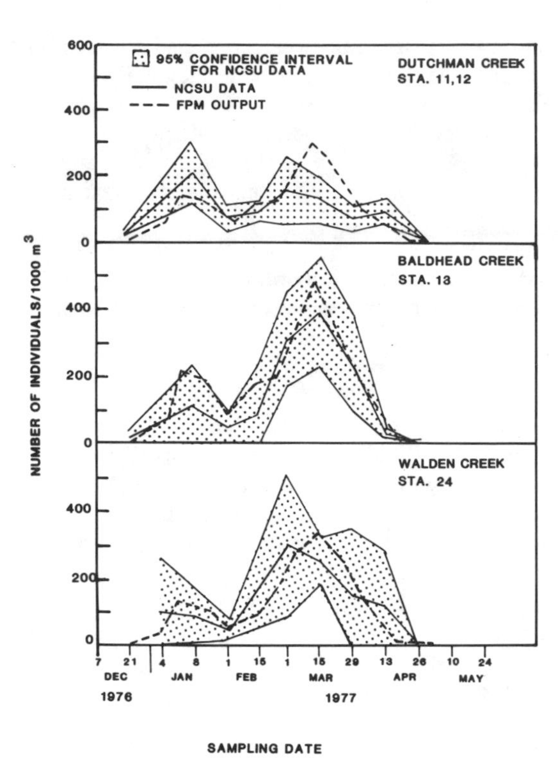

FIGURE 10.—Concentrations of spot and 95% confidence intervals in tidal creeks of selected Cape Fear River segments. Dutchman and Baldhead creeks are in river segment 2 (Figure 5); Walden creek is in river segment 3 adjacent to a power plant intake canal. NCSU = North Carolina State University; FPM = fish population model.

Evaluation of Input Parameters for Atlantic Croaker

As for spot, the rate of Atlantic croaker recruitment from ocean to river was determined from oceanic larval densities in the vicinity of the estuarine plume and the associated ocean exchange rates. For Atlantic croaker model runs, it was also assumed that not all larvae and juveniles flushed into the coastal sector were lost to the ocean. Some Atlantic croakers were returned to the estuary by recirculation flows or behavioral mechanisms.

Once inside the estuary, Atlantic croakers were concentrated near the bottom at all times. The vertical density splits, computed from the data collected in 1976–1978 (Table 5), were the same for all years, averaging 4:96 during the day and 25:75 at night for surface:bottom concentrations.

The March 1979 special study used a Tucker trawl to provide additional data regarding the vertical distribution of Atlantic croaker (Figure 12). The division of density profiles (Figure 12) into upper and lower layers gave percentage ratios of 22:78 during the day and 26:74 at night. These ratios were input into the FPM so that, after every time step, Atlantic croaker concentrations in the upper and lower layers could be adjusted to conform with the field observations.

Other life cycle parameters, such as natural mortality rates and larval growth rates, were also required by the FPM. Because Atlantic croaker larvae were not found in any substantial numbers at the marsh sampling stations, their natural mortality rates could not be directly estimated. Although the upriver main stem is the primary nursery for Atlantic croakers, it is regarded as a relatively "open" system, with some migration occurring most of the time, thus making it difficult to estimate the natural mortality rate from abundance data from this area. For purposes of the model runs, we assumed that the natural mortality rates of Atlantic croaker and spot were similar. Accordingly, natural mortality rates of 0.08, 0.06,

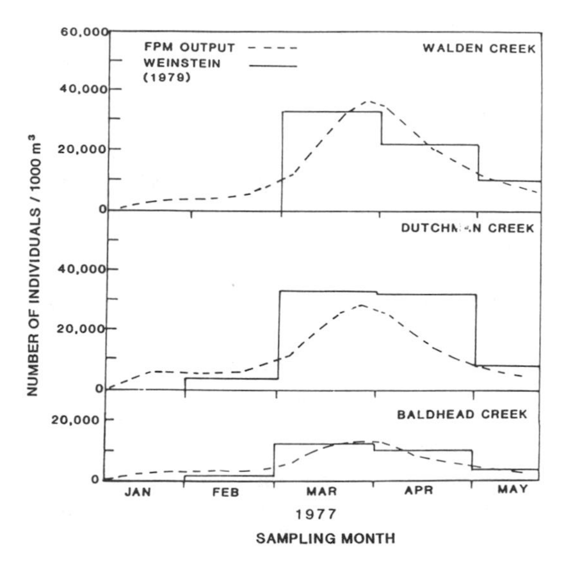

FIGURE 11.—Concentrations of spot in high-marsh tributaries of three tidal creeks. Dutchman and Baldhead creeks are in Cape Fear River segment 2 (Figure 5); Walden Creek is in river segment 3 adjacent to a power plant intake canal. FPM = fish population model.

and $0.04 \cdot d^{-1}$ were used for the 1977 calibration run for Atlantic croaker life stages 1, 2, and 3, respectively. The calibrated natural mortality rates were 0.08, 0.05, and $0.03 \cdot d^{-1}$ for life stages 1, 2, and 3 in 1978.

The growth rate of Atlantic croakers was estimated from the change in mean length of young of the year impinged at the power plant in various months. From March through September 1977, the average growth rate was computed as 0.32 $mm \cdot d^{-1}$. In the winter months, most Atlantic croakers were still in life stages 1 and 2 and their growth tended to be retarded because of cold temperatures. Therefore, their transfer to later life stages coincided with the actual appearance of those life stages at the power plant screens. Transfer rates between various life stages were then computed by dividing the length span of each life stage into the growth rate. These transfer rates were later adjusted during model calibration to yield life stage distributions consistent with field observations.

Atlantic croakers did not accumulate in the marsh areas in substantial numbers during 1976–1978. This is consistent with the observation that they were concentrated in the lower water column, because the principal pathway to the marsh areas was via the transport of water in the upper layer to the tributary creeks on flood tide. Thus, lateral transport coefficients were not required for the Atlantic croaker runs.

Model Calibration for Atlantic Croaker

The FPM was calibrated with the 1976–1977 Atlantic croaker data collected by Copeland and Hodson (1980; Figures 13, 14). Although the model output was outside the 95% confidence interval of field observations in a few instances, the agreement was considered satisfactory overall.

Conclusions

This paper has described an approach to the modeling and simulation of the recruitment of spot and Atlantic croaker in the Cape Fear estu-

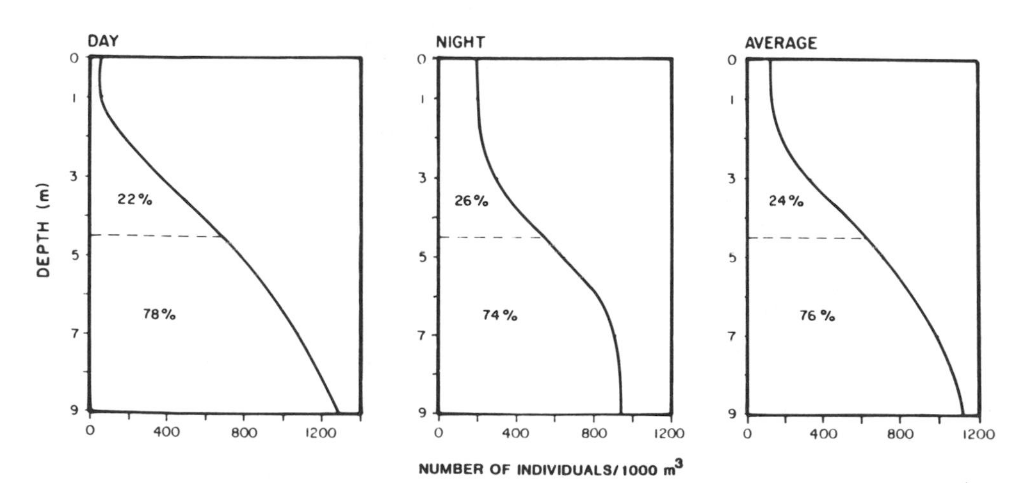

FIGURE 12.—Vertical density profiles for Atlantic croakers captured by Tucker trawl in the main channel of the Cape Fear River (March 1979). The broken horizontal lines represent the depth of no net water motion.

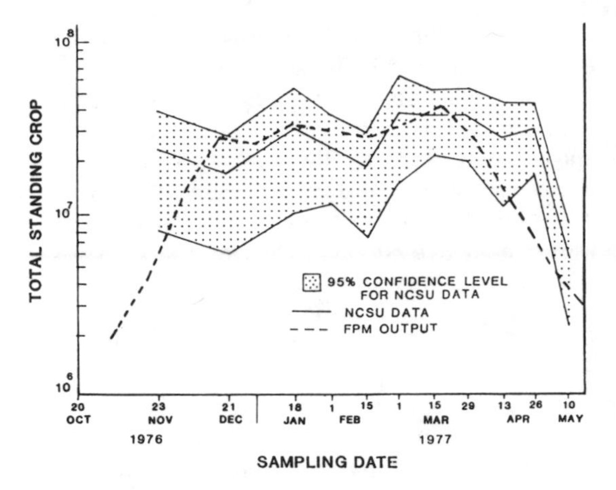

FIGURE 13.—Total standing crops and 95% confidence intervals for Atlantic croaker postlarvae in the main channel of the Cape Fear River estuary. NCSU = North Carolina State University; FPM = fish population model.

ary. Larvae and juveniles of both species were perceived to be transported into the estuary by the ocean exchange rate at the estuary mouth. Once inside the estuary, spot juveniles were transported by the lower-layer flow to the upriver nursery areas because they concentrated near the bottom during the day. At night, spot juveniles were evenly distributed in the water column and therefore could reach the tributary creek and marsh nursery areas via the flood tide into the creeks. Most Atlantic croaker juveniles, on the other hand, concentrated near the bottom at all times and were therefore found primarily in the upriver nursery areas only. Thus, physical mechanisms—represented by the ocean exchange rate, tidal flows, and net nontidal flows—and behavioral mechanisms—represented by diurnal vertical migration patterns—played major roles in the recruitment and distribution of these two species in the Cape Fear estuary.

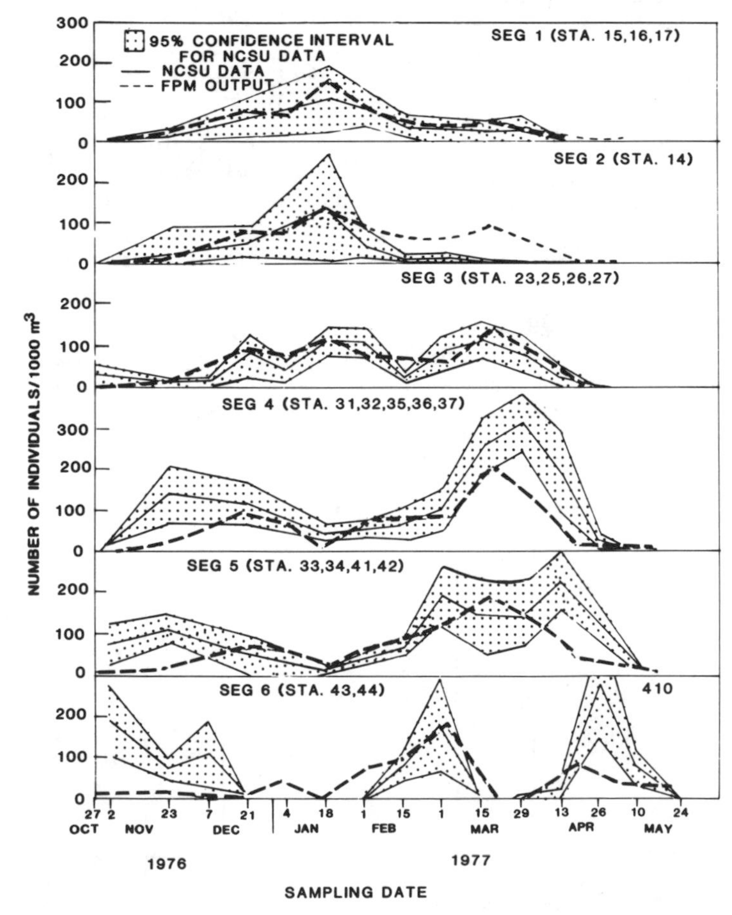

FIGURE 14.—Concentrations of Atlantic croakers and 95% confidence intervals in the main Cape Fear River channel of each model segment (SEG). NCSU = North Carolina State University; FPM = fish population model. Station locations (STA) are shown in Figure 5.

Construction activities at and around inlets would have the potential to interfere with recruitment rates, recirculation, behavioral cues, and other components of estuarine recruitment. The models described here have the potential to predict the outcome of inlet modification.

Acknowledgments

The development of the models and the collection of field data used in this study were funded by the Carolina Power & Light Company, Raleigh, North Carolina. We thank the reviewers for their comments, which improved the manuscript. M. Pinkham edited the final draft.

References

Abood, K. A. 1977. Evaluation of circulation in partially stratified estuaries as typified by the Hudson River. Doctoral dissertation. Rutgers University, New Brunswick, New Jersey.

Boehlert, G. W., and B. C. Mundy. 1988. Roles of behavioral and physical factors in larval and juvenile fish recruitment to estuarine nursery areas. American Fisheries Society Symposium 3:51–67.

Carolina Power & Light Company. 1980. Water chemistry report, volume 3. Brunswick Steam Electric Plant, Cape Fear studies. Final Report to the Carolina Power & Light Company, Raleigh, North Carolina.

Copeland, B. J., and R. G. Hodson. 1980. Larvae and post-larvae in the Cape Fear River estuary, NC, during operation, volume 7. Brunswick Steam Electric Plant, Cape Fear studies. Final Report to the Carolina Power & Light Company, Raleigh, North Carolina.

Miller, J. M. 1988. Physical processes and mechanisms of coastal migrations of immature marine fishes. American Fisheries Society Symposium 3:68–76.

Norcross, B. L., and R. F. Shaw. 1984. Oceanic and estuarine transport of fish eggs and larvae: a review. Transaction of the American Fisheries Society 113:153–165.

Sayre, W. W. 1968. Dispersion of mass in open-channel flow. Hydraulic Paper 3. Colorado State University, Fort Collins.

Weinstein, M. P. 1979. Shallow marsh habitats as primary nurseries for fishes and shellfish, Cape Fear River, North Carolina. U.S. National Marine Fisheries Service Fishery Bulletin 77:339–357.

Weinstein, M. P., and R. W. Davis. 1980. A comparison of the collection efficiency of seine and rotenone sampling. Estuaries 3:98–105.

Weinstein, M. P., R. G. Hodson, and L. R. Gerry. 1980a. Retention of three taxa of postlarval fishes in an intensively flushed tidal estuary, Cape Fear River, North Carolina. U.S. National Marine Fisheries Service Fishery Bulletin 78:419–436.

Weinstein, M. P., and M. F. Walters. 1981. Growth, survival and production in young-of-the-year populations of *Leiostomus xanthurus* Lacepede, residing in tidal creeks. Estuaries 4:185–197.

Weinstein, M. P., S. L. Weiss, and M. F. Walters. 1980b. Multiple determinants of community structure in shallow marsh habitats, Cape Fear River estuary, North Carolina. Marine Biology (Berlin) 58:227–243.

American Fisheries Society Symposium 3:132–148, 1988

Distribution of Fish Eggs and Larvae and Patterns of Water Circulation in Narragansett Bay, 1972–1973

DONALD W. BOURNE[1] AND JOHN J. GOVONI[2,3]

Marine Research, Incorporated, 141 Falmouth Heights Road
Falmouth, Massachusetts 02540, USA

Abstract.—About 42 species of ichthyoplankton, belonging to 28 families, were identified in some 6,900 plankton samples taken throughout Narragansett Bay, Rhode Island, between June 1972 and August 1973. One hundred sixty stations were established in 10 sampling sectors. Usually there were statistically significant population differences among these sectors, though confidence limits varied considerably among sectors and taxa. Species abundances varied from north to south in the bay, and some of the heaviest spawning occurred in the northern, most polluted part. Six or seven species accounted for 95% of all individuals. Between-year variation in numbers of a single species sometimes exceeded 100%. Limited sampling at discrete depths indicated that about 75% of the ichthyoplankton were in the upper half of the water column. Gravitationally forced water movement is apparently an important determinant of ichthyoplankton distribution and retention in Narragansett Bay, a partially mixed estuary.

Estuaries are beneficial, sometimes essential, for the reproduction of many fishes. Spawning takes place in them or the early life periods are spent there. The main advantage may be an exceptionally rich food supply. Young fish can grow rapidly to a size at which they can resist the dispersive powers of coastal water circulation. This provokes interest in the adaptations that have evolved among the very youngest life stages of coastal, estuarine, and anadromous fishes. How do they prolong access to this inshore abundance of food despite the net seaward transport of an estuary's water? Some fish produce adherent, demersal eggs and larvae that stay close to the bottom (Blaxter 1969). Others migrate vertically over the tidal cycle to reduce the time spent in ebbing water (Graham 1972). Counter-current circulation has been considered a mechanism that might retain or carry drifting, planktonic eggs or larvae into estuaries (see Weinstein et al. 1980).

Narragansett Bay (Rhode Island), New England's largest estuary, furnishes spawning and nursery grounds for many marine and estuarine fishes. The bay's principal freshwater sources are the Providence River in the northwest and the Taunton River in the northeast; lesser contributions come from the Warren–Barrington and Lees rivers. To the south, Narragansett Bay discharges into Rhode Island Sound through three channels: West Passage, East Passage and the Sakonnet River (Figure 1). The bay's area at mean low water is 325 km^2 and its mean depth 8.8 m. The maximum depth, in lower East Passage, is 57.3 m below mean low water (Hicks 1959). Mean depth is 17.7 m in East Passage and about 7.6 m in the other two channels.

The bay and its main tributaries are flanked by industry, which is concentrated at Providence, Rhode Island, and Fall River, Massachusetts, but which also occurs in small concentrations at other places along the shore. Six electricity-generating plants draw their cooling water from the Taunton and Providence rivers and discharge heated effluent at or near their connection with the bay. By far the largest of these is the New England Power Company plant located on Brayton Point, at the west side of the Taunton River's mouth.

The basis of this account is an intensive survey of the Narragansett Bay ichthyoplankton that began in June 1972 and ended in August 1973, and that resulted in about 6,900 plankton samples. The New England Power Company was the sponsor, in compliance with various state and federal licensing and environmental requirements that largely determined the survey's design. The survey consisted mainly of oblique plankton tows, though there was limited sampling of discrete depths at two stations in connection with power plant entrainment studies. Except for a survey of Cape Cod Bay between 1974 and 1976 (Bourne 1978; Scherer 1984), we know of no other ich-

[1]Present address: Post Office Box 282, Waquoit, Massachusetts 02536, USA.

[2]Present address: National Marine Fisheries Service, Southeast Fisheries Center, Beaufort Laboratory, Beaufort, North Carolina 28516, USA.

[3]Correspondence and reprint requests.

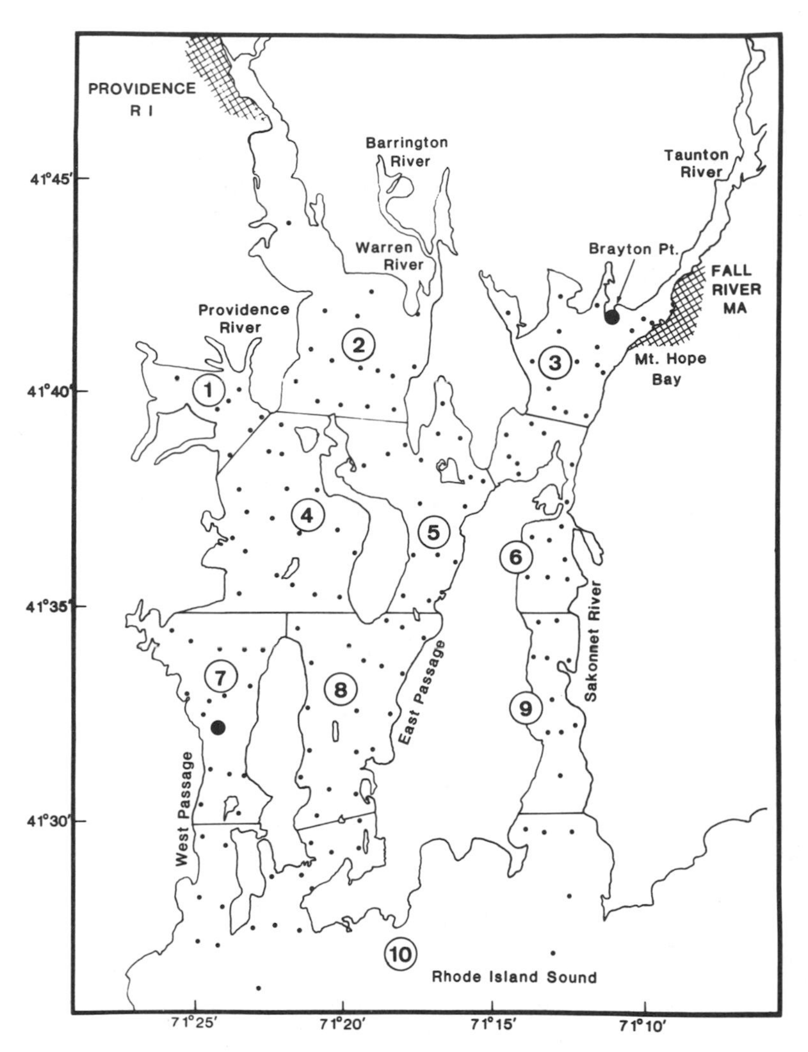

FIGURE 1.—Narragansett Bay with ichthyoplankton survey stations (small dots), vertical distribution stations (large, closed circles), and sectors (line dividers and encircled numbers). RI is Rhode Island; MA is Massachusetts.

thyoplankton sampling program of comparable extent in New England coastal waters.

Methods

Geographic survey.—To compare ichthyoplankton density among different parts of the bay, the bay was divided into 10 sectors, among which 160 sampling stations were apportioned (Figure 1). There were three sectors each in the upper, middle, and lower parts of the bay. A Rhode Island Sound sector (sector 10) stretched across the bay's entrances and had no definite offshore boundary. Its purpose was to characterize ichthyoplankton in the sound more or less qualitatively. Sampling began on 22 June 1972. Each sampling series of 160 stations was completed in 2 d (except, occasionally, when weather made a third day necessary). Four vessels operated simultaneously, each in a different group of sectors. The first tow on a sampling day was made at least 1 h after dawn and the last at least 1 h before sundown. Sampling series were carried out weekly through August 1972 and then monthly through February 1973. In March 1973 there were two series. Weekly sampling resumed in April and continued through August 30.

A series consisted of an oblique tow at each station with paired, bridleless, 60-cm-diameter bongo nets made with 505-μm Nitex mesh (Colton et al. 1980). The minimum length of tow was established, by experiment, at 4 min, generally long enough to produce useful numbers of fish eggs or larvae of the principal species. At the shallower stations in the bay, a single oblique tow took less time than the established minimum, so the net was repeatedly lowered each time it reached the surface and hauled back, until at least 4 min had passed. At deep stations, a single tow took as much as 7 min to complete. With the weights and gear used, wire angle averaged about 30° and haulback speed about 0.2 m/s. At an average vessel speed of 1.3 m/s, the speed of the net through the water was about 1.5 m/s, and the net had an oblique angle of about 4° from the horizontal. Variations in wind and sea conditions undoubtedly led to substantial deviations from these averages. Depth of gear was monitored with a pressure-activated transducer (ENDECO) mounted at the net and connected by conducting cable to a deck readout. Flowmeters (General Oceanics), for use in computing the quantity of water sampled in each tow, were mounted in the net mouth midway between the center and the rim. Samples were fixed in a 10% solution of formaldehyde in seawater, buffered with sodium borate.

Sample analysis.—A sample of ichthyoplankton was taken from one of the two nets on each bongo frame. The eggs and larvae were either counted completely under a dissecting microscope or they were subsampled when dominant species were so abundant that complete counting was unnecessary for the desired level of statistical confidence. Subsamples of the whole were taken from a partitioned plankton splitter (adapted from Motoda 1959). Eggs and larvae were identified to species in most cases, counted, and standardized to numbers/100 m^3, the population density. For an index of ichthyoplankton production in each sector, mean population densities of eggs and larvae were weighted by the half-tide volume of a sector, expressed in cubic meters.

Population differences among sectors.—The analysis was based on population densities of the eggs and larvae of abundant species—or the larvae alone if the eggs of a species were demersal rather than planktonic. For Atlantic menhaden *Brevoortia tyrannus*, anchovies *Anchoa* spp., and the labrid fishes, both eggs and larvae were sampled. There were two anchovy species (bay anchovy *A. mitchilli* and striped anchovy *A. hepsetus*) and two labrid species (cunner *Tautogolabrus adspersus* and tautog *Tautoga onitis*). The elongated eggs of the two anchovies are readily distinguishable, but the young larvae are not; hence, all the anchovy larvae are listed simply as *Anchoa* spp. In the case of the labrids, it is the eggs that cannot be distinguished readily by eye. Two other common species used for the analysis produce demersal eggs: winter flounder *Pseudopleuronectes americanus* and American sand lance *Ammodytes americanus*; only their larvae were counted.

For the Atlantic menhaden, anchovies, and labrids, counts of eggs and larvae at each station were averaged over 10-week periods in the summer of 1972 and the summer of 1973 when sampling series were accomplished weekly. For winter flounder and American sand lance larvae, similar station counts were averaged for 8 weeks in the spring of 1973 when sampling series were again accomplished weekly. These averages were then aggregated by sector and averaged again to provide sector means for each taxon.

Inasmuch as the abundances of eggs and larvae in the samples were far from normally distributed, a nonparametric statistical method, the Mann-Whitney *U*-test (Siegel 1956), was used to test for significant differences between pairs of sector populations.

Comparison of day and night samples.—Differences between densities of fish eggs and larvae estimated from samples collected during the day and during the night were assessed by comparing a group of samples from 21 stations in sectors 4 and 7. Samples were collected both day and night at these stations on two occasions in July 1973.

Vertical distribution.—Vertical distributions of fish eggs and larvae were assessed at two stations, one each in sectors 3 and 7 (Figure 1); our modification of a Tucker trawl (Tucker 1951; Clarke 1969) was used to sample discrete depths. This net was sent down closed. At the sampling depth, it was opened by a messenger sent down the wire to trip a mechanical release (General Oceanics). A second messenger closed it again for haulback. The net's mouth dimensions were 1 m across on top and bottom and 1.4 m long on the sides, which rake back from the top at 45°. Because of this diagonal cut of the side panels, the net's projected mouth area was about 1 m^2, depending on towing speed and the weights used at the bottom of the gear.

In 1972, day and night sampling for vertical distribution was carried out in the Brayton Point ship channel, near the mouth of the Taunton River in upper Mt. Hope Bay (sector 3) in water about 12 m deep, and under the Jamestown Bridge (sector 7) in water about 19 m deep. At each

TABLE 1.—List of species by family whose eggs or larvae were found in Narragansett Bay during the June 1972–August 1973 survey.

Family / species
Anguillidae
Anguilla rostrata
Ophichthidae
Clupeidae
Alosa spp.
Clupea harengus harengus
Brevoortia tyrannus
Engraulidae
Anchoa mitchilli
A. hepsetus
Osmeridae
Osmerus mordax
Lophiidae
Lophius americanus
Gadidae
Enchelyopus cimbrius
Gadus morhua
Melanogrammus aeglefinus
Merluccius bilinearis
Urophycis spp.
Pollachius virens
Atherinidae
Menidia menidia
Gasterosteidae
Gasterosteus aculeatus
Syngnathidae
Syngnathus fuscus
Pomatomidae
Pomatomus saltatrix
Sparidae
Stenotomus chrysops
Sciaenidae
Cynoscion regalis
Menticirrhus saxatilis
Labridae
Tautoga onitis
Tautogolabrus adspersus
Stichaeidae
Ulvaria subbifurcata
Pholidae
Pholis gunnellus
Ammodytidae
Ammodytes americanus
Gobiidae
Gobiosoma ginsburgi
Scombridae
Scomber scombrus
Stromateidae
Peprilus triacanthus
Triglidae
Prionotus spp.
Cottidae
Myoxocephalus spp.
Agonidae
Aspidophoroides monopterygius
Cyclopteridae
Liparis atlanticus
Bothidae
Scophthalmus aquosus
Paralichthys dentatus
P. oblongus
Pleuronectidae
Pseudopleuronectes americanus
Limanda ferruginea
Glyptocephalus cynoglossus
Hippoglossoides platessoides
Soleidae
Trinectes maculatus
Tetraodontidae
Sphoeroides maculatus

station, samples were taken at the surface, just over the bottom, and at two midwater levels evenly spaced between surface and bottom. In 1973, there was another series of day and night samples at the Brayton Point station; duplicate tows were made at each depth.

Sampling error.—Error in the density estimate at a single station is expected from three main sources: patchiness in the distribution of the plankton (tow effects); inconsistencies in field sampling technique between stations (station effects); and subsampling in the laboratory (subsampling effects). To determine the joint effects of these errors, we collected three replicate samples at each of five stations (one station each in sectors 3, 4, 7, 8, and 9 in July 1973). Each of the 15 samples was split into 16 aliquots, in every one of which all eggs and larvae were counted. Tow effects were nested with station effects in a nested, two-way analysis of variance with subsampling effects. Because population densities varied considerably between stations, we transformed density estimates to common logarithms so that variability among a station's replicate samples would be roughly equivalent among the five stations. From this, we obtained an unbiased estimate of variance for densities of eggs and larvae in a single sample (Afifi and Azen 1974).

Splitting samples did not appreciably increase error for the aggregate counts of all eggs or larvae in a single sample. The estimated variances were 0.036 for egg densities and 0.043 for larval densities when a whole sample was counted. For a one-sixteenth subsample (aliquot) the corresponding estimates were 0.015 for eggs and 0.025 for larvae. Confidence limits for egg abundances, calculated from the log-

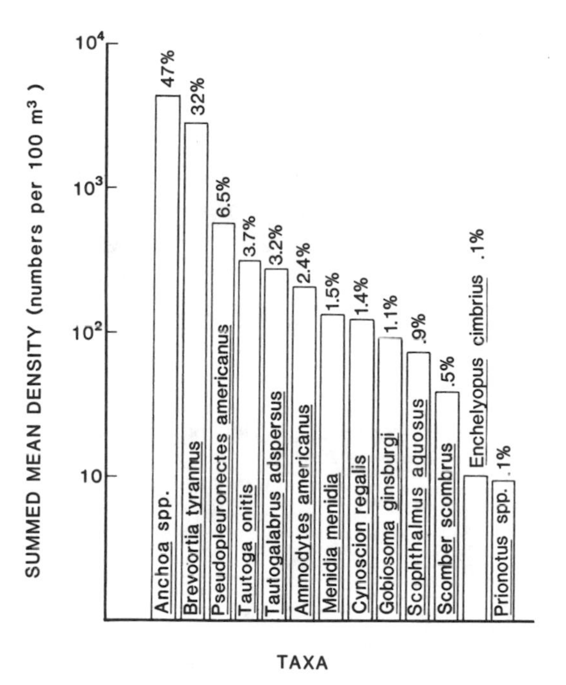

FIGURE 2.—Summed mean densities of the commonest fish larvae from sectors 1–9 (Figure 1) of Narragansett Bay, 1973, with the percentage contributed by each.

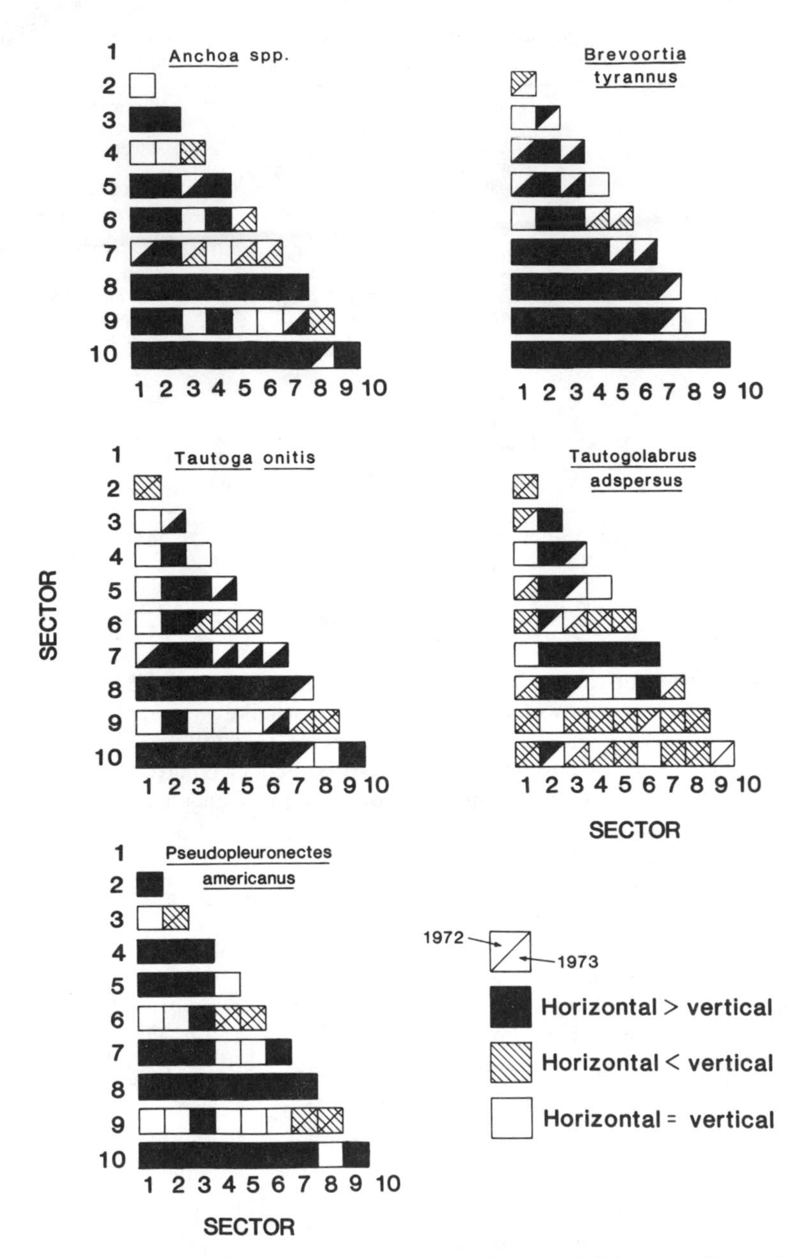

FIGURE 3.—Sector comparisons of the densities of the most abundant fish larvae collected from 22 June to 24 August 1972, and 19 June to 30 August 1973 in each of 10 sectors of Narragansett Bay. For each taxon, sector densities listed horizontally were significantly greater than those listed vertically if sections are solid black, and significantly less than vertical densities if sections are cross-hatched, as judged by the Mann–Whitney U-test ($P < 0.05\%$). Open sections indicate no significant differences.

transformed variances and expressed as percent error of the mean, were 79–136% in a whole sample and 71–140% in an aliquot. For larvae, the confidence limits were 66–151% for a whole sample and 59–168% for an aliquot.

Results

Ichthyoplankton

Ichthyoplankton of about 42 species, in 28 families, were taken during the survey (Table 1).

TABLE 2.—Abundances of selected taxa of fish eggs and larvae sampled in each sector (Figure 1) of Narragansett Bay, 22 June 1972–30 August 1973. The top number on each data pair is mean density during the period of occurrence, in number of eggs or larvae per 100 m^3 of water sampled. The lower number of each pair is an index of total annual production, in units of 10^8, obtained by weighting density by sector volume. Production could not be calculated for sector 10 because volume could not be delimited. An open data cell means that the taxon was not collected or that its production could not be calculated; a + sign indicates that the taxon was collected, but that its density was less than 1/100 m^3 or that its production was less than 10^3/year.

Life stage	Year	Sector 1	2	3	4	5	6	7	8	9	10
					Brevoortia tyrannus						
Egg	1972	356	139	438	306	544	455	264	122	401	12
		11	44	83	134	206	66	86	92	32	
	1973	846	628	2,301	138	412	944	32	163	399	63
		26	201	431	60	156	137	105	123	32	
Larva	1972	1,840	1,729	944	76	723	691	474	192	458	658
		56	552	189	339	273	100	156	145	37	
	1973	654	610	595	95	175	288	15	31	96	10
		20	195	113	41	66	42	5	24	7	
					Clupea harengus harengus						
Larva	1972	1	1	+	1	+	+	+	+	1	1
		+	+	+	+	+	20	+	+	+	
					Anchoa mitchilli						
Egg	1972	2,989	134	156	1,000	334	420	1,142	951	198	228
		90	50	30	437	126	61	374	718	159	
	1973	348	497	296	2,180	454	469	1,733	513	1,281	279
		105	159	56	958	172	68	568	387	103	
					***Anchoa* spp.**						
Larva	1972	717	356	105	267	132	123	123	20	124	8
		22	114	20	117	50	18	40	15	10	
	1973	1,637	678	140	456	80	184	351	48	221	34
		49	217	26	200	30	27	115	36	117	
					Enchelyopus cimbrius						
Egg	1972		+	+	+			+			
			+	+	+			+			
	1973	75	75	94	48	112	104	52	72	46	30
		2	24	18	21	42	15	17	54	4	
Larva	1972	1	+	+	+	+	+	+	+	+	+
		+	+	+	+	+	+	+	+	+	
	1973	1	1	1	1	1	1	1	2	2	1
		+	+	+	+	+	+	+	+	+	
					Menidia menidia						
Larva	1972	166	32	26	10	15	14	20	2	2	+
		5	1	5	4	6	2	6	1	+	
	1973	41	6	38	3	4	26	+	+	4	+
		1	2	7	1	2	4	+	+	+	
					***Stenotomus* and *Cynoscion* spp.**						
Egg	1972	476	155	97	356	125	291	284	110	430	45
		14	50	18	156	47	42	93	83	34	
	1973	356	222	366	299	219	687	135	148	791	77
		1	7	70	131	83	99	44	112	64	
					Cynoscion regalis						
Larva	1972	68	22	11	18	2	11	9	1	30	4
		2	7	2	8	1	2	3	1	2	
	1973	23	21	20	7	9	15	3	3	15	1
		1	7	4	3	4	2	1	2	1	

TABLE 2.—Continued.

Life stage	Year	Sector 1	2	3	4	5	6	7	8	9	10
						Menticirrhus saxatilis					
Egg	1972	238	40	30	31	30	18	18	6	39	1
		7	13	6	14	11	3	6	4	3	
	1973	55	30	22	11	14	16	1	3	7	1
		2	10	4	5	5	2	+	2	1	
Larva	1972	+	+	+	+	+	+	+	+	+	+
		+	+	+	+	+	+	+	+	+	
	1973		+	+	+	+	+	+	+	+	+
			+	+	+	+	+	+	+	+	
						Labridae					
Egg	1972	592	396	289	403	471	446	674	551	979	650
		18	127	55	176	178	65	221	416	79	
	1973	1,295	806	796	684	881	1,132	797	1,009	2,062	1,153
		39	257	151	299	333	164	261	761	166	
						Tautoga onitis					
Larva	1972	74	60	50	30	25	32	31	10	53	8
		2	19	9	13	10	5	10	7	4	
	1973	56	58	30	33	18	46	9	13	41	9
		2	18	6	10	7	7	3	10	3	
						Tautogolabrus adspersus					
Larva	1972	26	46	27	16	15	31	8	10	131	27
		1	15	5	7	6	4	3	8	11	
	1973	22	32	18	15	19	39	6	20	88	30
		1	10	3	6	7	6	2	15	7	
						Pholis gunnellus					
Larva	1973	+	+	+	+	+	+	+	+	+	+
		+	+	+	+	+	+	+	+	+	
						Ammodytes americanus					
Larva	1973	9	8	10	7	8	28	6	6	107	6
		+	3	2	3	3	4	2	5	9	
						Gobiosoma ginsburgi					
Larva	1972	3	4	8	2	1	10	+	+	6	+
		+	1	2	1	+	2	+	+	+	
	1973	15	26	12	5	2	12	1	+	12	+
		1	8	2	2	1	2	+	+	1	
						Scomber scombrus					
Egg	1973	1	2	3	11	15	3	13	141	220	652
		+	1	1	5	6	1	4	106	18	
Larva	1973	+	1	2	2	3	3	2	15	10	22
		+	+	1	1	+	+	1	11	1	
						Peprilus triacanthus					
Egg	1972	2	2	1	11	4	2	8	14	6	12
		+	1	+	5	2	+	3	11	1	
	1973	30	32	8	85	110	47	126	325	380	111
		1	10	2	37	42	7	42	245	31	
Larva	1972			+			+	+	+	+	+
				+			+	+	+	+	
	1973	1	1	1	1	+	2	+	1	1	1
		+	+	+	+	+	+	+	1	+	

TABLE 2.—Continued.

Life stage	Year	Sector 1	2	3	4	5	6	7	8	9	10
						Prionotus spp.					
Egg	1972	80	37	39	96	75	57	100	110	292	46
		2	12	7	42	28	08	33	83	24	
	1973	146	71	44	114	131	178	130	122	554	64
		4	23	8	50	50	26	43	92	44	
Larva	1972	2	1	3	1	1	1	1	1	1	1
		+	+	1	1	+	+	1	+	1	
	1973	1	2	1	1	+	2	+	+	2	+
		+	1	+	+	+	+	+	+	+	
						Myoxocephalus spp.					
Larva	1973	+	+	+	1	1	1	1	1	2	1
		+	+	+	+	+	+	+	1	+	
						Scophthalmus aquosus					
Egg	1972		+	+						+	+
			+	+						+	
	1973	260	168	142	121	112	189	88	60	323	39
		8	53	27	53	42	27	29	45	26	
Larva	1972	7	13	10	7	6	5	3	2	17	2
		+	4	2	3	2	1	1	2	1	
	1973	9	16	6	5	8	9	3	3	12	2
		+	5	1	2	3	1	1	2	1	
						Pseudopleuronectes americanus					
Larva	1973	107	57	91	34	34	71	28	11	96	12
		3	18	17	15	13	10	9	8	8	

Six or seven taxa generally accounted for 95% or more of all the planktonic fish eggs. Bay anchovy and Atlantic menhaden eggs made up nearly 80% of the total. Four of the 13 most abundant taxa of larvae (Figure 2) are hatched from demersal eggs.

Sector Comparisons

The highly polluted waters (USEPA 1971) of the Providence River (sector 2) and Mt. Hope Bay (sector 3) do not appear to be a barrier to fish spawning, though survival of the eggs and larvae may be diminished. The Providence River was outstandingly productive for many fishes (Table 2). In upper Mt. Hope Bay, productivity for most species was either greater than those in the other nine sectors (about one-half of the cases) or not significantly different (about one-third of the cases; Figure 3).

Various spawners favor particular sectors of the bay. Atlantic menhaden, for example, spawned mainly in the northern sectors, as indicated by the density of their eggs (Figure 4), whereas winter flounder larvae, hatched from demersal eggs, were most commonly found in shoal water, in or near coves and small bays; they were rare in the deep waters of the East Passage.

In most cases, for the abundant ichthyoplankton species (Atlantic menhaden, cunner, tautog, anchovies, and winter flounder), sampling intensity was sufficient to reveal differences in density (nonoverlapping confidence limits) among sectors within a given year and, in several cases, from one year to the next (Figure 5).

Annual Variation

Substantial variations were found in the production and apparent survival of eggs and larvae from the first to the second year of our survey. This was true not only for migratory species spawning there such as weakfish *Cynoscion regalis* and Atlantic menhaden, but also for residents such as tautog, cunner, and windowpane *Scophthalmus aquosus*. Abundance of most species declined from 1972 to 1973 (Table 2; Figure

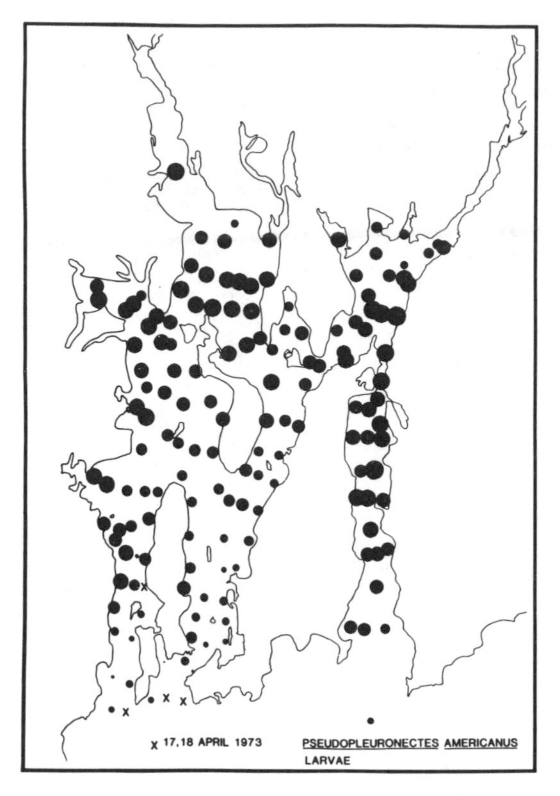

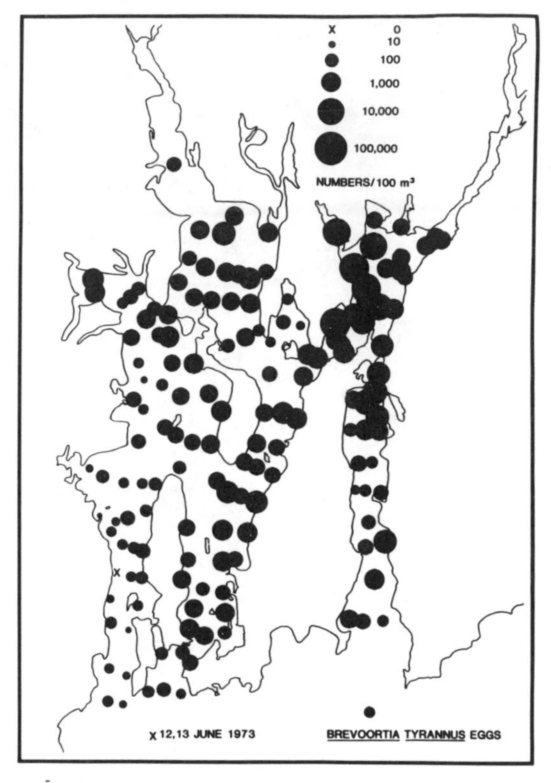

FIGURE 4.—Peak densities (numbers/100 m^3) of *Pseudopleuronectes americanus* larvae and *Brevoortia tyrannus* eggs in Narragansett Bay, 1973.

5). Of the truly abundant forms, only anchovy larvae increased their population density.

Day–Night Comparisons

More larvae were captured at night than during the day (Table 3). The distribution of larvae over the sampling area was evidently more random at night. An *F*-test of the log-transformed data for equality of variances showed this difference in the evenness of distribution to be significant (Table 3). The 95% confidence limits for difference between day and night time density estimates indicated that daytime sampling requires about four times the intensity of night sampling to achieve the same precision (Table 4).

Vertical Distribution

Closing-net samples showed eggs to be distinctly stratified even in the relatively shallow, wind-mixed waters of upper Mt. Hope Bay. Most eggs were in the upper water column (Table 5), although those of a few species such as fourbeard rockling *Enchelyopus cimbrius* and searobins *Prionotus* spp. showed a tendency to sink. In shallow parts of the bay, sea robin eggs were found in the channels. Certain egg taxa showed differences in vertical distribution from one time to another on the same date. This probably reflected local changes in water density, though these changes are very difficult to measure in an area so complex hydrographically.

Larvae as a group were more abundant in the upper water column (Table 5). Two species of fish larvae tended to concentrate in the upper water column by day, dispersing more evenly after dark: Atlantic menhaden and, especially, Atlantic silverside *Menidia menidia*. Seaboard goby *Gobiosoma ginsburgi*, winter flounder, and anchovy larvae tended to move downward by day and upward by night. Anchovy larvae, in particular, showed the most defined diel vertical pattern at the relatively deep West Passage station (Figure 6). At night, swarms of these larvae could be seen at the surface. Depth distributions of young anchovies in the early afternoon (1325–1540 hours) of 14 August 1972 and the following morning (0944–1043 hours) were similar, even though the first series was taken on a falling and the second on a rising tide.

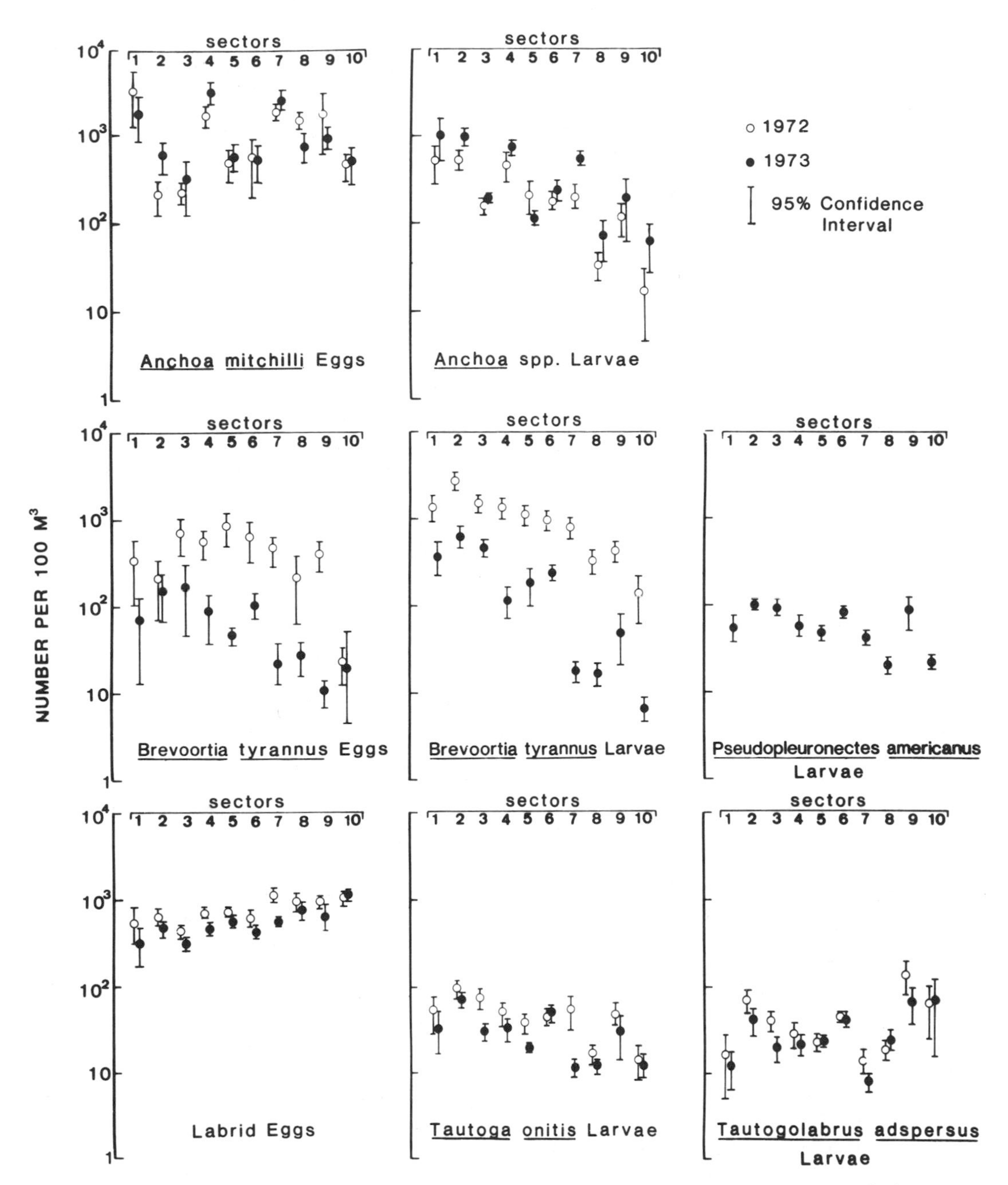

FIGURE 5.—Arithmetic mean sector densities and confidence intervals for larvae of *Pseudopleuronectes americanus* (11–13 April to 30 May 1973) and for eggs, larvae, or both of *Anchoa* spp., *Brevoortia tyrannus*, labrids, *Tautaga onitis*, and *Tautoglabrus adspersus* (22–23 June to 23–24 August 1972; 10–20 June to 20–30 August 1973) in Narragansett Bay.

Discussion

Ichthyoplankton

The ichthyoplankton of Narragansett Bay (Table 1) reflects the faunal diversity of southern New England coastal waters, where one finds a rough faunal boundary at the Cape Cod, Massachusetts, peninsula for many families and species of marine fishes (Azarovitz and Grosslein 1982, 1987). The bay has more species and families in its

TABLE 3.—Geometric mean densities (number/100 m^3) and 95% confidence limits (in parentheses) of total fish eggs and larvae sampled during day and at night at 21 stations in sectors 4 and 7 of Narragansett Bay on two occasions (1, 2) in July 1983, and F-statistics (log_{10}) of the equality of variances. Asterisks denote $P < 0.05$* and $P < 0.01$**.

Sampling occasion and life stage	Density		
	Day	Night	F
(1) Egg	7,239 (5,033–10,411)	5,480 (3,810–7,881)	1.198
(1) Larva	1,439 (1,088–1,903)	2,204 (1,667–2,915)	4.748*
(2) Egg	3,239 (1,651–6,355)	6,227 (3,174–12,218)	1.922
(2) Larva	1,896 (1,440–2,496)	4,509 (3,426–5,935)	20.30**

TABLE 4.—Geometric means and 95% confidence limits (in parentheses) of the differences between densities of fish eggs and larvae (numbers/100 m^3) sampled during the day and at night at 21 stations in sectors 4 and 7 of Narragansett Bay on two occasions in July 1983.

Life stage	9–11 July	17–18 July
Egg	0.544 (0.757–1.054)	1.28 (2.15–3.59)
Larva	1.06 (1.49–2.09)	1.63 (2.32–3.31)

ichthyoplankton than estuaries to the north and fewer than estuaries to the south. North of Cape Cod, along the Maine coast, 18 or 19 families and about 26 species are represented in the estuarine ichthyoplankton (Chenoweth 1973; Hauser 1973). In Delaware Bay, to the south, 26 families and 56 species have been reported (Scotton et al. 1973). In Cape Cod Bay, 34 families and 60 species are represented in the ichthyoplankton (Bourne 1978; Scherer 1984), but some middle Atlantic forms would probably rarely, if ever, be found north of Cape Cod were it not for the Cape Cod Canal, through which there is a large tidal flow. Narragansett Bay had, with few exceptions, the same ichthyoplankton as the Weweantic River estuary in nearby Massachusetts, south of Cape Cod (Lebida 1969), and the Mystic River estuary in Connecticut (Pearcy and Richards 1962).

The faunal composition and relative abundance of ichthyoplankton do not appear to have changed appreciably in Narragansett Bay since it was sampled by Herman (1958, 1963). The monthly mean sector densities from our survey compare well with Herman's data, when his are adjusted to numbers/100 m^3 (Table 6; Figure 7). The seasonality of egg abundance that he measured in 1957–1958 from weekly samples at four stations is remarkably similar to the abundance we found at 160 stations during our 1972–1973 survey. Herman's egg data show virtually the same rank order as our 1973 data for the 10 most abundant taxa. Eggs of Atlantic mackerel *Scomber scombrus*, however, which ranked ninth in abundance in our 1973 collections, did not appear at all on Herman's list. Except for the labrids, Herman's average egg densities for abundant species over the 1958 season were lower than ours for 1973, which may largely reflect natural annual variation. In addition, his four station locations were not in areas where we found high densities of bay anchovy and Atlantic menhaden eggs.

A striking difference between Herman's results and ours was in the comparatively large number of larvae that he caught in the late fall and winter. The fall catch was due to the abundant fall spawning of Atlantic menhaden that took place in 1958 but not in 1972. Herman reported that the January and March peaks in his records of larval abundance were due principally to sculpins *Myoxocephalus* spp. Sculpins were not present in 1972 in the numbers they were in 1958.

Other differences between Herman's (1958) data and ours may be attributable to gear. Our maximum densities of eggs were of the same order as our larval densities. Herman's were not. He found more than a hundred times as many larvae as he did eggs at the seasonal peak. The relative scarcity of winter flounder larvae in Herman's collections and the dominance of sculpins—larvae that do not figure prominently in the other surveys mentioned here—indicate that Herman's 1-mm mesh size retained the large and heavy-bodied sculpin larvae while allowing many of the more slender winter flounder, anchovy, and Atlantic menhaden larvae to pass through.

TABLE 5.—Distribution (%) of total eggs and larvae among Tucker trawl samples at each four depths in the Brayton Point Ship Channel, March–July 1973.

Stratum	Percent of total eggs, day plus night	Percent of total larvae	
		Day	Night
Surface	47	55	30
3 m	25	30	43
6 m	13	10	18
9 m	15	5	9

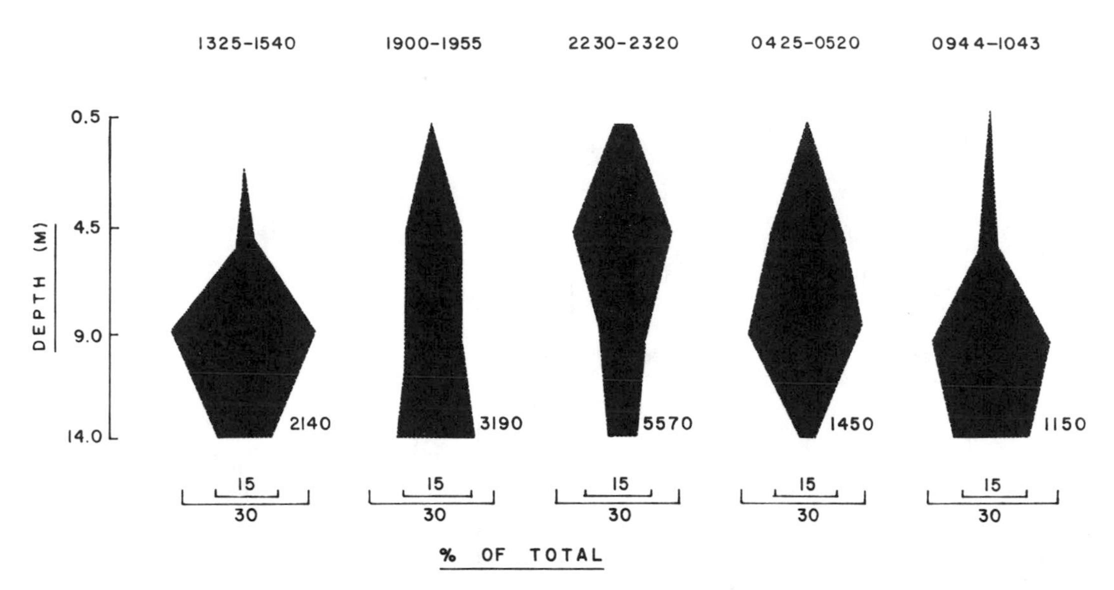

FIGURE 6.—The vertical distribution of *Anchoa* spp. larvae in West Passage, 14–15 August 1972. Collection hours are along the top, and the number next to each image is the number of larvae collected in an entire vertical series.

Among the dominant taxa of larvae collected in Narragansett Bay (Figure 2), anchovies, Atlantic menhaden, and cunner are also among the top six most abundant larvae in the contiguous waters of Long Island and Block Island sounds (Perlmutter 1939; Merriman and Sclar 1952; Wheatland 1956; Richards 1959). Winter flounder and American sand lance larvae appear on all but Perlmutter's list, no doubt because he collected only between May and October.

TABLE 6.—Mean densities of fish eggs (numbers/100 m^3) for all stations over the period of occurrence and percent contribution to total egg abundance from Herman's (1958) Narragansett Bay survey and the present study.

Taxon	Present study: Mean density	Present study: % of total	Herman (1958): Mean density	Herman (1958): % of total
Labrids	1,051	30	1,260	62
Anchoa mitchilli	876	25	217	11
Brevoortia tyrannus	651	18	90	4
Cynoscion and *Stenotomus* spp.	358	10	217	11
Prionotus spp.	166	5	100	5
Scophthalmus aquosus	162	5	49	2
Peprilus triacanthus	127	4	39	2
Enchelyopus cimbrius	75	2	29	1
Scomber scombrus	46	1	0	0
Menticirrhus saxatilis	18	<1	34	2

Water Circulation and Ichthyoplankton Distribution

Circulation in Narragansett Bay is due mainly to strong tidal currents (Levine and Kenyon 1975), with gravitational (induced by horizontal pressure gradients) and wind-driven currents superimposed (Hess 1976). In the bay's headwaters, current velocities are slower than in the lower bay; slack water in the upper bay may last as long as 3 h (McMaster 1960). The bay's waters are only partially stratified and there is considerable vertical mixing (Pilson 1985). There is typically no sharp halocline and no evidence of any large, isolated water mass (Pilson 1985). A salinity difference of only 2‰ from surface to bottom is common. The tides may be more important in turbulent mixing of the bay's water than they are in its net transport (Gordon and Spaulding 1987), and east–west density differences (Hess 1976) also help to mix the water vertically. In the bay's northern reaches, the bottom water is fresher than at the mouth, while the surface water is at least slightly saltier than water from the tributary rivers sampled only a little distance northward. To some extent, fish eggs and small larvae are carried passively upward or downward just as the salt and fresh waters are.

There is an overall southward flow of low-salinity water in the upper water column and a northward counterflow of saltier water at deeper

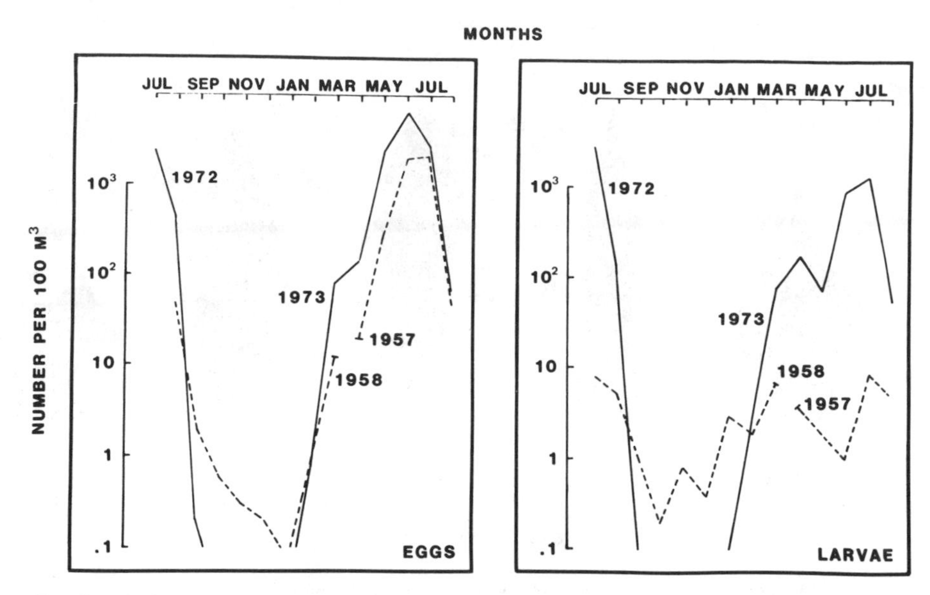

FIGURE 7.—Comparison of total densities of fish eggs and larvae collected in the present study (solid lines) and Herman's (1963) 1957–1958 survey of Narragansett Bay (broken lines). Herman's data have been converted to numbers/100 m³.

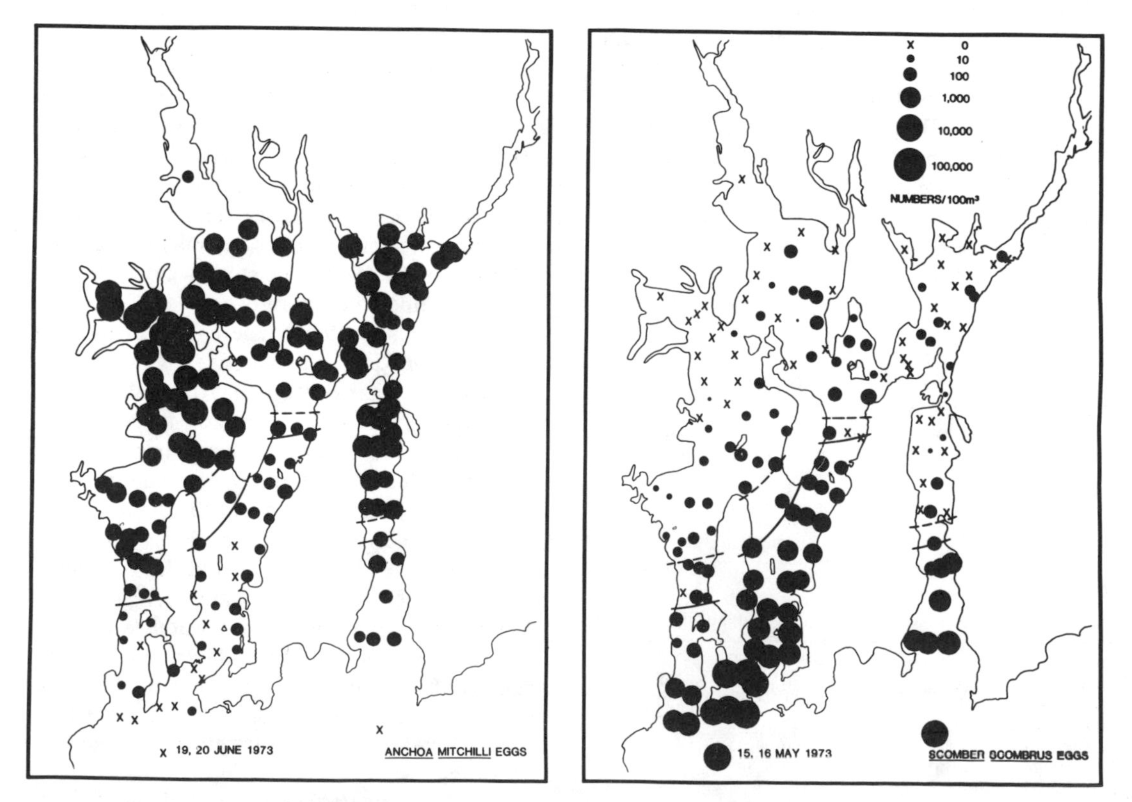

FIGURE 8.—Peak densities (numbers/100 m³) of *Anchoa mitchilli* and *Scomber scombrus* eggs in Narragansett Bay, 1973. Isohalines (31‰) are from Hicks (1959); solid contours represent surface and broken contours bottom salinities.

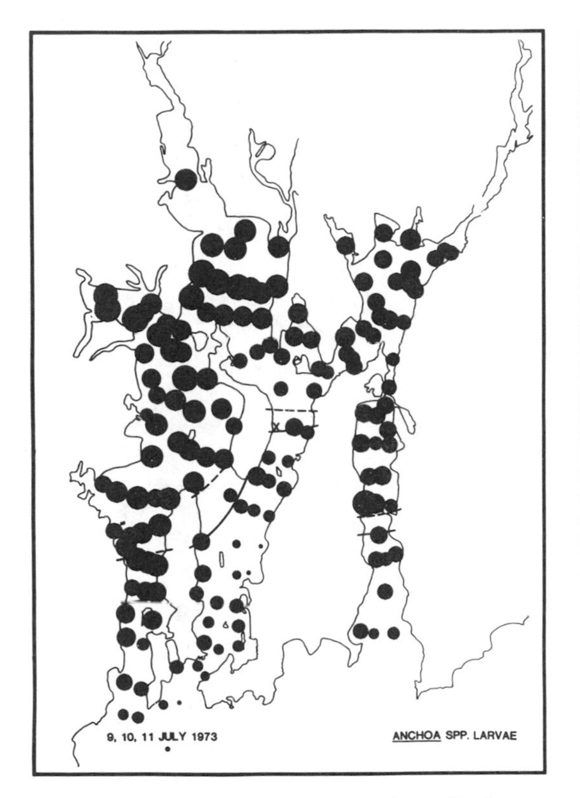

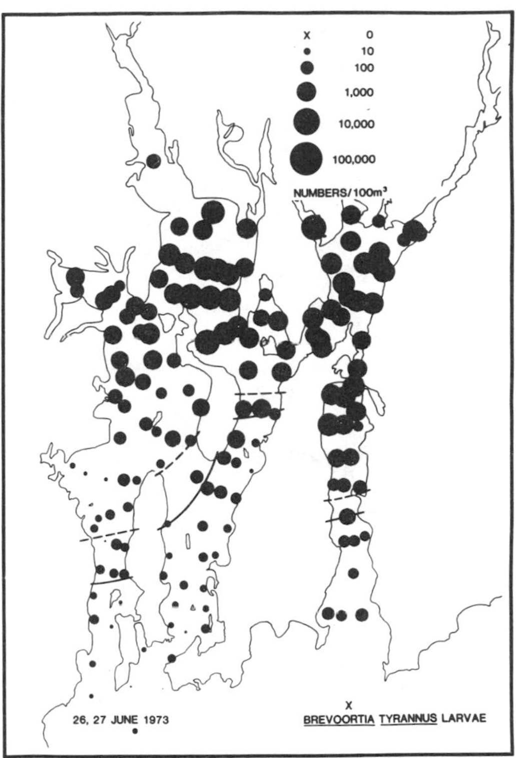

FIGURE 9.—Peak densities (numbers/100 m³) of *Anchoa* spp. and *Brevoortia tyrannus* larvae in Narragansett Bay, 1973. Isohalines (31‰) are from Hicks (1959); solid contours represent surface and broken contours bottom salinities.

levels (Hicks 1959). Owing in part to bottom depths, higher salinity water flows into the bay through the East Passage and lower salinity water flows out of the bay through the West Passage and Sakonnet River (Hicks 1959). At times, however, wind-induced currents can disrupt this basic flow pattern (Weisberg 1976; Weisberg and Sturgis 1976; Gordon and Spalding 1987). The result of wind stress may be quite different flow patterns with large-scale eddies in some sectors of the bay and a reversal of flow such that high-salinity water flows into the bay through the West Passage and Sakonnet River and low salinity water flows out through the East Passage. These reversals will episodically occur when the winds blow from the south and southwest, as they typically do in the southern New England spring and summer (Gordon and Spaulding 1987). Most of a year's ichthyoplankton is produced at these seasons.

Vertical distribution of eggs may partially reflect spawning periodicity over a day–night cycle as proposed by Kendall and Naplin (1981). For fish species that spawn over a fairly wide range of salinities, different groups of eggs spawned in different parts of an estuary and at different times might vary distinctly in buoyancy, which for gadid eggs—to take an example—is said to decrease with development, the initial value being determined at fertilization (Sundnes et al. 1964). Even so, the overall vertical distribution of eggs in the bay showed them to be more abundant in surface waters than in deep water. Williams (1968), working in Long Island Sound in waters of quite uniform salinity, found that pelagic eggs of seven species studied (all abundantly represented in our collections as well) were also most plentiful toward the surface, though some showed a far more pronounced gradient with depth than others.

The retention of fish eggs and very small larvae that lack the ability to move vertically, and that thus behave as passive particles (Boicourt 1982; Fortier and Leggett 1982) must be related to the residence time of water. Residence time of water in Narragansett Bay is 26 d on the average, 40 d in late summer and 10 d in late spring and early summer (Pilson 1985), far longer than the time required for egg incubation and yolk-sac absorption for most fishes spawned in the bay (see Theilacker and Dorsey 1980 for a review).

Some effects of the bay's circulation on ichthyoplankton can be seen in the different distributions of bay anchovy and Atlantic mackerel eggs. Bay anchovy spawn well up in the bay, whereas Atlantic mackerel spawn outside it or in the deep reaches of the lower East Passage. Atlantic mackerel eggs spawned in Rhode Island Sound appear to penetrate much farther up into East Passage than into the West Passage and the Sakonnet River (Figure 8). This conforms with Hick's (1959) analysis of the bay's general circulation, in which he noted the relatively high salinity at both the surface and bottom in East Passage compared with salinity in the other two outlets, both much shallower than East Passage. He calculated that the flow of salty bottom water into East Passage exceeded the flows into West Passage and the Sakonnet River by factors of 2.8 and 7.2, respectively. Salinities taken from the present study, but agreeing generally with Hicks's, varied seasonally from 23.0 to 27.3‰ in the northern part of the bay (sectors 1–3) and from 26.6 to 31.5‰ in the southern part (sectors 7–9). Hicks's (1959) 31‰ isohalines roughly coincide with the main concentration of Atlantic mackerel eggs (Figure 8), though some eggs are carried north all the way to the bay's headwaters. This transport of eggs spawned outside the bay, however, can be interrupted by episodic southerly winds that can move water, and eggs, into the bay through the West Passage and Sakonnet River and out of the Bay in the East Passage. The response depends upon wind velocity and direction and water depth (Gordon and Spaulding 1987). The distribution of bay anchovy eggs spawned in the upper bay also is generally bounded by the 31‰ isohalines (Figure 8), but these eggs are mainly to the north, in opposition to the Atlantic mackerel egg distribution.

Light is established as a principal influence on the vertical distribution of many fish larvae (Ahlstrom 1959). The presence of a thermocline or halocline may also affect vertical distribution (Smith et al. 1978; Kendall and Naplin 1981; Sameoto 1982). Larval Atlantic cod *Gadus morhua* position themselves vertically in response to changes in hydrostatic pressure (Dannevig and Hansen 1959). Our sampling for vertical distribution was not sufficient to sort out the various determinants in Narragansett Bay, though there is

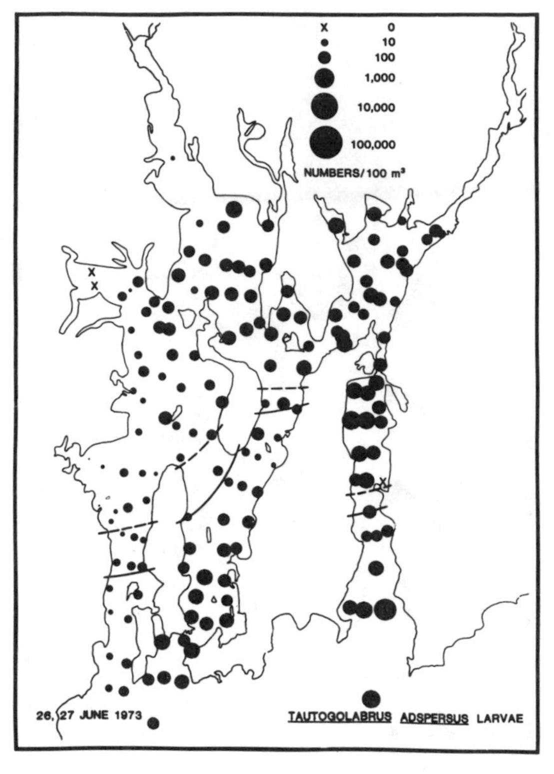

FIGURE 10.—Peak densities (numbers/100 m^3) of *Tautogolabrus adspersus* larvae in Narragansett Bay, 1973. Isohalines (31‰) are from Hicks (1959); solid contours represent surface and broken contours bottom salinities.

a strong indication that, for certain species, light is dominant.

Some fish larvae have been thought to penetrate or maintain themselves within partially mixed estuaries such as Narragansett Bay by migrating vertically bottomward on the flood tide, to ride the deep flow of salty water up the estuary, and then, on the ebb, moving to a level in the water column where seaward flow is minimal (Graham 1972; Weinstein et al. 1980; Fortier and Leggett 1982). Microscale turbulence is the likely stimulus for this response to tidal currents (McCleave and Kleckner 1982), responses to light notwithstanding. This behavior might occur among some Narragansett Bay fish larvae. Atlantic menhaden, for instance, spawn in the upper bay and a distributional chart of their larvae (Figure 9) shows them largely confined to the half of the bay north of the 31‰ isohaline. Anchovy larvae show a similar though less well-defined pattern (Figure 9). As evidenced by their diel vertical distribution, both species can and do move vertically into or through the stratum of no net horizontal motion: this—if we assume that the stratum of no horizontal motion coincides with the maximum salinity gradient—is at about 2.7 m in the Providence River, 1.8 m in Mt. Hope Bay, 5.2 m in the Sakonnet River, and, on the average, at 6.2 m in the West Passage and 8.7 m in the East Passage (calculated from Hicks 1959). Depending upon the timing of this movement, these two species might accomplish net maintenance of a horizontal position within the estuary.

The pattern of concentration in the upper bay does not prove that the suggested behavior occurs. Atlantic menhaden and anchovy larvae may simply hatch from eggs spawned in the upper bay whence they are carried slowly southward in a net seaward flow, their diminishing population densities reflecting their dispersal into ever-larger volumes of water. With our data, we cannot settle on either interpretation. It is, however, worth noting that cunner larvae, which showed little evidence of vertical movement, were more evenly distributed north to south in the bay (Figure 10).

References

Afifi A. A., and S. P. Azen. 1974. Statistical analysis: a computer oriented approach. Academic Press, New York.

Ahlstrom, E. H. 1959. Vertical distribution of pelagic fish eggs and larvae off California and Baja California. U.S. Fish and Wildlife Service Fishery Bulletin 60:106–146.

Azarovitz, T. R., and M. D. Grosslein. 1982. Fish distribution. MESA (Marine Ecosystems Analysis) New York Bight Atlas Monograph 15.

Azarovitz, T. R., and M. D. Grosslein. 1987. Fishes and squids. Pages 315–346 *in* R. H. Backus and D. W. Bourne, editors. George's Bank. MIT Press, Cambridge, Massachusetts.

Blaxter, J. H. S. 1969. Development: eggs and larvae. Pages 177–252 *in* W. S. Hoar and D. J. Randall, editors. Fish physiology, volume 3. Academic Press, New York.

Boicourt, W. C. 1982. Estuarine larval retention mechanisms of two scales. Pages 445–452 *in* V. S. Kennedy, editor. Estuarine comparisons. Academic Press, New York.

Bourne, D. W. 1978. Entrainment investigations and Cape Cod Bay ichthyoplankton studies, March 1970 and June 1972 and March 1974–July 1977. Summary Report to Boston Edison Company, Boston.

Chenoweth, S. B. 1973. Fish larvae of the estuaries and coast of Maine. U.S. National Marine Fisheries Service Fishery Bulletin 71:105–113.

Clarke, M. R. 1969. A new midwater trawl for sampling discrete depth horizons. Journal of the Marine Biological Association of the United Kingdom 49:945–960.

Colton, J. B., J. R. Green, R. R. Byron, and J. L. Frisella. 1980. Bongo net retention rates as affected by towing speed and mesh size. Canadian Journal of Fisheries and Aquatic Sciences 37:606–623.

Dannevig, A., and S. Hansen. 1959. Faktorer av betydning for fiskeeggenes og fiskeyngelens oppvekst. Fiskeridirektoratets Skrifter, Serie Havundersokelser 10:1–36.

Fortier, L., and W. C. Leggett. 1982. Fickian transport and the dispersal of fish larvae in estuaries. Canadian Journal of Fisheries and Aquatic Sciences 39:1150–1163.

Gordon, R. B., and M. L. Spaulding. 1987. Numerical simulations of the tidal- and wind-driven circulation in Narragansett Bay. Estuarine, Coastal and Shelf Science 24:611–636.

Graham, J. J. 1972. Retention of larval herring within the Sheepscot estuary of Maine. U.S. National Marine Fisheries Service Fishery Bulletin 70:299–321.

Hauser, W. J. 1973. Larval fish ecology of the Sheepscot River–Montsweag Bay estuary, Maine. Doctoral dissertation. University of Maine, Orono.

Herman, S. S. 1958. Planktonic fish eggs and larvae of Narragansett Bay. Master's thesis. University of Rhode Island, Kingston.

Herman, S. S. 1963. Planktonic fish eggs and larvae of Narragansett Bay. Limnology and Oceanography 8:103–109.

Hess, K. W. 1976. A three-dimensional numerical model of the estuary circulation and salinity in Narragansett Bay. Estuarine and Coastal Marine Science 4:325–338.

Hicks, S. D. 1959. The physical oceanography of Narragansett Bay. Limnology and Oceanography 4:316–327.

Kendall, A. W., and M. A. Naplin. 1981. Diel-depth distribution of summer ichthyoplankton in the Middle Atlantic Bight. U.S. National Marine Fisheries Service Fishery Bulletin 79:705–726.

Lebida, R. 1969. The seasonal abundance and distribution of eggs, larvae, and juvenile fishes in the Weweantic River estuary, Massachusetts, 1966. Master's thesis. University of Massachusetts, Amherst.

Levine, E. R., and K. E. Kenyon. 1975. The tidal energetics of Narragansett Bay. Journal of Geophysical Research 80:1683–1688.

McCleave, J. D., and R. C. Kleckner. 1982. Selective tidal stream transport in the estuarine migration of glass eels of the American eel (*Anguilla rostrata*). Journal du Conseil, Conseil International pour l'Exploration de la Mer 40:262–271.

McMaster, R. L. 1960. Sediments of Narragansett Bay system and Rhode Island Sound, Rhode Island. Journal of Sedimentary Petrology 30:249–274.

Merriman, D., and R. C. Sclar. 1952. Hydrographic and biological studies of Block Island Sound. The pelagic fish eggs and larvae of Block Island Sound. Bulletin of the Bingham Oceanographic Collection, Yale University 13(3):165–219.

Motoda, S. 1959. Devices of simple plankton apparatus. Memoirs of the Faculty of Fisheries Hokkaido University 7:73–94.

Pearcy, W. G., and S. W. Richards. 1962. Distribution

and ecology of fishes of the Mystic River estuary, Connecticut. Ecology 43:248–259.

Perlmutter, A. 1939. An ecological survey of young fish and eggs identified from tow-net collections. Pages 11–71 *in* A biological survey of the salt waters of Long Island, 1938. Salt water survey 15. New York State Conservation Department, Supplement to Annual Report 28, Albany.

Pilson, M. E. Q. 1985. On the residence time of water in Narragansett Bay. Estuaries 8:2–14.

Richards, S. W. 1959. Pelagic fish eggs and larvae of Long Island Sound. Bulletin of the Bingham Oceanographic Collection, Yale University 17(1):95–124.

Sameoto, D. 1982. Vertical distribution and abundance of the Peruvian anchovy, *Engraulis ringens*, and sardine, *Sardinops sagax*, larvae during November 1977. Journal of Fish Biology 21:171–185.

Scherer, M. D. 1984. Ichthyoplankton of Cape Cod Bay. Lecture Notes on Coastal and Estuarine Studies 11:151–190.

Scotton, L., R. E. Smith, N. S. Smith, K. S. Price, and D. P. DeSylva. 1973. Pictorial guide to fish larvae of Delaware Bay with information and bibliographies useful for the study of fish larvae. University of Delaware, College of Marine Studies, Delaware Bay Report Series 7, Newark.

Siegel, S. 1956. Non-parametric statistics for the behavioral sciences. McGraw-Hill, New York.

Smith, W. G., J. D. Sibunka, and A. Wells. 1978. Diel movements of larval yellowtail flounder, *Limanda ferruginea*, determined from discrete depth sampling. U.S. National Marine Fisheries Service Fishery Bulletin 76:167–178.

Sundnes, G., H. Leivestad, and O. Iverson. 1964. Buoyancy determinations of eggs from the cod (*Gadus morhua* L.). Journal du Conseil, Conseil International pour l'Exploration de la Mer 29:249–252.

Theilacker, G. H., and K. Dorsey. 1980. Larval fish diversity, a summary of laboratory and field research. IOC (Intergovernmental Oceanographic Commission) Workshop Report 28:105–142. (United Nations Educational, Scientific and Cultural Organization, Paris.)

Tucker, G. H. 1951. Relation of fishes and other organisms to the scattering of underwater sound. Journal of Marine Research 10:215–278.

USEPA (U.S. Environmental Protection Agency). 1971. Report on pollution of the interstate waters of Mt. Hope Bay and its tributary waters. USEPA, Boston.

Weinstein, M. P., S. L. Weiss, R. G. Hodson, and L. R. Gerry. 1980. Retention of three taxa of postlarval fishes in an intensively flushed tidal estuary, Cape Fear River, North Carolina. U.S. National Marine Fisheries Service Fishery Bulletin 78:419–436.

Weisberg, R. H. 1976. The non-tidal flow in the Providence River of Narragansett Bay: a stochastic approach to estuarine circulation. Journal of Physical Oceanography 6:721–734.

Weisberg, R. H., and W. Sturgis. 1976. Velocity observations in the West Passage of Narragansett Bay: a partially mixed estuary. Journal of Physical Oceanography 6:345–354.

Wheatland, S. B. 1956. Oceanography of Long Island Sound 1952–54. VII. Pelagic fish eggs and larvae. Bulletin of the Bingham Oceanographic Collection, Yale University 15(7):234–413.

Williams, G. C. 1968. Bathymetric distribution of planktonic fish eggs in Long Island Sound. Limnology and Oceanography 13:382–385.

American Fisheries Society Symposium 3:149–162, 1988

Null Hypotheses, Models, and Statistical Designs in the Study of Larval Transport

DAVID R. COLBY

U.S. National Marine Fisheries Service
Southeast Fisheries Center, Beaufort Laboratory
Beaufort, North Carolina 28516, USA

Abstract.—In larval transport investigations, the broad range of relevant spatial and temporal scales, the imprecision of measurements of organism abundance, and the potentially complex motions of water masses are particular concerns in developing research designs. These concerns underscore a need both for collaboration between biologists and physicists, and for rapid adoption of emerging technologies for measuring relevant processes at appropriate scales in the field. Many branches of statistics have application to larval transport investigations. Requirements for accurate description imply a need for greater emphasis on statistical estimation procedures and attention to sources of bias, and accordingly less emphasis on statistical hypothesis testing, especially with observational data. Strong null hypotheses should be advanced that are consistent with common experience and reflect current knowledge. They then provide an invaluable framework against which to compare empirical data. In larval transport research, advection–diffusion models, physical models, explicit conceptual models, and stochastic models all can serve as null hypotheses. Random-walk, Markov-chain models of the movements and expected vertical distribution of larvae in the water column have merit for studies of vertical migration as a transport mechanism. They also can be used to model the accumulation of larvae along shorelines and other boundaries and may help explain layering of oceanic plankton. Spatial and temporal scales, target and sampling populations, independent replication, and sampling gear bias are among the concepts that require attention in the design of studies of the flux of larvae through an inlet, or of a particular transport mechanism. Because, in larval transport studies, a variable of interest may change in response to many external factors, careful examination of patterns of covariation may provide more insight than will a focus on average values. New statistical approaches, such as the bootstrap, offer promising alternatives to traditional statistical methodologies, and they underscore the increasing application of computers to all phases of larval transport research, from research design and data acquisition to data analysis and stochastic modeling.

One conclusion drawn at a recent North Atlantic Treaty Organization Advanced Research Institute on mechanisms of migration by fishes was that "lack of statistical rigor may be one of the greatest weaknesses in previous work" (McCleave et al. 1984b). This lack is understandable, given the potential spatiotemporal complexity of fish migration, the range of scales involved, the relative imprecision and unknown accuracy of techniques for measuring fish abundance, the relatively low resolution of many environmental data sets, and the difficult logistics and high costs associated with gathering such data. The biophysical nature of fish migration and larval transport points up the value of collaboration between biologists and hydrographers (Crawford and Carey 1985) and the desirability of rapid application of emerging technologies for measuring relevant processes at appropriate spatiotemporal scales.

Fish migration research is not the only area within ecology that has been criticized for a lack of statistical rigor, nor is the problem simply nonapplication of statistical concepts to the design, analysis, and interpretation of research. Problems frequently have arisen, though, because design concepts have been misunderstood and, consequently, analytical procedures have been misapplied (Underwood 1981; Hurlbert 1984).

In what follows, I discuss certain statistical concepts as they relate to investigation of larval transport through inlets. I will address these concepts primarily in regard to fishes that are spawned on the continental shelf off the southeastern United States and move landward through inlets into estuaries (Weinstein 1981; Miller et al. 1984). This context was chosen because it is one with which I am familiar, and because Oregon Inlet, connecting the Atlantic Ocean with Pamlico Sound, North Carolina, has been proposed for modification by installation of large jetties, which is of immediate interest because of the potential effect of the jetties on larval transport.

The difficulty of research into larval transport becomes more apparent when we consider what would be required for an adequate description of, for example, the transport of larval Atlantic menhaden *Brevoortia tyrannus* from spawning sites on the edge of the continental shelf to low-salinity waters at the head of an estuary. The temporal scales extend from the minute-by-minute changes in tidal current velocities, and perhaps in the vertical elevation of the larva in the water column, to the several months between the spawning of the first egg and the arrival of the last larva at the estuarine nursery ground. The spatial scales extend from the millimeters or centimeters relevant to a larva's response to boundaries, such as the water surface or ocean floor, to the 10s of kilometers required in describing the larva's transport route. Further, a description must take into account that the water mass in which the larva swims is itself moving through space in complex and difficult-to-predict ways (Harris 1980).

In order to provide an accurate, comprehensive, and easily assimilated understanding of the transport of Atlantic menhaden, then, a descriptive model should simulate a temporal sequence of synoptic "frames," like a motion picture, showing how the larvae move through time and space. One would want to be able to "zoom" up to observe the entire population as it dispersed through time, and to "zoom" down to observe larval behavior on small spatial and short temporal scales. All this implies a need for sampling synoptically and repeatedly at a range of spatial scales, and measuring not merely fish densities but a series of environmental variables as well. It seems likely that the data required to develop such a descriptive model will only be obtained in a piecemeal fashion over a long period of time, if at all. However, several researchers have demonstrated that remarkable insights into larval transport phenomena can be gained with rather limited resources if ingenious sampling designs and novel analytical approaches are used.

Hypothesis Testing versus Estimation

Platt (1964) argued that science advances most rapidly through the systematic application of an analytical inductive method he termed "strong inference." The scientist, Platt contended, should approach any problem by first devising a set of alternative hypotheses, then designing and carrying out "crucial" experiments that will permit exclusion of one or more of the hypotheses. Repeated application of these steps produces a branching logical tree and rapid progress in understanding. It is not irrelevant that he chose to illustrate the power of the method with examples from experimental molecular biology and high-energy physics.

Quinn and Dunham (1983) rejected "strong inference" as an appropriate model for investigating ecological phenomena, arguing that "postulated ecological causes or relationships can rarely be strictly disproven, although they may often be shown to be unimportant or improbable." They go on to point out that, because patterns and processes in natural communities generally result from many causal factors that act simultaneously and interactively, the objective of research is usually to evaluate the relative importance of the different causal factors, rather than to distinguish between several mutually exclusive alternative hypotheses.

At first glance, testing hypotheses through the formal methodology of statistical analysis is attractive because it appears to offer an objective basis for separating signal from background noise. However, there is the inevitable trade-off between failure to detect that a null hypothesis is false (type-II error) and rejection of the null hypothesis when it is true (type-I error). Everything else being equal, if the investigator decides to reduce the probability of type-I error by specifying a priori a lower value of alpha, there is a simultaneous decrease in the power of the test to detect "real" differences. The magnitude of a difference required for "statistical significance," other things being equal, is a function of replication; at low levels of replication, the researcher may be in a very weak position to detect treatment effects, even when they are of substantial magnitude. At the other extreme, high levels of replication may allow the researcher to detect treatment effects that are of such low magnitude that they are, for all intents and purposes, trivial.

Ross (1985) has reminded us that hypothesis testing is framed in terms of risk, the money or other resources that will be lost as a result of a given decision. As such, Ross argued, "it is in fact irrelevant to science which is concerned with making sense, not conserving resources."

Box et al. (1978) introduced the concept of statistical hypothesis testing in a particularly ingenious way. They showed how a time series can be used as a "relevant reference set" for testing a hypothesis about, in their example, the yield of a chemical following modification of an industrial process. That is, their test employs the actual measured yields of the chemical prior to modification of the process as a basis of comparison rather than a mathematical distribution such as Student's t. Their test requires no assumptions about statistical independence of successive values or about random sampling, unlike many standard techniques that are often incorrectly applied to temporal or spatial sequences of data. Box et al. also discussed the advantage of confidence interval estimation over hypothesis testing: "Significance testing in general has been a greatly overworked procedure, and in many cases where significance statements have been made it would have been better to provide an interval within which the value of the parameter would be expected to lie."

The appeal of statistical hypothesis testing is that it leads to relatively unequivocal decisions. Yet empirical data never conform exactly to the mathematical distributions underlying parametric theory, and so tests are to a greater or lesser extent approximations. The confidence limits approach, on the other hand, reminds the researcher that he has measured some average or pattern with a certain level of imprecision. The reason why the theory associated with estimation would seem to be more relevant to larval transport research is that the taking of measurements out in nature is fundamentally a procedure to enable accurate description. Whatever control is employed (e.g., depth of a sampling device) is nominal at best and therefore evidence bearing on a hypothesis is necessarily associative or correlative in principle—and this remains true whether hypothesis-testing statistics are computed or not.

Statistical hypothesis testing is most appropriate in an experimental context wherein treatments can be randomly assigned to experimental units. In larval transport research, statistical hypothesis testing seems to be most useful in the analysis of controlled experiments comparing, for example, behavioral or physiological responses of a species to different levels of a stimulus (e.g., Kelly et al. 1982). Interpretation of statistical tests of hypotheses becomes increasingly problematical as one moves from relatively simple phenomena, in which univariate hypotheses are tested with rigidly controlled experiments and treatments are randomly assigned to experimental units, towards complex multivariate phenomena measured in a natural system, in which several uncontrolled factors may simultaneously influence several response variables and treatments are not (or can not be) randomly assigned to experimental units.

The inadequacy of statistical hypothesis testing of ecological data is well illustrated by a recent paper by Millard and Lettenmaier (1986). They described a procedure for optimizing the statistical power of monitoring programs designed to detect ecological change following an environmental perturbation such as activation of a new wastewater discharge system. They admit the ambiguity implicit in the interpretation of tests of hypotheses for before–after designs "even if the results of the ANOVA [analysis of variance] indicate a change in organism density after the occurrence of the event, the change can reasonably be attributed to the event only if it is known that no other changes occurred in the environment that might also have affected the density." They then conclude: "A better approach, therefore, is to utilize control stations." But the "control station" approach simply adds additional assumptions about spatial uniformity, aside from wastewater discharge, between the control and potentially perturbed stations. Thus, a strong element of faith remains within the hypothesis test, and the skeptic can question the purpose of computing exact ratios of mean squares and P-values.

Perhaps ecologists have been oversold on the desirability of, and necessity for, formal statistical hypothesis testing because of the emphasis this topic usually receives in courses on applied statistics and in applied statistical textbooks (e.g., Snedecor and Cochran 1967; Winer 1971; Steel and Torrie 1980; Sokal and Rohlf 1981). It is true that experimentation both in the laboratory and in the field can provide profound insights into ecological phenomena, but it is also true that ecology is an observational science, necessarily concerned with accurate description of uncontrolled phenomena out in the natural world. It is for the latter reason that the statistical training of ecologists deserves more emphasis on the statistical theory of sampling and survey design. This branch of statistics focuses, for example, on the efficiency (in terms of costs and precision) of alternative sampling designs for estimation of parameters, and on ways of quantifying and dealing with bias from various sources (e.g., Francis 1984; Hankin 1984; Schweigert et al. 1985). Ecologists, perhaps, have not sufficiently appreciated that the validity of parameter estimates based upon sampling depend crucially upon the way in which the sample units to be actually measured are selected from among those in the population. The term "scientific sample" denotes a sample in which the measured sample units are selected with known (not necessarily equal) probability.

Multivariate analysis (Pielou 1984), time-series analysis (Bloomfield 1976), spectral analysis (Denman 1975), and circular-distribution analysis (Batschelet 1981) are other branches of applied statistics that seem particularly useful in larval transport research.

Weak versus Strong Null Hypotheses

Whether one chooses to employ statistical hypothesis testing or not, one still should formulate null hypotheses. The role of null hypotheses and so-called "null models" in community ecology has received a great deal of attention in recent years (Strong et al. 1984). Although their value for addressing certain ecological phenomena is debatable (Quinn and Dunham 1983), properly formulated null hypotheses can provide a necessary basis of reference against which empirical data may be compared (Strong 1980).

Too often, biologists have been content with weak or trivial null hypotheses that specify, for example, that mean values are equal or that a regression or correlation coefficient is zero, even when experience indicates such a null hypothesis is false, foolish, or absurd. Trivial null hypotheses of this form are intellectually unsettling if for no other reason than that they imply *nothing* is known about the phenomenon in question. To quote Ross (1985): "If knowledge were accumulating the incumbent hypothesis would hardly ever be that of no correlation or no effect. Yet this is what is tested routinely, as if all that had gone before counted for nought."

The null hypothesis for a specific case should specify the current state of knowledge of the investigator with regard to a phenomenon. For example, abundant scientific evidence and common experience confirm that larval Atlantic menhaden spawned on the continental shelf are subsequently found within North Carolina estuaries. The question then is not whether transport occurs, but rather: "Are more Atlantic menhaden transported to the estuaries than one would expect based on random dispersal?" Atlantic menhaden are highly fecund (Nelson et al. 1977), and it is conceivable that their abundance in estuaries has obscured the possibilities (1) that the vast majority of each cohort fails to reach estuaries because they disperse to inhospitable environs and subsequently die, and (2) that variation in annual cohort strength in estuaries simply reflects variation in annual patterns of circulation on the continental shelf. If we are to assess the relative effectiveness of transport, we must account for those elements of the spawn that do not attain their nursery ground, as well as those that do (Leggett 1984). This would require an extensive program to monitor the currents, eggs, and larvae on the outer continental shelf and the Gulf Stream for several months, in each of several years. In any case, it is apparent that many larval fishes are carried far beyond their normal ranges (Markle et al. 1980; Flierl and Wroblewski 1985).

Models provide the grist for specifying nontrivial null hypotheses. Fortier and Leggett (1982) have skillfully employed advection–diffusion models in larval transport research and DeAngelis and Yeh (1984) have demonstrated the utility of biased random-walk models for simulating transport. Miller et al. (1984) have described seasonal circulation patterns over the continental shelf off North Carolina. A null hypothesis appropriate for investigation of cross-shelf transport off North Carolina might, therefore, consist of a three-dimensional, biased, random-walk model that incorporates the circulation patterns described by Miller et al. (1984) for the time between spawning on the outer shelf and recruitment of larvae to the estuary. If mortality rates were incorporated into the model, the model could be used to provide location- and depth-specific estimates of larval density, which could then be compared with empirical measurements of larval density and current vectors at depth (Legendre and Demers 1984).

Scaled-down physical models also have application to larval transport research. For example, the construction of a scaled-down physical replica of Oregon Inlet permitted simulation of current patterns and larval transport through the inlet, both with and without the proposed large jetties in place (Hollyfield and Frankensteen 1980). These simulations could provide guidance in the allocation of sampling effort in time and space as well as a framework against which to compare empirical measurements of transport through the inlet.

Vertical Migration Models

By moving vertically in the water column, an organism may (1) enter an estuary or move landward or both, (2) maintain its approximate horizontal position within an estuary and avoid being flushed seaward, or (3) enhance its movement seaward (Rogers 1940; Pearcy 1962; Pearcy and Richards 1962; Graham and Davis 1971; Graham 1972; Able 1978; Weinstein et al. 1980; Fortier and Leggett 1982; Rothlisberg 1982). In some cases, the organism makes regular vertical movements in phase with the tide; in others, it attains a partic-

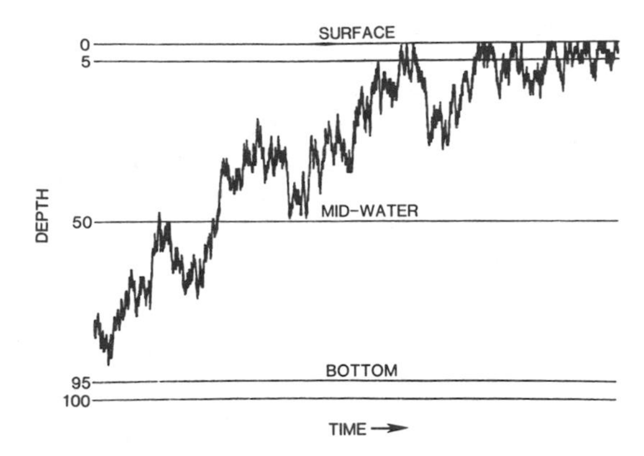

FIGURE 1.—The simulated track of an organism moving about the water column over time. The water column has an arbitrary depth of 100 units. The track represents 5,000 stochastic steps of a random walk; the organism has a 50% probability of moving either up or down, or, if it is a boundary, a 50% probability of either staying at or moving from the boundary.

ular stratum in the water column and stays there. These considerations suggest that stochastic models of vertical migration may be useful as null hypotheses in studies of vertical migration as a larval transport mechanism.

As a first model for examining the behavior of a hypothetical organism moving randomly in the vertical plane, let us consider a simple random-walk Markov chain (Parzen 1960; Hillier and Lieberman 1980; Legendre and Legendre 1983) in which the organism, with equal probability, moves either one unit up or one unit down during each time interval. Think of the water column as subdivided horizontally into several "slices" of unit thickness. On entering the uppermost slice, bounded by the surface film, the organism, during the next time interval (or step), either remains at the surface with probability p, or moves one unit downward with probability $q = 1 - p$. The corresponding specification applies to the organism if it enters the lowermost slice as well. A pseudorandom number generator on a computer was used to simulate the track of one such hypothetical organism through time for the case in which $p = q = 0.50$ (Figure 1). The transition matrix for this model, and others to be considered, will have as many rows (and columns) as there are slices making up the water column (Table 1).

One could use Monte Carlo simulations to track the organism through time and, by keeping account of how much time it spent in each slice, get a picture of its expected vertical distribution. But there is an easier, analytical, approach. Parzen (1960) showed that a Markov chain of this form will achieve a statistical equilibrium after a large number of steps (movements of the organism). One can, therefore, directly compute the stationary probabilities for each slice of the water column. For the model discussed above, where $p = q = 0.5$, the stationary probabilities are equal, meaning that, over the long term, the organism will spend an equal amount of time in each slice.

The attractiveness of the bottom and surface can be varied by varying the transition probabilities, p and q, in the top and bottom rows of the transition matrix (Table 1). Recall that q is the probability that in the next time interval the organism moves from the top or bottom slice towards midwater. If the water column is arbitrarily subdivided into 100 slices, the stationary probabilities (π) are:

$\pi_1 = \pi_{100} = (2 + 196q)^{-1}$ for the surface and bottom slices, and $\pi_2, \pi_3, \ldots, \pi_{99} = (2q)(2 + 196q)^{-1}$ for intermediate slices. So for any speci-

TABLE 1.—Transition matrix for a random-walk Markov chain in which an organism has an equal probability of moving one unit up or one unit down during each successive time interval. If it reaches the surface it will remain there during the next time interval with probability p, or move down one unit with probability $q = 1 - p$. Similarly, if it reaches the bottom it will remain there during the next time interval with probability p, or move upwards one unit with probability q.

p	q	0	0	0	•	•	•						
0.5	0	0.5	0	0	•	•	•						
0	0.5	0	0.5	0	•	•	•						
0	0	0.5	0	0.5	•	•	•						
•	•	•	•	•				•	•	•	•	•	•
•	•	•	•	•				•	•	•	•	•	•
•	•	•	•	•				•	•	•	•	•	•
					•	•	•	0.5	0	0.5	0	0	0
					•	•	•	0	0.5	0	0.5	0	0
					•	•	•	0	0	0.5	0	0.5	0
					•	•	•	0	0	0	0.5	0	0.5
					•	•	•	0	0	0	0	q	p

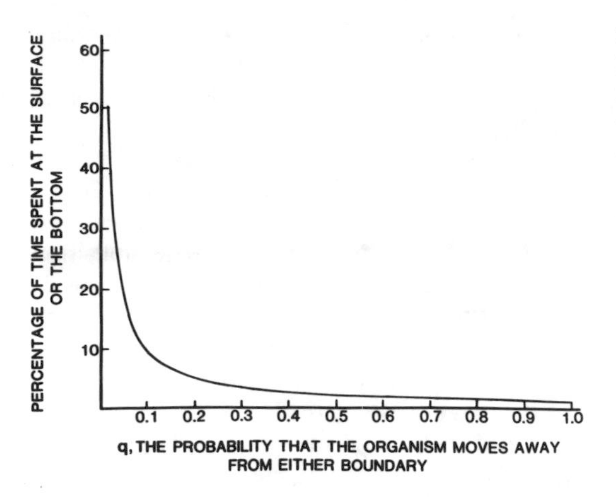

FIGURE 2.—Relationship between the expected amount of time an organism would occupy the surface or bottom "slices" of the water column and the probability that, on encountering either boundary, the organism moves toward midwater, based on a random-walk, Markov-chain model.

fied value of q, the proportion of time spent in the boundary slices will be $2(2 + 196q)^{-1}$. Only when q drops below about 0.3 does the proportion of time spent in the two boundary slices appreciably increase (Figure 2).

In the above model, the organism does not "recognize" the surface or bottom, and consequently modify its behavior, until it is within either the upper or lower one-hundredth of the water column. In nature, however, water velocity decreases as one approaches a fixed boundary (Vogel 1981), and it is worthwhile to consider a model in which the larvae take advantage of the slower velocities of the boundary layer in the lowermost portion of the water column to make greater progress against a current, or, at least, to reduce the rate at which they would otherwise be swept seaward on a falling tide. In such a model, it would be assumed that the larvae detect the lower current velocities of the boundary layer and respond by decreasing the probability that they will move upward in the water column away from the slower velocities adjacent to the bottom. A simple random-walk Markov chain to model the hypothesized boundary layer transport mechanism would have transition matrix probabilities that reflect logarithmically decreasing water velocities down through the five lowermost "slices" that make up the boundary layer, but that here, for simplicity, are held constant in the uppermost 95 slices (Table 2). Above the boundary layer, then, the organism is equally likely to move up or down, but once it enters the boundary layer its probability of moving up sharply decreases. The stationary probabilities for this model indicate that one would expect to find the hypothetical larval fish in the lowest (100th) slice about 98.8% of the time. Additional realism could be incorporated into this admittedly oversimplified model by making provision for variation in current velocities throughout the water column.

The attractiveness of this particular model for me arises from my observation of thousands of very young sockeye salmon *Oncorhynchus nerka* swimming steadily upstream through a stretch of rapids by remaining very close to the cobble bottom of a river in western Alaska. Several published studies have suggested the importance of the boundary layer to small fish likely to be swept downriver (Weinstein et al. 1980) or offshore by current (Brewer et al. 1984; Brewer and Kleppel 1986) and, therefore, lend indirect support to the boundary layer model. At very high current velocities, however, the converse situation may also apply because of bottom scouring. A correlation between catch rates of small plaice *Pleuronectes platessa* and amounts of resuspended matter in the water column led Rijnsdorp

TABLE 2.—Transition matrix for a random-walk Markov chain in which the transition probabilities for an organism reflect an approximately logarithmic decrease in water velocity in a hypothetical lower boundary layer.

0.5	0.5	0	0	0	•	•	•							
0.5	0	0.5	0	0	•	•	•							
0	0.5	0	0.5	0	•	•	•							
•	•	•	•	•				•	•	•	•	•	•	•
•	•	•	•	•				•	•	•	•	•	•	•
•	•	•	•	•				•	•	•	•	•	•	•
					•	•	•	0	0.5	0	0	0	0	0
					•	•	•	0.5	0	0.5	0	0	0	0
					•	•	•	0	0.5	0	0.5	0	0	0
					•	•	•	0	0	0.18	0	0.82	0	0
					•	•	•	0	0	0	0.07	0	0.93	0
					•	•	•	0	0	0	0	0.03	0	0.97
					•	•	•	0	0	0	0	0	0.01	0.99

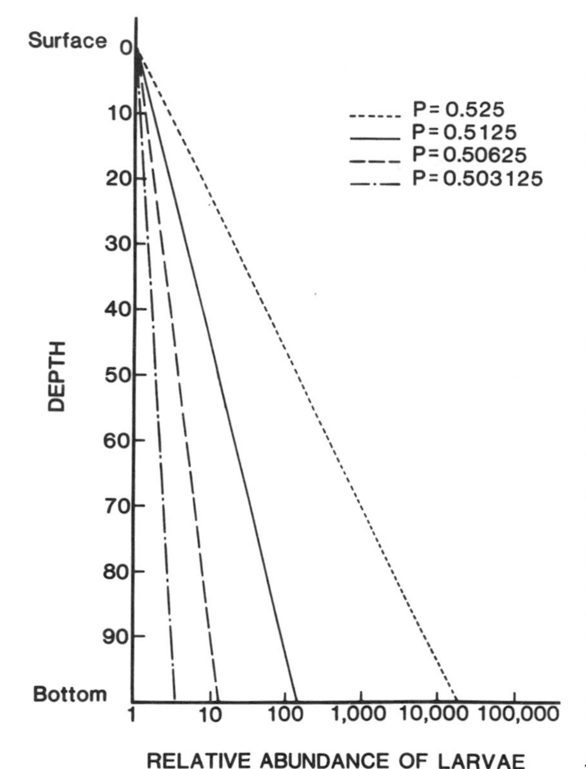

FIGURE 3.—Stationary vertical distributions of larvae in the water column that apply for different values of p; p is the probability that, in midwater, an organism will move downward, and $1.0 - p = q$ is the probability that it will move upward. The stationary probability for the ith horizontal slice of water is $(p/q)^j\ [1 - (p/q)]\ [1 - (p/q)^{100}]^{-1}$; $j = 0, 1, 2, \ldots, 99$. Relative abundance is expressed as multiples of the number of larvae at the surface.

et al. (1985) to suggest that the fish were being swept off the bottom at the time of peak tidal flows.

Finally, let us consider a model that reveals how only a very subtle vertical preference on the part of the organism implies a substantial departure from a uniform distribution. A solution presented by Parzen (1960) for a random-walk model with retaining barriers can be used to model the vertical distribution of a larva that has a greater tendency to swim upwards than downwards, or vice versa, perhaps because of light or temperature preferences, predator avoidance, or feeding. Even if the organism has only a slightly greater probability of moving downward than upward, it will spend far more time in the lower part of the water column (Figure 3). For example, if the probability of the organism moving downwards is only 0.503125 versus 0.496875 for moving upwards, the model implies that the organism should be about five times more abundant near the bottom than near the surface. Slightly increasing the probability of moving downward from 0.503125 to 0.5125 sharply increases the departure from a uniform distribution (Figure 3).

In order to use a random-walk Markov chain to model the vertical migration or distribution of a particular larval fish or invertebrate, it would be necessary to incorporate information on the depth of the water column and the organism's potential rate of movement vertically, i.e., to introduce specific scales to the model. In a more general sense, if we assume a larval fish moves forward at the rate of 1 body length/s, then we can examine the relationship between the minimum time required to move 10 m up or down in the water column, and the angle of declination of the fish's path from the horizontal (Figure 4). For example, a 20-mm larval fish tilting downward at an angle of 4° would descend 10 m in about 2 h. This sort of approach only provides a rough idea of what is possible, given certain assumptions. By way of comparison, empirical measurements of free-living, 8.1–10.0-mm yellowtail flounder *Limanda ferruginea* showed that they may move vertically as much as 25 m in 3 h (Smith et al. 1977). Furthermore, many marine invertebrate larvae can move vertically at rates of 10–50 m/h (Mileikovsky 1973; Righter 1973).

I have described these random-walk Markov-chain models in terms of the vertical dimension, with the surface and bottom or, for example, surface and pycnocline as boundaries, but they can serve in the horizontal plane as well to model phenomena such as the accumulation of organisms along a frontal system. Because, in general, the current will decrease as one approaches the lateral boundaries of an estuary or tidal creek, the boundary layer model could be conceptually rotated 90° to simulate the aggregation of larvae along shorelines.

Statistical Design Considerations

I next discuss certain statistical design concepts as they apply to measuring larval transport through coastal inlets. It is assumed that the purpose of such a study would be to either measure the flux of larvae or to test hypotheses about mechanisms employed by larvae to effect their transport.

An initial consideration is that of scales (Haury et al. 1977; Harris 1980; Gagnon and LaCroix 1982). An organism may move through a medium that is itself moving, and questions of appropriate

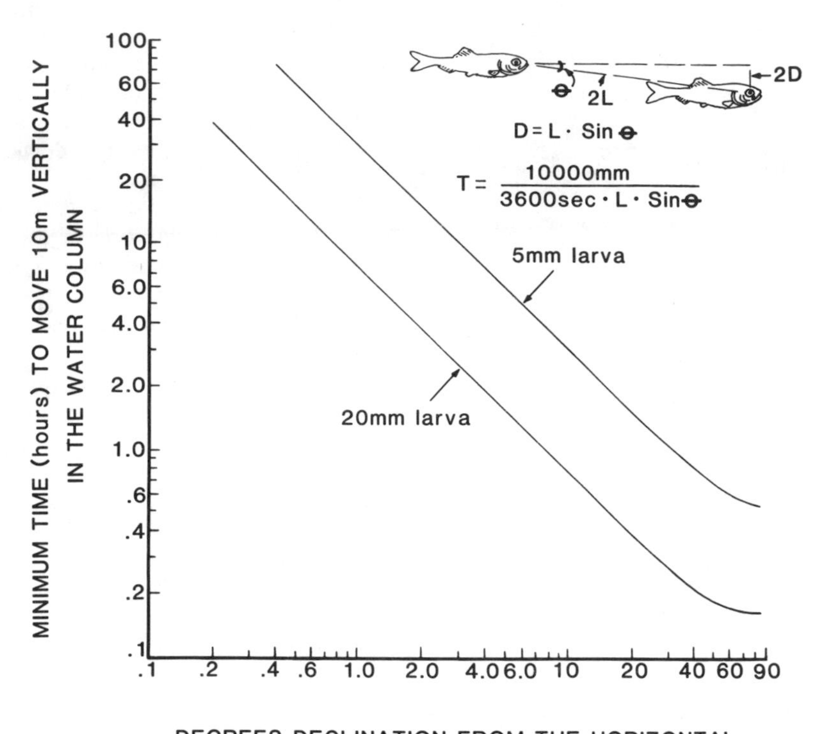

FIGURE 4.—Relationship between the rate of vertical movement (T) and the angle of declination from the horizontal for a larval fish moving forward at one body length (*L*) per second.

temporal and spatial scales for the design arise immediately. If, for example, the hypothesis involves the larva's response to physical features having scales on the order of 10s of centimeters, the design and the sampling device must provide for replicate sampling on the order of centimeters. Similarly, if the system under study involves any periodic source of variability (e.g., tide) then the design must provide for a sampling frequency at least four times the frequency of the periodic influence in order to avoid a biased and overly simplistic representation of the system (Platt and Denman 1975; Kelly 1976). If episodic, short-term events are hypothesized to have major effects on transport, the sampling period must be long enough to encompass several such events, as well as nonevent periods for comparison. If the purpose of the research is to compare flux of larvae across a cross-section of an inlet under different environmental conditions, one needs to be aware of sources of bias that may arise from flux computation methods based on the product of average velocity times average concentration (density) (see Boon 1978).

A second consideration is the care that must be exercised to ensure that the population of sample units from which a sample will be selected (sampling population), coincides with the population of sampling units for which the investigator wishes to make inferences (target population) (Cochran 1977). In larval transport research, a sampling unit might be a certain volume of water in a specified time interval. The target population might then be specified as all those volumes of water contained within some geographical boundaries during the specified time interval. However, if all of the sample units are not accessible to the sampling device, the sampling population will not coincide with the target population, and inferences about the target population will require extrapolation. The magnitude of the bias will depend upon the extent to which the excluded sample units differ in terms of the dependent variable(s) from those in the sampling population.

A third statistical concept is that of replication. The validity of many types of statistical inferences is grounded in independent replication at the design level(s) at which inferences will be made.

This means that, if inferences are to be made about the vertical distribution of a particular organism in a particular estuary during a particular time interval, samples must be taken repeatedly and independently throughout the water column within that time interval and location. If inferences are to be more general about the vertical distribution of this species in estuaries, for example, samples must be taken from more than one estuary. If one wants to make inferences about the consistency of an annual pattern, one must replicate at the level of years, etc. The more varied external environmental factors are during the period, the stronger one's inferences are; i.e., the fewer the caveats that need to be attached to the inferences.

The question of how much replication is adequate is not easily answered even though formulae for estimating sample size requirements are available (e.g., Cochran 1977; Cohen 1977; Green 1979). A conventional procedure is to estimate the variation likely to be encountered either by doing some preliminary sampling or by making calculations from existing data. In theory, this approach has much to recommend it, but, because sample variances themselves have variances, estimates of required sample size may be far from the mark. For example, the distribution of sample variances from a negative binomial distribution, often an appropriate model for positively skewed distributions of catch rates, shows that estimated sample size requirements should be viewed as rough approximations (Figure 5).

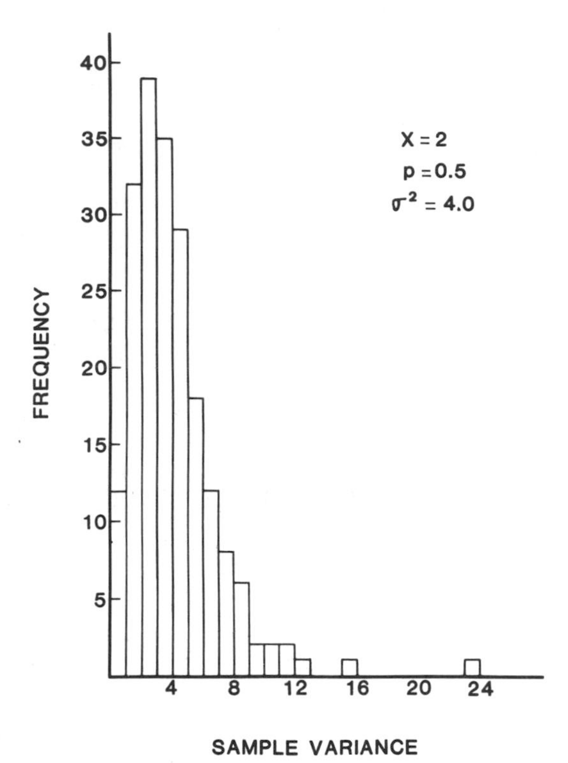

FIGURE 5.—The frequency distribution of sample variances for 200 random samples of size 10 from a negative binomial distribution; X is the Bernoulli success parameter, σ^2 is the population variance; p is the Bernoulli probability parameter.

Replication levels often are determined not by the inherent variability of the variable(s) being measured, but by constraints imposed by the availability of sampling equipment, personnel, and other resources. When this is the case, estimates of inherent variability can be used to estimate the likely precision of parameter estimates. This information then can be used to refine the design in terms of allocation of total sampling effort. At this stage, decisions can be made about what information to sacrifice and what variables to sample at higher intensity. In extreme cases, there may be such an imbalance between required precision and available resources for sampling that it makes no sense to continue beyond the initial planning phase. The patchiness of ichthyoplankton means, in general, that relatively high levels of replication will be required to estimate mean densities with even a moderate level of precision. To develop a description of the movements of young plaice over the tidal cycle in a 100 × 1,000-m area, Kuipers (1973) took "about a thousand" samples with a small trawl.

Some of these statistical considerations will now be addressed, first in the context of an investigation of larval flux through an inlet, and later in the context of examining a hypothesis about a transport mechanism.

We might begin a design for a larval flux study by deciding on the three-dimensional space that is relevant. Inasmuch as inferences are to be made about transport through the inlet as a whole, the water within the entire cross-section of the inlet must be included within the sampling population. Unless the water is homogeneous with respect to chemical and physical properties and organism densities, the efficiency of the design would be increased by subdividing the cross-section into two or more strata, portions of the cross-section that are expected to be relatively homogeneous. Note that there is no requirement that all the sampling units making up a stratum be contiguous, although this may usually be the case.

If we wish to obtain a broader picture of circulation patterns and larval densities, we would extend the sampling seaward and landward. Because larvae can attain more than intermittent (tidal) residency in the estuary only if they move from the water mass being advected in and out on the tidal cycle, we might especially wish to extend the sampling landward to where turbulent diffusion and active swimming (Fortier and Leggett 1982) permit the larvae to enter estuarine water. Further, it would be desirable to extend the sampling seaward from the inlet to include any areas that might be expected to accumulate larvae because of hydrography (Tanaka 1985).

The temporal dimension of the design will depend upon the null hypothesis as well. For example, if vertical migration is the hypothesized mechanism of transport, stratification of the design in the temporal dimension would be advantageous because one could then disproportionally allocate sampling effort to those times when the mechanism is predicted to be most operational. On the other hand, if the hypothesis involves the importance of irregular, sporadic events, the sampling must extend over a sufficiently long period to include several of these events. In the sense that the flux of larvae through an inlet is like the movement of a fluid through a transparent pipe, irregularities (bubbles in the fluid) may be especially enlightening about the rate of transport. Thus, if the sampling frame extends from seaward of the inlet to well up in the estuary, it may be feasible to track irregularities in larval densities (or cohorts, if they can be identified; see Graham and Townsend 1985) through time and space and thereby estimate rates of movement.

After stratifying the sampling frame into what seems to be relatively homogeneous sets of sample units, one is faced with decisions about sampling gears, concurrent measurements of chemical and physical variables, replication levels, and sampling frequency. All of these will depend upon particular systems, organisms, and hypotheses, so it is difficult to generalize beyond the suspicion that designs for the study of transport mechanisms will be far easier to implement than designs for measuring flux (Imberger et al. 1983). When those attempting to measure the flux of sediments, particulate matter, or nutrients in the estuary throw up their arms in dispair (Kjerfve et al. 1978), there seems no cause for optimism for those interested in accurately measuring the flux of actively swimming fishes and crustaceans.

On the other hand, if the hypothesis concerns a transport mechanism, then relatively simple designs can be very effective. In a series of papers, Fortier and Leggett (1982, 1983, 1984, 1985) demonstrated the insights to be gained from relatively simple designs that involve sampling several depth strata at frequent intervals for periods of several days. The high frequency and concurrent sampling of fish larvae, current velocities and directions, salinities, and temperature allowed Fortier and Leggett to employ spectral analysis and cross-correlation analysis to relate characteristics of the fish to characteristics of the water mass from which they were collected. They then placed these findings in the context of other information on the circulation of the St. Lawrence estuary and biology of the fish to give us some of our clearest evidence of transport mechanisms for larval fishes.

Averages and Short-Term Events

If one is sampling from a single statistical distribution, it makes sense to focus on measures of central tendency such as the arithmetic or geometric mean. But if one is dealing with a dynamic variable that continually changes in response to several external factors, it makes more sense to focus upon that variable's pattern of change and its relationship to the patterns of change of environmental factors. The spatial and temporal heterogeneity of zooplankton patches, for example was recently underscored by Omori and Hamner (1982), who reported that swarms of some zooplankters can have 1,000 times the average population density. Where that level of heterogeneity is common, the meaning of an average is elusive, and the planktologist's preoccupation with patchiness probably reflects an implicit recognition of this. In this regard, Mackas et al. (1985) pointed out that it is thus only the exceptional organism that lives out its life under "average conditions."

Several authors have argued that more attention should be concentrated on measuring transport during what may be relatively ephemeral events, such as the passage of cold fronts (Weinstein 1981; Yoder 1983; Shaw et al. 1985). It is during brief, anomalous environmental conditions that we may gain our greatest insights into transport mechanisms, either because the mechanisms only function episodically or because environmental extremes induce extreme variation in transport rates and thereby increase the chances of detecting underlying mechanisms. Because of the inherent unpredictability of such events, however, one may be forced to sample for protracted periods in order to observe enough anomalies to draw conclusions.

Additional Considerations

Ontogenetic changes in the physiology and behavior of larvae imply that the efficiency of a given sampling gear may change substantially during the period of transport from the outer continental shelf to estuarine waters. A slowly towed bongo net may accurately measure the density of larvae that are less than a week old and well up in the water column, but it may give very biased estimates of older larvae that are more concentrated in a boundary layer or are strong enough swimmers to avoid the net. If an organism's transport through an inlet is largely contingent on its acquisition of swimming competence, sampling techniques must change from the outer continental shelf to the estuary to maintain comparable capture efficiencies. As Omori and Hamner (1982) argued, too great an emphasis on standardization of sampling gear can easily lead to inaccurate measurements of plankton densities. They pointed out that the question should determine the sampling technique, not the other way around. They went on to state: "If we all sample the same way, we are constrained to examine our data from the same point of view, models will be further refined, and the data will take on a fascinating precision . . . which is mistaken for true and accurate description."

Longhurst (1984) described how new sampling instruments have changed our concepts of the pattern of biota in the upper part of the water column: "Two principal findings have emerged: first, that some of the observed horizontal patchiness was an artifact because horizontal layers do not always lie quite flat, so that their peaks and troughs may be recorded as spurious patchiness in the horizontal plane; second, and much more importantly, that the layering of biota and related variables is of a repeatability and complexity previously unsuspected."

Concluding Remarks

The theory of experimental design, sample survey design, and such parametric procedures for statistical inference as the analysis of variance, have been around for decades. However, the increasing availability of digital computers in recent years has made feasible the development of an entirely new methodology for assessing statistical error. These new methods extend, and to some extent replace, older analytical methodologies based upon assumed mathematical distributions with nonparametric approaches that utilize the power of digital computers to rapidly manipulate, resample, and calculate. These new methods both avoid assumptions about underlying distributions and provide for more comprehensive examination of the characteristics of data (Efron and Gong 1983). These "computer-intensive" methods include the bootstrap, the jackknife, cross-validation, random subsampling, and balanced repeated replication (Efron 1982; Diaconis and Efron 1983). They share in common the feature that a large number of "fake" data sets are generated by sampling with replacement from an original data set. The desired statistic (e.g., mean, median, regression coefficient, principal component coefficient, etc.) is computed for each fake data set, and the resulting distribution of these values is then used in assessing the probable error of the corresponding estimate computed from the original data. Kimura and Balsiger (1985) have provided an example of how the bootstrap can be used to improve fish survey design, and Heltshe and Forrester (1985) used the jackknife to examine the statistical properties of Brillouin's and Simpson's indices of diversity. It seems likely that these new methods will find a great deal of application in all areas of science, including marine ecology.

Acknowledgments

I thank C. W. Krouse for assistance in computer programming, J. L. Fulford for typing the manuscript, and H. R. Gordy for preparation of the figures. The manuscript was improved by suggestions from A. B. Powell, J. J. Govoni, J. W. Whitsett, G. W. Thayer, F. A. Cross, and two anonymous reviewers.

References

Able, K. W. 1978. Ichthyoplankton of the St. Lawrence estuary: composition distribution and abundance. Journal of the Fisheries Research Board of Canada. 35:1518–1531.

Batschelet, E. 1981. Circular statistics in biology. Academic Press, New York.

Bloomfield, P. 1976. Fourier analysis of time series: an introduction. Wiley, New York.

Boon, J. D., III. 1978. Suspended solids transport in a salt marsh creek—an analysis of errors. Pages 147–160 *in* B. Kjerfve, editor. Estuarine transport processes. University of South Carolina Press, Columbia.

Box, G. E. P., W. G. Hunter, and J. S. Hunter. 1978. Statistics for experimenters: an introduction to design, data analysis and model building. Wiley, New York.

Brewer, G. D, and G. S. Kleppel. 1986. Diel vertical distribution of fish larvae and their prey in near-shore waters of southern California. Marine Ecology Progress Series 27:217–226.

Brewer, G. D., G. S. Kleppel, and M. Dempsey. 1984. Apparent predation on ichthyoplankton by zooplankton and fishes in near-shore waters of southern California. Marine Biology (Berlin) 80:17–28.

Cochran, W. G. 1977. Sampling techniques, 2nd edition. Wiley, New York.

Cohen, J. 1977. Statistical power analysis for the behavioral sciences. Academic Press, New York.

Crawford, R. E., and C. G. Carey. 1985. Retention of winter flounder larvae within a Rhode Island salt pond. Estuaries 8:217–227.

DeAngelis, D. L., and G. T. Yeh. 1984. An introduction to modeling migratory behavior of fishes. Pages 445–469 *in* McCleave et al. (1984a).

Denman, K. L. 1975. Spectral analysis: a summary of the theory and techniques. Canada Fisheries and Marine Service Technical Report 539.

Diaconis, P., and B. Efron. 1983. Computer-intensive methods in statistics. Scientific American 248(5):116–130.

Efron, B. 1982. The jackknife, the bootstrap and other resampling plans. Society for Industrial and Applied Mathematics, Philadelphia.

Efron, B., and G. Gong. 1983. A leisurely look at the bootstrap, the jackknife, and cross-validation. American Statistician 37:36–48.

Flierl, G. R., and J. S. Wroblewski. 1985. The possible influence of warm core gulf stream rings upon shelf water larval fish distribution. U.S. National Marine Fisheries Service Fishery Bulletin 83:313–330.

Fortier, L., and W. C. Leggett. 1982. Fickian transport and the dispersal of fish larvae in estuaries. Canadian Journal of Fisheries and Aquatic Sciences 39:1150–1163.

Fortier, L., and W. C. Leggett. 1983. Vertical migrations and transport of larval fish in a partially mixed estuary. Canadian Journal of Fisheries and Aquatic Sciences 40:1543–1555.

Fortier, L., and W. C. Leggett. 1984. Small-scale covariability in the abundance of fish larvae and their prey. Canadian Journal of Fisheries and Aquatic Sciences 41:502–512.

Fortier, L., and W. C. Leggett. 1985. A drift study of larval fish survival. Marine Ecology Progress Series 25:245–257.

Francis, R. I. C. C. 1984. An adaptitive strategy for stratified random trawl surveys. New Zealand Journal of Marine and Freshwater Research 18:59–71.

Gagnon, M., and G. LaCroix. 1982. The effects of tidal advection and mixing on the statistical dispersion of zooplankton. Journal of Experimental Marine Biology and Ecology 56:9–22.

Graham, J. J. 1972. Retention of larval herring within the Sheepscot estuary of Maine. U.S. National Marine Fisheries Service Fishery Bulletin 70:299–305.

Graham, J. J., and C. W. Davis. 1971. Estimates of mortality and year class strength of larval herring in western Maine, 1964–67. Rapports et Procès-Verbaux des Réunions, Conseil International pour l'Exploration de la Mer 160:147–152.

Graham, J. J., and D. W. Townsend. 1985. Mortality, growth and transport of larval Atlantic herring *Clupea harengus* in Maine coastal waters. Transactions of the American Fisheries Society 114:490–498.

Green, R. H. 1979. Sampling design and statistical methods for environmental biologists. Wiley, New York.

Hankin, D. G. 1984. Multistage sampling designs in fisheries research: applications in small streams. Canadian Journal of Fisheries and Aquatic Sciences 41:1575–1591.

Harris, G. P. 1980. Temporal and spatial scales in phytoplankton ecology. Mechanisms, methods, models, and management. Canadian Journal of Fisheries and Aquatic Sciences 37:877–900.

Haury, L. R., J. A. McGowan, and P. H. Wiebe. 1977. Patterns and processes in the time–space scales of plankton distributions. Pages 277–327 *in* J. H. Steele, editor. Spatial pattern in plankton communities. Plenum, New York.

Heltshe, H. F., and N. C. Forrester. 1985. Statistical evaluation of the jacknife estimate of diversity when using quadrat samples. Ecology 66:107–111.

Hillier, F. S., and G. J. Lieberman. 1980. Introduction to operations research, 3rd edition. Holden-Day, Oakland, California.

Hollyfield, N. W., and E. D. Frankensteen. 1980. Oregon Inlet larval transport sensitivity study. U.S. Army Corps of Engineers, Wilmington, North Carolina.

Hurlbert, S. H. 1984. Pseudoreplication and the design of ecological field experiments. Ecological Monographs 54:187–211.

Imberger, J., and seven coauthors. 1983. The influence of water motion on the distribution and transport of materials in a salt marsh estuary. Limnology and Oceanography 28:201–214.

Kelly, J. C. 1976. Sampling the sea. Pages 361–387 *in* D. H. Cushing and J. J. Walsh, editors. The ecology of the seas. Saunders, Philadelphia.

Kelly, P., S. D. Sulkin, and W. F. Van Heukelem. 1982. A dispersal model for larvae of the deep sea red crab *Geryon quinquedens* based upon behavioral regulation of vertical migration in the hatching stage. Marine Biology (Berlin) 72:35–43.

Kimura, D. K., and J. W. Balsiger. 1985. Bootstrap methods for evaluating sablefish pot index surveys. North American Journal of Fisheries Management 5:47–56.

Kjerfve, B. K., R. Dyer, and J. R. Schubel. 1978. Epilogue: where do we go from here? Pages 319–324 *in* B. Kjerfve, editor. Estuarine transport processes. University of South Carolina Press, Columbia.

Kuipers, B. 1973. On the tidal transport of young plaice (*Pleuronectes platessa*) in the Wadden Sea. Netherlands Journal of Sea Research 6:376–388.

Legendre, L., and S. Demers. 1984. Towards dynamic biological oceanography and limnology. Canadian Journal of Fisheries and Aquatic Sciences 41:2–19.

Legendre, L., and P. Legendre. 1983. Numerical ecology. Elsevier Science Publishing New York.

Leggett, W. C. 1984. Fish migrations in coastal and estuarine environments: a call for new approaches to the study of an old problem. Pages 159–178 *in* McCleave et al. (1984a).

Longhurst, A. 1984. Heterogeneity in the ocean—implications for fisheries. Rapports et Procès-Verbaux des Réunions Conseil International pour l'Exploration de la Mer 185:268–282.

Mackas, D. L., K. L. Denman, and M. R. Abbot. 1985. Plankton patchiness: biology in the physical vernacular. Bulletin of Marine Science 37:652–674.

Markle, D. F., W. B. Scott, and A. C. Kohler. 1980. New and rare records of Canadian fishes and the influence of hydrography on resident and nonresident Scotian Shelf ichthyofauna. Canadian Journal of Fisheries and Aquatic Sciences 37:49–65.

McCleave, J. D., G. P. Arnold, J. J. Dodson, and W. H. Neill, editors. 1984a. Mechanisms of migration in fishes. Plenum, New York.

McCleave, J. D., F. R. Harden Jones, W. C. Leggett, and T. G. Northcote. 1984b. Fish migration studies: future directions. Pages 545–554 *in* McCleave et al. (1984a).

Mileikovsky, S. A. 1973. Speed of active movement of pelagic larvae of marine bottom invertebrates and their ability to regulate their vertical position. Marine Biology (Berlin) 23:11–17.

Millard, S. P., and D. P. Lettenmaier. 1986. Optimal design of biological sampling programs using the analysis of variance. Estuarine, Coastal and Shelf Science 22:637–656.

Miller, J. M., J. P. Reed, and L. J. Pietrafesa. 1984. Patterns, mechanisms and approaches to the study of migration of estuarine-dependent fish larvae and juveniles. Pages 209–225 *in* McCleave et al. (1984a).

Nelson, W. R., M. C. Ingham, and W. E. Schaaf. 1977. Larval transport and year-class strength of Atlantic menhaden, *Brevoortia tyrannus*. U.S. National Marine Fisheries Service Fishery Bulletin 75:23–41.

Omori, M., and W. M. Hamner. 1982. Patchy distribution of zooplankton: population assessment and sampling problems. Marine Biology (Berlin) 72:193–200.

Parzen, E. 1960. Modern probability theory and its application. Wiley, New York.

Pearcy, W. G. 1962. Ecology of and estuarine population of winter flounder, *Pseudopleuronectes americanus* (Walbaum). Part 3. Bulletin of the Bingham Oceanographic Collection, Yale University 18(1):16–38.

Pearcy, W. G., and S. W. Richards. 1962. Distribution and ecology of fishes of the Mystic River estuary, Connecticut. Ecology 43:248–259.

Pielou, E. C. 1984. The interpretation of ecological data: a primer on classification and ordination. Wiley, New York.

Platt, J. R. 1964. Strong inference. Science (Washington, D.C.) 146:347–353.

Platt, T., and K. L. Denman. 1975. Spectral analysis in ecology. Annual Review of Ecology and Systematics 6:189–210.

Quinn, J. F., and A. E. Dunham. 1983. On hypothesis testing in ecology and evolution. American Naturalist 122:602–617.

Righter, G. 1973. Field and laboratory observations of the diurnal vertical migration of marine gastropod larvae. Netherlands Journal of Sea Research 7:126–134.

Rijnsdorp, A. D., M. van Stralen, and H. W. van der Veer. 1985. Selective tidal transport of North Sea plaice larvae *Pleuronectes platessa* in coastal nursery areas. Transactions of the American Fisheries Society 114:461–470.

Rogers, H. M. 1940. Occurrence and retention of plankton within the estuary. Journal of the Fisheries Research Board of Canada 5:164–171.

Ross, J. 1985. Misuse of statistics in social sciences. Nature (London) 318:514.

Rothlisberg, P. C. 1982. Vertical migration and its effect on dispersal of penaeid shrimp larvae in the Gulf of Carpentaria, Australia. U.S. National Marine Fisheries Service Fishery Bulletin 80:541–554.

Schweigert, J. F., C. W. Haegele, and M. Stocker. 1985. Optimizing sampling design for herring spawn surveys in the Strait of Georgia, B.C. Canadian Journal of Fisheries and Aquatic Sciences 42:1806–1814.

Shaw, R. F., W. J. Wiseman, Jr., R. E. Turner, L. J. Rouse, Jr., R. E. Condrey and F. J. Kelly, Jr. 1985. Transport of larval gulf menhaden *Brevoortia patronus* in continental shelf waters of western Louisiana: a hypothesis. Transactions of the American Fisheries Society 114:452–460.

Smith, W. G., J. D. Sibunka, and A. Wells. 1977. Diel movements of larval yellowtail flounder, *Limanda ferruginea,* determined from discrete depth sampling. U.S. National Marine Fisheries Service Fishery Bulletin 76:167–178.

Snedecor, G. W., and W. G. Cochran. 1967. Statistical methods, 6th edition. Iowa State University Press, Ames.

Sokal, R. R., and F. J. Rohlf. 1981. Biometry: the principles and practice of statistics in biological research, 2nd edition. Freeman, San Francisco.

Steel, R. G. D., and J. H. Torrie. 1980. Principles and procedures of statistics, 2nd edition. McGraw-Hill, New York.

Strong, D. R. 1980. Null hypotheses in ecology. Synthesis 43:167–171.

Strong, D. R., D. Simberloff, L. G. Abele, and A. B. Thistle, editors. 1984. Ecological communities: conceptual issues and the evidence. Princeton University Press, Princeton, New Jersey.

Tanaka, M. 1985. Factors affecting the inshore migration of pelagic larval and demersal juvenile red sea bream *Pagrus major* to a nursery ground. Transac-

tions of the American Fisheries Society 114:471–477.
Underwood, A. J. 1981. Techniques of analysis of variance in experimental marine biology and ecology. Oceanography and Marine Biology: an Annual Review 19:513–605.
Vogel, S. 1981. Life in moving fluids. Willard Grant Press, Boston.
Weinstein, M. 1981. Plankton productivity and the distribution of fishes on the southeastern U.S. Continental Shelf. Science (Washington, D.C.) 214:351–352.
Weinstein, M. P., S. L. Weiss, R. G. Hodson, and L. R. Gerry. 1980. Retention of three taxa of postlarval fishes in an intensively flushed estuary, Cape Fear River, North Carolina. U.S. National Marine Fisheries Service Fishery Bulletin 78:419–436.
Winer, B. J. 1971. Statistical principles of experimental design, 2nd edition. McGraw-Hill, New York.
Yoder, J. A. 1983. Statistical analysis of the distribution of fish eggs and larvae on the southeastern U.S. Continental Shelf with comments on oceanographic processes that may affect larval survival. Estuarine, Coastal and Shelf Science 17:637–650.

American Fisheries Society Symposium 3:163–165, 1988

Epilogue

MICHAEL P. WEINSTEIN

Lawler, Matusky & Skelly Engineers, One Blue Hill Plaza, Pearl River, New York 10965, USA

On the second day of the workshop an attempt was made to devise a generic study plan of sufficient sensitivity to assess the impact of jetty (or other) construction on larval fish and shellfish transport through inlets. It should be emphasized that the summary of the workshop below does not necessarily reflect individual opinions or a group consensus; rather it is simply a synthesis that is based upon the collective experience evidenced by the workshop participants. The methodology devised was tempered by the need to be cost-conscious in designing any field program.

Initially, workshop participants were divided into two groups of manageable size to discuss the status of our knowledge of larval transport and to outline the elements of a field monitoring program. A purposeful attempt was made to segregate workshop members by geography (e.g., people who worked in the Gulf of Mexico region, or the North Carolina "contingent") and to bring together physical, oceanographic, and biological disciplines. Later that afternoon, a spokesperson from each group reported on that group's activity to the participants as a whole. An open discussion was then led by the workshop moderator and the results were recorded. The entire day's discussion was taped, and the transcripts have been used to guide this written summary, which focuses on the afternoon's discussion.

Several study elements were quickly identified. Two built on previously recognized difficulties in sampling larvae and the "scale" of the problem:

- time scales of measurements as they affect long-term (e.g., seasonal) and short-term (e.g., tidal, diel, meteorological, age-dependent) estimates of larval abundance and distribution;
- spatial scales that are appropriately referenced in the vertical as well as the horizontal dimension (inside, alongshore, and outside of the inlet) as they affect abundance and distribution estimates.

Confounding our ability to establish spatiotemporal patterns of larval distribution is the typically patchy distribution of larvae in the water column and, in certain instances, their association with boundary layers that are difficult to sample. Further, the need to characterize a continuous and highly dynamic process (larval recruitment) from a limited series of "snapshots" (sampling events) of the system is particularly difficult from a biological standpoint.

From a physical oceanographic standpoint, there are two equally challenging problems:

- how to determine the boundaries of the water mass that is influenced by the inlet under various external forcing conditions, i.e., the volume of water entrained by the inlet under varying conditions of wind, weather, freshwater runoff, longshore currents, etc.;
- how to determine the efficacy of the estuarine "pump" at delivering ocean water and organisms to the estuarine side of the inlet.

As the discussion progressed, it rapidly became clear that a program of modeling and field verification would be a potentially viable approach to the overall situation and would facilitate interfacing the physical oceanographic data with the biological data. It then became a question of the time scale for sampling, the sampling sequence (e.g., synoptic versus sequential sampling of physical and biological variables) and, of course, the intensity of sampling. For simplicity of approach, the null hypothesis was posed that the larvae would act as purely passive particles and would be simulated as such. Because the larvae are typically aggregated in distribution, it was suggested that the analytical framework should incorporate the distribution of larvae according to a negative binomial. The consensus was that this could be done.

Much discussion ensued on the kinds and frequency of biological sampling that would be required. A constant concern with high background variability in catches pervaded these discussions, and it was decided that, in general, fewer sampling stations with greater replication was desirable. It was also suggested that a program develop in two basic steps. Effort and money first would be concentrated on developing a model that describes the physical oceanographic processes, with the assumption that the "passive" organisms would be transferred entirely by hydrodynamic processes. "Biological" money and effort would then be expended in two phases. The model would be calibrated to characterize the spatial distribution of larvae, particularly in the vertical dimension (to be done synoptically with physical sampling). Next, a more intensive sampling program for larvae would be undertaken at a limited number of stations (with greater replication) specifically to address background variability.

Several questions emerged during the discussion, the answers to which would partially constitute the data base to be developed.

- What are the pertinent physical oceanographic features of the nearshore system that control the volume (and configuration) of the water mass entrained into the inlet?
- From where does that volume originate?
- How does the volume change through time (seasonally and in the short term due to winds, freshwater flow, or other meteorological conditions)?
- What is the spatial and temporal distribution of larvae and postlarvae in this volume that is entrained into the estuary (for purposes of calibrating the model)?
- What is the nature of the transfer process from this volume that keeps larvae and postlarvae in the estuary and prevents them from being flushed out (estuarine spawners, too)?
- Where should stations be established in terms of optimal design for determining flux and net transport of larvae and water to the estuary?
- What is the role of the boundary layer, where particles (organisms) may concentrate?

Stated as an endpoint, we wished to determine the fraction of larvae being retained in the estuary from each oceanic water parcel (containing "new" larvae) that is influenced (entrained) by the inlet. The system may be viewed as three components coupled in time: (1) oceanic water parcels moving near the inlet, (2) inlet passage, and (3) estuarine retention of some fraction of a given oceanic parcel (and larvae contained therein). The tidal excursion was the obvious time step for the modeling effort.

Once a fully calibrated physical model was achieved, the premise of passive larvae could be tested. Significant departure from the distribution expected from passivity alone would be taken as evidence that an active component of larval flux was operative. If this were the case, the hydrodynamic model could be modified to account for larval behavior as a component of the recruitment process. Finally, the adjusted model could be run with and without the presence of jetties to ascertain the impact on the recruitment process.

The generic approach to studying the effects of inlet modification on larval transport can be summarized in five steps.

(1) Develop a hydrodynamic model with input from a field program to collect data on physical processes, the three-dimensional distribution of larvae, and the source of larvae up to several kilometers from shore. Important factors to consider are

- longshore flow, determined by wind-induced pressure gradients from 0 to about 100 km offshore;
- longshore flow in the coastal boundary layer from the shore out to about the 10-m depth contour or to about 10 km offshore;
- longshore flow to the outer bar;
- cross-shelf flow;
- Stokes drift;
- tides;
- inlet plume dynamics.

Information on topography, season of recruitment, boundaries, and specific variables potentially influencing the physical or biological response of the system (including temperature, salinity, river runoff, tide, and wind) was also judged important. The larval fish or shellfish sampling program would be designed to determine the distribution and extent of patches of larvae and would be tested for goodness of fit against a negative binomial distribution. This effort would initially emphasize the spatial distribution of sampling stations at the expense of excessive replication, but would be of sufficient intensity to detect aggregation. The program would be conducted synoptically with physical data collection.

(2) Provide verification of the physical model, by determining the fate of a parcel of water with dye studies.

(3) Simulate the spatial distribution of larval "patches" in the model with and without jetties.

(4) Conduct, if necessary, additional field sampling of larvae to verify the model and to test the hypothesis that larvae act as passive particles. The decision to conduct step (4) will depend on the desired level of precision and accuracy that is defined by steps (1–3). The workshop participants were in agreement that

- verification at this step must be serious and rigorous (at a predetermined level of sensitivity);
- simulations should include several different scenarios for circulation within the estuary;
- field sampling at step (4) must include sufficient replication to achieve a prespecified level of confidence and will necessarily be expensive;
- levels of difference must be specified a priori in order to move to step (5).

(5) If there is a mismatch between model prediction and observed distribution(s), and the physical model seems to be correct, the existence of behavioral components should be considered.

From a practical standpoint, the results of the workshop have several implications for the planning of future engineering projects at coastal inlets. The first three steps of the proposed generic study plan (i.e., hydrodynamic model development, model verification via dye studies, and model simulation of larval distributions with and without jetties) would likely be feasible from a funding perspective. A moderate investment of funds might suffice to provide the environmental planner with some basic insights into jetty (or other) construction impacts. Steps (4) (field verification of larval distribution and testing of the passive particle hypothesis) and (5) (assessment of the importance of the behavioral component in the analysis), however, would require funding at a level difficult to justify relative to the costs of construction and the perceived value of the project. Nevertheless, the recommended generic approach should provide environmental planners with options for evaluating impacts of jetty projects on a case-by-case basis.